Springer Undergraduate Mathematics Series

Springer

London
Berlin
Heidelberg
New York
Barcelona
Hong Kong
Milan
Paris
Santa Clara
Singapore
Tokyo

Advisory Board

Professor P.J. Cameron *Queen Mary and Westfield College*
Dr M.A.J. Chaplain *University of Dundee*
Dr K. Erdmann *Oxford University*
Professor L.C.G. Rogers *University of Bath*
Dr E. Süli *Oxford University*
Professor J.F. Toland *University of Bath*

Other books in this series

Analytic Methods for Partial Differential Equations *G. Evans, J. Blackledge and P. Yardley*
Applied Geometry for Computer Graphics and CAD *D. Marsh*
Basic Linear Algebra *T.S. Blyth and E.F. Robertson*
Basic Stochastic Processes *Z. Brzeźniak and T. Zastawniak*
Elements of Logic via Numbers and Sets *D.L. Johnson*
Elementary Number Theory *G.A. Jones and J.M. Jones*
Groups, Rings and Fields *D.A.R. Wallace*
Hyperbolic Geometry *J.W. Anderson*
Introduction to Laplace Transforms and Fourier Series *P.P.G. Dyke*
Introduction to Ring Theory *P.M. Cohn*
Indroductory Mathematics: Algebra and Analysis *G.Smith*
Introductory Mathematics: Applications and Methods *G.S. Marshall*
Measure, Integral and Probability *M. Capińksi and E. Kopp*
Multivariate Calculus and Geometry *S. Dineen*
Topologies and Uniformities *I.M. James*
Vector Calculus *P.C. Matthews*

G. Evans, J. Blackledge and P. Yardley

Numerical Methods for Partial Differential Equations

With 55 Figures

 Springer

Gwynne A. Evans, MA, Dphil, DSc
Jonathan M. Blackledge, BSc, PhD, DIC
Peter D. Yardley, BSc, PhD

Faculty of Computing Sciences & Engineering, De Montfort University,
The Gateway, Leicester LE1 9BH, UK

Cover illustration elements reproduced by kind permission of:
Aptech Systems, Inc., Publishers of the GAUSS Mathematical and Statistical System, 23804 S.E. Kent-Kangley Road, Maple Valley, WA 98038, USA. Tel: (206) 432 - 7855 Fax (206) 432 - 7832 email: info@aptech.com URL: www.aptech.com
American Statistical Association: Chance Vol 8 No 1, 1995 article by KS and KW Heiner 'Tree Rings of the Northern Shawangunks' page 32 fig 2
Springer-Verlag: Mathematica in Education and Research Vol 4 Issue 3 1995 article by Roman E Maeder, Beatrice Amrhein and Oliver Gloor 'Illustrated Mathematics: Visualization of Mathematical Objects' page 9 fig 11, originally published as a CD ROM 'Illustrated Mathematics' by TELOS: ISBN 0-387-14222-3, German edition by Birkhauser: ISBN-3-7643-5100-4.
Mathematica in Education and Research Vol 4 Issue 3 1995 article by Richard J Gaylord and Kazume Nishidate 'Traffic Engineering with Cellular Automata' page 35 fig 2. Mathematica in Education and Research Vol 5 Issue 2 1996 article by Michael Trott 'The Implicitization of a Trefoil Knot' page 14.
Mathematica in Education and Research Vol 5 Issue 2 1996 article by Lee de Cola 'Coins, Trees, Bars and Bells: Simulation of the Binomial Process page 19 fig 3. Mathematica in Education and Research Vol 5 Issue 2 1996 article by Richard Gaylord and Kazume Nishidate 'Contagious Spreading' page 33 fig 1. Mathematica in Education and Research Vol 5 Issue 2 1996 article by Joe Buhler and Stan Wagon 'Secrets of the Madelung Constant' page 50 fig 1.

ISBN 3-540-76125-X Springer-Verlag Berlin Heidelberg New York

British Library Cataloguing in Publication Data
Evans, G.
    Numerical methods for partial differential equations. - (Springer undergraduate mathematics series)
    1. Differential equations, Partial - Numerical solutions
    I. Title II. Blackledge, J. III. Yardley, P.
    515.3'53
ISBN 354076125X

Library of Congress Cataloging-in-Publication Data
Evans, Gwynne.
    Numerical methods for partial differential equations / G. Evans, J. Blackledge and P. Yardley.
        p. cm. – (Springer undergraduate mathematics series)
    Includes bibliographical references and index.
    ISBN 3-540-76125-X (alk. paper)
    1. Differential equations, Partial–Numerical solutions.
I. Blackledge, J.M. (Jonathan M.) II. Yardley, P. (Peter), 1948-   .
III. Title. IV. Series.
QA377. E947 1999   99-36323
515'.353–dc21   CIP

Typesetting by Focal Image Ltd, London.
Printed and bound at the Athenæum Press Ltd., Gateshead, Tyne & Wear
12/3830-543210 Printed on acid-free paper SPIN 10555798

*To our past and present students without
whom this work would not have been developed*

Preface

The subject of partial differential equations holds an exciting and special position in mathematics. Partial differential equations were not consciously created as a subject but emerged in the 18th century as ordinary differential equations failed to describe the physical principles being studied. The subject was originally developed by the major names of mathematics, in particular, Leonard Euler and Joseph-Louis Lagrange who studied waves on strings; Daniel Bernoulli and Euler who considered potential theory, with later developments by Adrien-Marie Legendre and Pierre-Simon Laplace; and Joseph Fourier's famous work on series expansions for the heat equation. Many of the greatest advances in modern science have been based on discovering the underlying partial differential equation for the process in question. James Clerk Maxwell, for example, put electricity and magnetism into a unified theory by establishing Maxwell's equations for electromagnetic theory, which gave solutions for problems in radio wave propagation, the diffraction of light and X-ray developments. Schrödinger's equation for quantum mechanical processes at the atomic level leads to experimentally verifiable results which have changed the face of atomic physics and chemistry in the 20th century. In fluid mechanics, the Navier-Stokes' equations form a basis for huge number-crunching activities associated with such widely disparate topics as weather forecasting and the design of supersonic aircraft.

Inevitably the study of partial differential equations is a large undertaking, and falls into several areas of mathematics. At one extreme the main interest is in the existence and uniqueness of solutions, and the functional analysis of the proofs of these properties. At the other extreme lies the applied mathematical and engineering quest to find useful solutions, either analytically or numerically, to these important equations which can be used in design and construction. In both this text, and the companion volume, 'Analytic methods for partial differential equations' (Springer, 1999), the emphasis is on the practical solution rather than the theoretical background, though this important work is recognised by pointers to further reading. The approach is based on courses given by the authors at De Montfort University.

This numerical volume is in six chapters. Chapter 1 contains mathematical preliminaries, mainly areas of linear algebra, solution of linear equations, iterative solution and characteristics.

The solution of the heat equation is in Chapter 2, with stability considerations leading to the coverage of implicit methods. Important considerations of consistency and convergence are also made in this chapter.

The wave equation and the method of characteristics is the material for Chapter 3, with Laplace's equation and elliptic problems being considered in Chapter 4. This latter chapter is concluded with a brief mention of more advanced topics such as methods for sparse linear systems including the multigrid methods for Laplace's equation.

Chapter 5 is in the nature of an introduction to finite element methods as applied to ordinary differential equations. Here the various methods are applied to the same exercises and hence a comparison can be made of the solutions, giving a built-in check of the validity of the codes. At the same time, the reader is developing skills for this alternative approach to the numerical solution of partial differential equations, making concepts in the next chapter more familiar.

The finite element method for partial differential equations is covered in Chapter 6. The advantages of a preliminary look at the topic for ordinary differential equations now becomes apparent, as there is inevitably considerable detail in even the simplest solution for partial differential equations which can be very disconcerting to the uninitiated reader. The principles can easily be lost in the mass of matrix elements, and the preliminary approach will make some of the fundamental ideas familiar before such detail is encountered. Again the method is applied to many of the same problems which have been earlier solved by finite differences, agreement of solutions being obtained, and comparisons made.

Most sections have a set of exercises, and fairly full solutions can be found in the Appendix. As in the companion volume, 'Analytic methods for partial differential equations', exceptions occur with some introductory sections and where two successive sections are required to complete a topic. In this numerical volume, the solutions have been completed for the most part using small programs and guidance as to these codes is given in the Appendix without writing the algorithms in a specific computer language which was thought to be fraught with problems. Instead careful specifications are given to allow individual readers to easily generate code in a convenient environment. In the case of the finite element chapter, all the exercises are at the end of the chapter, as it was considered impractical to complete part of a solution and then return to the problem for completion later.

In the last stages of the preparation, the completed manuscript was read by Endre Süli (Oxford University), and we are very grateful for his general remarks and detailed comments.

We would like to express our thanks to Susan Hezlet who was our first point of contact with Springer-Verlag. She was instrumental in steering this book through to its conclusion, though the final stage is in the capable hands of David Ireland. We are also to grateful for the continued support of De Montfort University, Leicester, and particularly the technical staff who kept our computer systems running for the duration of the writing work.

Contents

1

Background Mathematics

1.1 Introduction

The companion volume of this book, 'Analytic methods for partial differential equations', is concerned with solution of partial differential equations using classical methods which result in analytic solutions. These equations result when almost any physical situation is modelled, ranging from fluid mechanics problems, through electromagnetic problems to models of the economy. Some specific and fundamental problems were highlighted in the earlier volume, namely the three major classes of linear second order partial differential equations. The heat equation, the wave equation and Laplace's equation will form a basis for study from a numerical point of view for the same reason as they did in the analytic case. That is, the three equations are the canonical forms to which any quasi-linear second order equation may be reduced using the characteristic transformation.

The history of the numerical solution of partial differential equations is much more recent than the analytic approaches, and the development of the numerical approach has been heavily influenced by the advent of high speed computing machines. This progress is still being seen. In the pre-computer days, the pressures of war were instrumental in forcing hand-worked numerical solutions to problems such as blast waves to be attempted. Large numbers of electro-mechanical machines were used with the 'programmer' controlling the machine operators. Great ingenuity was required to allow checks to be made on human error in the process. The relaxation method was one of the results of these processes.

Once reliable and fast electronic means were available, the solution of more

1

and more complex partial differential equations became feasible. The earliest method involved discretising the partial derivatives and hence converting the partial differential equation into a difference equation. This could either be solved in a step-by-step method as with the wave equation, or required the solution of a large set of sparse linear algebraic equations in the case of Laplace's equation. Hence speed was not the only requirement of the computing machine. Storage was also crucial. At first, matrix blocks were moved in and out of backing store to be processed in the limited high speed store available. Today, huge storage requirements can be met relatively cheaply, and the progress in cheap high speed store and fast processing capability is enabling more and more difficult problems to be attempted. Weather forecasting is a very well-known area in which computer power is improving the accuracy of forecasts: admittedly now combined with the knowledge of chaos which gives some degree of forecast reliability.

In the chapters which follow the numerical solution of partial differential equations is considered, first by using the three basic problems as cases which demonstrate the methods. The finite difference method is considered first. This is the method which was first applied in the early hand-computed work, and is relatively simple to set up. The area or volume of interest is broken up into a grid system on which the partial derivatives can be expressed as simple differences. The problem then reduces to finding the solution at the grid points as a set of linear algebraic equations. Hence some attention is paid to solving linear algebraic equations with many elements in each row being zero. The use of iterative solutions can be very effective under these circumstances.

A number of theoretical considerations need to be made. Firstly, it needs to be established that by taking a finer and finer grid, the difference equation solution does indeed converge to the solution of the approximated partial differential equation. This is the classical problem of accuracy in numerical analysis. However real computers execute their arithmetic operations to a finite word length and hence all stored real numbers are subject to a rounding error. The propagation of these errors is the second main theme of numerical analysis in general, and partial differential equations in particular. This is the problem of numerical stability. Do small errors introduced as an inevitable consequence of the use of finite word length machines grow in the development of the solution? Questions of this sort will be considered as part of the stability analysis for the methods presented. There will be exercises in which the reader will be encouraged to see just how instability manifests itself in an unstable method, and how the problem can be circumvented.

Using finite differences is not the only way to tackle a partial differential equation. In 1960, Zienkiewicz used a rather different approach to structural problems in civil engineering, and this work has developed into a completely separate method of solution. This method is the finite element method (Zienkiewicz, 1977). It is based on a variational formulation of the partial differential equation, and the first part of the description of the method requires some general ways of obtaining a suitable variational principle. In many problems there is a natural such principle, often some form of energy conservation. The problem is then one of minimising an integral by the choice of a dependent function. The classic method which then follows is the Rayleigh–Ritz method. In the finite element method, the volume over which the integral is taken is split up into a set of elements. These may be triangular or prismatic for example. On each element a simple solution form may be assumed, such as a linear form. By summing over each element, the condition for a minimum reduces to a large set of linear algebraic equations for the solution values at key points of the element, such as the vertices of the triangle. Again the use of sparse linear equation solvers is required.

This first chapter is concerned with some of the mathematical preliminaries which are required in the numerical work. For the most part this chapter is quite independent of the equivalent chapter in the first volume, but the section on classification reappears here for completeness.

1.2 Vector and Matrix Norms

The numerical analysis of partial differential equations requires the use of vectors and matrices both in setting up the numerical methods and in analysing their convergence and stability properties. There is a practical need for measures of the 'size' of a vector or matrix which can be realised computationally, as well as be used theoretically. Hence the first section of this background chapter deals with the definition of the norm of a vector, the norm of a matrix and realises some specific examples.

The norm of vector $\mathbf{x}$ is a real positive number, $||\mathbf{x}||$, which satisfies the axioms:

(i) $||\mathbf{x}|| > 0$ if $\mathbf{x} \neq \mathbf{0}$ and $||\mathbf{x}|| = 0$ if $\mathbf{x} = \mathbf{0}$;

(ii) $||c\mathbf{x}|| = |c|\,||\mathbf{x}||$ for a real or complex scalar c; and

(iii) $||\mathbf{x} + \mathbf{y}|| \leq ||\mathbf{x}|| + ||\mathbf{y}||$.

If the vector $\mathbf{x}$ has components $x_1, \ldots, x_n$ then there are three commonly used norms:

(i) The one-norm of $\mathbf{x}$ is defined as

$$||\mathbf{x}||_1 = |x_1| + |x_2| + \cdots + |x_n| = \sum_{i=1}^{n} |x_i|. \qquad (1.2.1)$$

(ii) The two-norm of $\mathbf{x}$ is

$$||\mathbf{x}||_2 = (|x_1|^2 + |x_2|^2 + \cdots + |x_n|^2)^{\frac{1}{2}} = \left[\sum_{i=1}^n |x_i|^2\right]^{\frac{1}{2}}. \qquad (1.2.2)$$

(iii) The infinity norm of $\mathbf{x}$ is the maximum of the moduli of the components or

$$||\mathbf{x}||_\infty = \max_i |x_i|. \qquad (1.2.3)$$

In a similar manner the norm of a matrix A is a real positive number giving a measure of the 'size' of the matrix which satisfies the axioms

(i) $||A|| > 0$ if $A \neq 0$ and $||A|| = 0$ if $A = 0$;

(ii) $||cA|| = |c|\,||A||$ for a real or complex scalar c;

(iii) $||A + B|| \leq ||A|| + ||B||$; and

(iv) $||AB|| \leq ||A||\,||B||$.

Vectors and matrices occur together and so they must satisfy a condition equivalent to (iv), and with this in mind matrix and vector norms are said to be *compatible* or *consistent* if

$$||A\mathbf{x}|| \leq ||A||\,||\mathbf{x}||, \quad \forall x \in \mathcal{R}^n (\text{or } \mathcal{C}^n).$$

There is a class of matrix norms whose definition depends on an underlying vector norm. These are the subordinate matrix norms. Let A be an $n \times n$ matrix and $\mathbf{x} \in S$ where

$$S = \{(n \times 1) \text{ vectors} : ||\mathbf{x}|| = 1\};$$

now in general $||A\mathbf{x}||$ varies as $\mathbf{x}$ varies $(\mathbf{x} \in S)$. Let $\mathbf{x}_0 \in S$ be such that $||A\mathbf{x}||$ attains its maximum value. Then the norm of matrix A, subordinate to the vector norm $||.||$, is defined by

$$||A|| = ||A\mathbf{x}_0|| = \max_{||\mathbf{x}||=1} ||A\mathbf{x}||. \qquad (1.2.4)$$

The matrix norm that is subordinate to the vector norm automatically satisfies the compatibility condition since, if $\mathbf{x} = \mathbf{x}_1 \in S$, then

$$||A\mathbf{x}_1|| \leq ||A\mathbf{x}_0|| = ||A|| = ||A||\,||\mathbf{x}_1|| \quad \text{since} \quad ||\mathbf{x}_1|| = 1.$$

Therefore $||A\mathbf{x}|| \leq ||A||\,||\mathbf{x}||$ for any $\mathbf{x} \in \mathcal{R}^n$. Note that for all subordinate matrix norms

$$||I|| = \max_{||\mathbf{x}||=1} ||I\mathbf{x}|| = \max_{||\mathbf{x}||=1} ||\mathbf{x}|| = 1. \qquad (1.2.5)$$

The definitions of the subordinate one, two and infinity norms with $||\mathbf{x}|| = 1$ lead to:

- The one norm of matrix A is the maximum column sum of the moduli of the elements of A, and is denoted by $||A||_1$.

- The infinity norm of matrix A is the maximum row sum of the moduli of the elements of A, and is denoted by $||A||_\infty$.

- The two norm of matrix A is the square root of the spectral radius of $A^H A$ where $A^H = (\bar{A})^T$ (the transpose of the complex conjugate of A). This norm is denoted by $||A||_2$. The *spectral radius* of a matrix B is denoted by $\rho(B)$ and is the modulus of the eigenvalue of maximum modulus of B.

Hence for example if

$$A = \begin{pmatrix} -1 & 1 \\ 3 & -2 \end{pmatrix} \quad \text{then} \quad A^H A = A^T A = \begin{pmatrix} 10 & -7 \\ -7 & 5 \end{pmatrix}$$

has eigenvalues 14.93 and 0.067. Then using the above definitions

$$||A||_1 = 1 + 3 = 4, \ ||A||_\infty = 3 + 2 = 5, \ ||A||_2 = \sqrt{14.93} = 3.86.$$

Note that if A is real and symmetric

$$A^H = A \quad \text{and} \quad ||A||_2 = [\rho(A^2)]^{\frac{1}{2}} = [\rho^2(A)]^{\frac{1}{2}} = \rho(A) = \max_i |\lambda_i|.$$

A number of other equivalent definitions of $||A||_2$ appear in the literature. For example the eigenvalues of $A^H A$ are denoted by $\sigma_1^2, \sigma_2^2, \ldots \sigma_n^2$ and the σ_i are called the *singular values* of A. By their construction the singular values will be real and non-negative. Hence from the above definition

$$||A||_2 = \sigma_1$$

where

$$\sigma_1 = \max_{1 \leq i \leq n} \sigma_i.$$

For a symmetric A, the singular values of A are precisely the eigenvalues of A apart from a possible sign change, and

$$||A||_2 = |\lambda_1|,$$

where λ_1 is the largest absolute value eigenvalue of A. A bound for the spectral radius can also be derived in terms of norms. Let λ_i and $\mathbf{x}_i$ be corresponding eigenvalue and eigenvector of the $n \times n$ matrix A, then $A\mathbf{x}_i = \lambda_i \mathbf{x}_i$, and

$$||A\mathbf{x}_i|| = ||\lambda_i \mathbf{x}_i|| = |\lambda_i| \, ||\mathbf{x}_i||.$$

For all compatible matrix and vector norms

$$|\lambda_i| \, ||\mathbf{x}_i|| = ||A\mathbf{x}_i|| \leq ||A|| \, ||\mathbf{x}_i||.$$

Therefore $|\lambda_i| \leq ||A||$, $i = 1(1)n$. Hence $\rho(A) \leq ||A||$ for any matrix norm that is compatible with a vector norm.

A few illustrative exercises which are based on the previous section now follow.

EXERCISES

1.1 A further matrix norm is $||A||_E$ or the Euclidean norm, defined by

$$||A||_E^2 = \sum_{i,j} a_{ij}^2.$$

Prove that

$$||A||_2 \leq ||A||_E \leq n^{1/2}||A||_2$$

for an nth order matrix A. Verify the inequality for the matrix

$$\begin{bmatrix} 1.2 & -2.0 \\ 1.0 & 0.6 \end{bmatrix}.$$

1.2 Compute the four norms $||A||_1$, $||A||_2$, $||A||_\infty$ and $||A||_E$ for the matrix

$$\begin{bmatrix} 1 & 0 & 1 \\ 2 & 3 & 0 \\ 2 & 1 & 4 \end{bmatrix}$$

and find the characteristic polynomial of this matrix and hence its eigenvalues. Verify that

$$|\lambda_i| \leq ||A||$$

for $i = 1, 2, 3$.

1.3 Compute the spectral radius of the matrix

$$\begin{bmatrix} 9 & -2 & 1 \\ 4 & 5 & -2 \\ 1 & -3 & -5 \end{bmatrix}$$

and confirm that this value is indeed less than or equal to both $||A||_1$ and $||A||_\infty$.

1.4 For the solution of a set of linear algebraic equations

$$A\mathbf{x} = \mathbf{b}$$

the condition number is given by

$$\kappa = ||A||_2||A^{-1}||_2.$$

The input errors, such as machine precision, are multiplied by this number to obtain the errors in the solution x. Find the condition number for the matrix

$$\begin{bmatrix} 1 & 2 & 6 \\ 2 & 6 & 24 \\ 6 & 24 & 120 \end{bmatrix}.$$

1.5 Sketch the curve of $\kappa(A)$ as defined above, against the variable c for the matrix

$$\begin{bmatrix} 1 & 1/2 & 1/3 \\ 1/2 & 1/3 & 1/4 \\ 1/3 & 1/4 & c \end{bmatrix}.$$

Large values of κ will indicate where there will be high loss of accuracy in the solution of linear equations with the matrix A. For $c = 1/5$ the matrix is the 3×3 Hilbert matrix.

1.3 Gerschgorin's Theorems

The first of the theorems which gives bounds on eigenvalues is

Theorem 1.1

The largest of the moduli of the eigenvalues of the square matrix A cannot exceed the largest sum of the moduli of the elements along any row or any column. In other words $\rho(A) \le ||A||_1$, or $||A||_\infty$.

Proof

Let λ_i be an eigenvalue of the $n \times n$ matrix A and $\mathbf{x}_i$ be the corresponding eigenvector with components $v_1, v_2, \ldots, v_n$. Then $A\mathbf{x}_i = \lambda_i \mathbf{x}_i$ becomes in full

$$\begin{aligned} a_{11}v_1 + a_{12}v_2 + \cdots + a_{1n}v_n &= \lambda_i v_1 \\ a_{21}v_1 + a_{22}v_2 + \cdots + a_{2n}v_n &= \lambda_i v_2 \\ \vdots &= \vdots \\ a_{s1}v_1 + a_{s2}v_2 + \cdots + a_{sn}v_n &= \lambda_i v_s \\ \vdots &= \vdots \end{aligned}$$

Let v_s be the largest in modulus of $v_1, \ldots, v_n$, noting that $v_s \ne 0$. Select the sth equation and divide by v_s giving

$$\lambda_i = a_{s1}\left(\frac{v_1}{v_s}\right) + a_{s2}\left(\frac{v_2}{v_s}\right) + \cdots + a_{sn}\left(\frac{v_n}{v_s}\right).$$

Therefore

$$|\lambda_i| \le |a_{s1}| + |a_{s2}| + \cdots + |a_{sn}|$$

since

$$\left|\frac{v_i}{v_s}\right| \le 1, \quad i = 1, 2, \ldots, n.$$

If this is not the largest row sum then $|\lambda_i| <$ the largest row sum. In particular this holds for

$$|\lambda_i| = \max_{1 \leq s \leq n} |\lambda_s|.$$

Since the eigenvalues of A^T are the same as those for A the theorem is also true for columns.

The second theorem gives the approximate position of the eigenvalues of a matrix and is *Gerschgorin's circle theorem or first theorem, or Brauer's theorem.*

Theorem 1.2

Let P_s be the sum of the moduli of the elements along the sth row excluding the diagonal element a_{ss}. Then each eigenvalue of A lies inside or on the boundary of at least one of the circles $|\lambda - a_{ss}| = P_s$.

Proof

By the previous proof

$$\lambda_i = a_{s1}\left(\frac{v_1}{v_s}\right) + a_{s2}\left(\frac{v_2}{v_s}\right) + \cdots + a_{ss} + \cdots + a_{sn}\left(\frac{v_n}{v_s}\right).$$

Hence

$$|\lambda_i - a_{ss}| = \sum_{j \neq s} a_{sj}\frac{v_j}{v_s} = \sum_{j \neq s} |a_{sj}|.$$

The third theorem is *Gerschgorin's second theorem*:

Theorem 1.3

If p of the circles of Gerschgorin's circle theorem form a connected domain that is isolated from the other circles, then there are precisely p eigenvalues of the matrix A within this connected domain. In particular, an isolated Gerschgorin's circle contains one eigenvalue.

Proof

Split the matrix A into its diagonal elements D and off-diagonal elements C to give

$$A = \text{diag}(a_{ii}) + C = D + C \tag{1.3.1}$$

and then using P_s as defined in the circle theorem, the matrices $(D + \epsilon C)$ can be considered. For $\epsilon = 0$ the matrix D is returned whereas $\epsilon = 1$ gives A. However the characteristic polynomial of $(D + \epsilon C)$ has coefficients which are themselves polynomials in ϵ, and therefore the roots of this characteristic polynomial are continuous in ϵ. Hence eigenvalues traverse continuous paths as ϵ varies, by the circle theorem as, for any eigenvalue, the eigenvalues lie in circular discs with centres a_{ii} and radii ϵP_i.

Suppose the first s discs form a connected domain. The discs may be reordered if this is not the case. Then $(n-s)$ discs with radii $P_{s+1}, P_{s+2}, \ldots, P_n$ are isolated from the s with radii $P_1, P_2, \ldots, P_s$. This also applies to the discs of radii ϵP_i for all $\epsilon \in [0,1]$. When $\epsilon = 0$, the eigenvalues are $a_{11}, \ldots, a_{nn}$ and clearly the first s lie inside the domain corresponding to the first s discs, and the other $(n-s)$ lie outside. Hence by the continuity this state must continue through to $\epsilon = 1$ to prove the theorem.

When the eigenvalues λ_i of a matrix A are estimated by the circle theorem, the condition $|\lambda_i| \leq 1$ is equivalent to $||A||_\infty \leq 1$ or $||A||_1 \leq 1$, for we have $|\lambda - a_{ss}| \leq P_s$. Hence $-P_s \leq \lambda - a_{ss} \leq P_s$ so that $-P_s + a_{ss} \leq \lambda \leq P_s + a_{ss}$. Now λ will satisfy $-1 \leq \lambda \leq 1$, if $P_s - a_{ss} \leq 1$ and $P_s + a_{ss} \leq 1$ for $s = 1, \ldots, n$, as

$$||A||_\infty = \max_{1 \leq s \leq n} \sum_{j=1}^{n} |a_{sj}| = P_s + |a_{ss}| \leq 1. \qquad (1.3.2)$$

Now P_s is the sum of the moduli of the elements of A in the sth row (excluding a_{ss}), and a_{ss} may be positive or negative. Hence inequality (1.3.2) is equivalent to

$$\sum_{j=1}^{n} |a_{sj}| \leq 1, \quad s = 1, \ldots, n, \qquad (1.3.3)$$

or to $||A||_\infty \leq 1$ for rows. For any consistent pairs of matrix and vector norms,

$$|\lambda|\,||\mathbf{x}|| = ||\lambda \mathbf{x}|| = ||A\mathbf{x}|| \leq ||A||\,||\mathbf{x}||$$

and hence $|\lambda| \leq ||A||$. Hence

$$||A||_2^2 = \text{max eigenvalue of} \quad A^H A$$

$$\leq ||A^H A||_1 \leq ||A^H||_1 ||A||_1 = ||A||_\infty ||A||_1.$$

Hence $||A||_2^2 \leq ||A||_1 ||A||_\infty$, and so if both inequalities hold it follows automatically that $||A||_2 \leq 1$. Hence Gerschgorin's circle theorem can be used to establish conditions for stability which will be dealt with in Chapter 2.

[Note: It can be shown that if the off-diagonal elements of a real tri-diagonal matrix are one-signed then all its eigenvalues are real (Smith, 1978).]

The following exercises can now be attempted to complete the understanding of this section.

EXERCISES

1.6 Use Gerschgorin's theorems to investigate the regions in which the eigenvalues of the matrix

$$\begin{bmatrix} 2 & -1 & 0 & 0 \\ -1 & 2 & -1 & 0 \\ 0 & -1 & 2 & -1 \\ 0 & 0 & -1 & 2 \end{bmatrix}$$

lie, and confirm that they lie in the range $[0, 4]$.

1.7 Use Gerschgorin's theorems to investigate whether an estimate of the condition number κ for the matrix of exercise 1.4 can be found. For a symmetric matrix, κ reduces to the ratio of the eigenvalue of largest modulus to that with the smallest modulus. This exercise highlights the shortcomings of the theorems.

1.8 Consider the same problem as in exercise 1.7. but with the $n \times n$ matrix whose elements are defined by

$$a_{i,j} = (i + j - 1)!.$$

This is an example of a well-known class of very ill-conditioned matrices.

1.9 Use Gerschgorin's theorems to find bounds on the condition factor $\kappa(A)$ for the matrix

$$\begin{bmatrix} 1 & 0.2 & 0.3 & 0 \\ 1 & 8 & 1 & 0 \\ 0 & 1 & 10 & 4 \\ 0 & 0 & 4 & 100 \end{bmatrix}.$$

1.4 Iterative Solution of Linear Algebraic Equations

In general the use of difference methods for the solution of partial differential equations leads to an algebraic system $A\mathbf{x} = \mathbf{b}$ where A is a given matrix which is sparse and of large order. Direct methods for solving this system, such as Gaussian elimination (Evans, 1995) tend to be inefficient and it is more usual to use an iterative method. The problem with elimination methods is that the initially sparse matrix begins to fill-in as the process develops, and more and more of the originally zero elements now have to be processed. There is a consequential cost in both storage and processor time. Iterative methods do not alter the matrix structure and so preserve sparseness, though in some cases problems of convergence may become an issue.

Consider splitting the matrix A into three components in which L is lower triangular, D is diagonal and U is upper triangular. Let us suppose that

$$A = \begin{bmatrix} d_{11} & u_{12} & u_{13} & \cdots & u_{1n} \\ l_{21} & d_{22} & u_{23} & \cdots & u_{2n} \\ l_{31} & l_{32} & d_{33} & \cdots & u_{3n} \\ \vdots & & & & \vdots \\ l_{n1} & l_{n2} & l_{n3} & \cdots & d_{nn} \end{bmatrix}.$$

Then

$$A = L + D + U, \qquad (1.4.1)$$

with

$$L = \begin{bmatrix} 0 & & & \\ l_{11} & 0 & & \\ \vdots & \vdots & \ddots & \\ \cdots & \cdots & l_{n,n-1} & 0 \end{bmatrix},$$

$$D = \begin{bmatrix} d_{11} & & & \\ & \ddots & & \\ & & \ddots & \\ & & & d_{nn} \end{bmatrix},$$

$$U = \begin{bmatrix} 0 & u_{12} & \cdots & \cdots \\ & 0 & \ddots & \vdots \\ & & 0 & u_{n-1,n} \\ & & & 0 \end{bmatrix},$$

and $A\mathbf{x} = \mathbf{b}$ becomes

$$(L + D + U)\mathbf{x} = \mathbf{b} \qquad (1.4.2)$$

or

$$D\mathbf{x} = -(L + U)\mathbf{x} + \mathbf{b}. \qquad (1.4.3)$$

Assuming that D^{-1} exists, this leads to the iterative scheme

$$\mathbf{x}^{(r+1)} = -D^{-1}(L + U)\mathbf{x}^{(r)} + D^{-1}\mathbf{b}, \quad x^{(0)} \text{ given}, \qquad (1.4.4)$$

which is the *Jacobi method*.

The matrix formulation used in the above manipulations do not provide the easiest method of implementation. The obvious implementation is to rewrite the linear equations with the diagonal term on the left-hand side and then iterate equation by equation as in the example below. Consider the set of equations

$$\begin{bmatrix} 2 & 1 & 1 \\ -1 & 3 & 1 \\ 1 & 2 & -4 \end{bmatrix} \mathbf{x} = \begin{bmatrix} 4 \\ -5 \\ 6 \end{bmatrix}$$

which yields the iteration

$$
\begin{aligned}
2x_1^{(r+1)} &= 4 - x_2^{(r)} - x_3^{(r)} \\
3x_2^{(r+1)} &= -5 + x_1^{(r)} - x_3^{(r)} \\
-4x_3^{(r+1)} &= 6 - x_1^{(r)} - 2x_2^{(r)}.
\end{aligned}
$$

Starting with the zero vector for $\mathbf{x}^{(0)}$ gives the results in Table 1.1.

Table 1.1.

r	x_1	x_2	x_3
0	0.0	0.0	0.0
1	2.000	−1.6667	−1.5000
2	3.5833	−0.5000	−1.83333
3	3.16667	0.13889	−0.85417
4	2.35764	−0.32639	−0.63889
5	2.48264	−0.66782	−1.07378
6	2.87080	−0.48119	−1.21325
7	2.84722	−0.30531	−1.02289
8	2.66410	−0.37663	−0.94085
9	2.65874	−0.46501	−1.02228
	⋮		⋮
20	2.72244	−0.41584	−1.02716

Intuitively, faster convergence would be expected the greater in magnitude the diagonal elements are compared with the off-diagonal elements. This is known as *diagonal dominance* and is only weakly exhibited in this example. Hence there is quite slow convergence to the correct result of (2.72222, −0.4166667, −1.0277778).

An improvement to this method is obtained by using the newly calculated elements as soon as they are available. Hence the elements which multiply L in (1.4.4) are known and could also be placed on the left-hand side of the iteration to give

$$(L + D)\mathbf{x}^{r+1} = -U\mathbf{x}^r + \mathbf{b}, \tag{1.4.5}$$

giving the *Gauss–Seidel iteration*

$$\mathbf{x}^{r+1} = -(L + D)^{-1}U\mathbf{x}^r + (L + D)^{-1}\mathbf{b}. \tag{1.4.6}$$

Note that (1.4.5) can be written

$$D\mathbf{x}^{r+1} = -L\mathbf{x}^{r+1} - U\mathbf{x}^r + \mathbf{b} \tag{1.4.7}$$

or

$$\mathbf{x}^{r+1} = \mathbf{x}^r + D^{-1}(\mathbf{b} - L\mathbf{x}^{r+1} - U\mathbf{x}^r - D\mathbf{x}^r). \tag{1.4.8}$$

Hence the new approximation is given by the old approximation together with a displacement (or correction).

In a practical form the above example set up for Gauss–Seidel iteration has the form:

$$\begin{aligned} 2x_1^{(r+1)} &= 4 - x_2^{(r)} - x_3^{(r)} \\ 3x_2^{(r+1)} &= -5 + x_1^{(r+1)} - x_3^{(r)} \\ -4x_3^{(r+1)} &= 6 - x_1^{(r+1)} - 2x_2^{(r+1)} \end{aligned}$$

and the iterations are shown in Table 1.2.

Table 1.2.

r	x_1	x_2	x_3
0	0.0	0.0	0.0
1	2.000	−1.000	−1.5000
2	3.25	−0.08333	−0.72917
3	2.40625	−0.62153	1.2092
4	2.91536	−0.21918	−0.91706
5	2.60444	−0.49283	−1.09531
6	2.79406	−0.37021	−0.98659
7	2.67840	−0.44501	−1.05290
8	2.74895	−0.39938	−1.01245
9	2.70592	−0.42721	−1.03712
$\vdots$	$\vdots$		$\vdots$
20	2.72229	−0.41662	−1.027737

Here the better convergence of the Gauss–Seidel method over Jacobi's method can be seen. In the practical applications of these methods to partial differential equations, the nature of the finite differencing often yields diagonally dominant matrices which give quite rapid convergence without the need for storing full matrices. Methods of increasing the convergence rate are of considerable interest and include the following approach.

If successive displacements are all one-signed, as they usually are for the approximating difference equations of elliptic problems, it would seem reasonable to expect convergence to be accelerated if a larger (displacement correction) was given than is defined above. This leads to the *successive over relaxation* or *SOR iteration* defined by

$$\mathbf{x}^{r+1} = \mathbf{x}^r + \omega D^{-1}[\mathbf{b} - L\mathbf{x}^{r+1} - U\mathbf{x}^r - D\mathbf{x}^r] \qquad (1.4.9)$$

where ω, the acceleration parameter or relaxation factor, generally lies in the range $1 < \omega < 2$.

Thus, (1.4.9) becomes

$$(D + \omega L)\mathbf{x}^{r+1} = D\mathbf{x}^r + \omega \mathbf{b} - \omega U\mathbf{x}^r - \omega D\mathbf{x}^r \qquad (1.4.10)$$

which can be rewritten as

$$\mathbf{x}^{r+1} = (D + \omega L)^{-1}[(1 - \omega)D - \omega U]\mathbf{x}^r + \omega(D + \omega L)^{-1}\mathbf{b}. \qquad (1.4.11)$$

All the iterative schemes described so far have the form

$$\mathbf{x}^{r+1} = G\mathbf{x}^r + H\mathbf{b}. \qquad (1.4.12)$$

The rate of convergence of a scheme will be shown to be dictated by the magnitude of the dominant eigenvalue of the matrix G. Choosing the relaxation parameter in SOR suitably can result in savings in computational effort by a significant factor. In some special cases, optimal parameters can be found analytically, and the reader is referred to Young (1971) and Varga (1962) for further details. Usually experimentation and experience enable the user to obtain near optimal parameters.

The success of an iterative method depends on the rate of convergence. A point iterative method is one in which each component of $\mathbf{x}^r$ is calculated explicitly in terms of existing estimates of other components. A stationary iterative method is one in which $\mathbf{x}^r$ is calculated from known approximations by the same cycle of operations for all r. Jacobi, Gauss–Seidel and SOR are stationary iterative methods and have the form (1.4.12) where G is the iteration matrix and $H\mathbf{b}$ is a column of vectors of known values. Equation (1.4.12) was derived from the original equations $A\mathbf{x} = \mathbf{b}$ and hence the unique solution of $A\mathbf{x} = \mathbf{b}$ is the solution of

$$\mathbf{x} = G\mathbf{x} + H\mathbf{b}. \qquad (1.4.13)$$

Define

$$\boldsymbol{\Delta}^r = \mathbf{x}^{r+1} - \mathbf{x}^r \quad \text{and} \quad \mathbf{e}^r = \mathbf{x} - \mathbf{x}^r \qquad (1.4.14)$$

which leads recursively to

$$\mathbf{e}^r = G^r \mathbf{e}^0. \qquad (1.4.15)$$

Hence the iteration will converge if and only if $\lim_{r \to \infty} G^r = 0$. We assume that the eigenvalues of G are real and that an eigenvector basis exists. Taking the eigenvectors $\mathbf{v}_1, \ldots, \mathbf{v}_N$ to be arranged so that $\mathbf{v}_i$ has corresponding eigenvalue λ_i, where

$$|\lambda_i| \leq |\lambda_{i-1}|, \quad i = 2, \ldots, n,$$

we note that $\mathbf{e}^0$ can be expressed uniquely as a linear combination of the eigenvectors to give

$$\mathbf{e}^0 = \gamma_1 \mathbf{v}_1 + \gamma_2 \mathbf{v}_2 + \cdots + \gamma_n \mathbf{v}_n \qquad (1.4.16)$$

where $\gamma_i, \quad i = 1, \ldots, n$, are scalars. Then

$$
\begin{aligned}
G\mathbf{e}^0 &= G(\gamma_1 \mathbf{v}_1 + \gamma_2 \mathbf{v}_2 + \cdots + \gamma_n \mathbf{v}_n) \\
&= \gamma_1 \lambda_1 \mathbf{v}_1 + \gamma_2 \lambda_2 \mathbf{v}_2 + \cdots + \gamma_n \lambda_n \mathbf{v}_n \\
&= \lambda_1 \left(\gamma_1 \mathbf{v}_1 + \gamma_2 \frac{\lambda_2}{\lambda_1} \mathbf{v}_2 + \cdots + \gamma_n \frac{\lambda_n}{\lambda_1} \mathbf{v}_n \right)
\end{aligned}
$$

and hence

$$G^r e^0 = \lambda_1^r \left(\gamma_1 \mathbf{v}_1 + \gamma_2 \left(\frac{\lambda_2}{\lambda_1} \right)^r \mathbf{v}_2 + \cdots + \gamma_n \left(\frac{\lambda_n}{\lambda_1} \right)^r \mathbf{v}_n \right). \tag{1.4.17}$$

Letting $\rho = |\lambda_1|$, results in the definitions that λ_1 is called the dominant eigenvalue of G and ρ is the spectral radius. Hence the iteration will converge for arbitrary $\mathbf{x}^0$ if and only if the spectral radius ρ of G is less than one. If r is large, then (1.4.17) can be written

$$\mathbf{e}^r \simeq \lambda_1^r \gamma_1 \mathbf{v}_1.$$

Hence if the ith component in $\mathbf{e}^r$ is e_i^r and the ith component of $\mathbf{v}_1$ is v_{1i} then

$$|e_i^r| \simeq \rho^r |\gamma_1 v_{1i}|.$$

Ultimately, therefore the error in the approximation decreases by a factor $\sim 1/\rho$ with each iteration

$$\left(\frac{|e_i^r|}{|e_i^{r+1}|} \simeq \frac{1}{\rho} \right).$$

Suppose that we require to continue the iteration until no component of $\mathbf{e}^r$ exceeds E. We then require $|e_i^r| < E$, $i = 1, \ldots, n$. Set

$$m = \max_{1 \leq i \leq n} |\gamma_1 v_{1i}|.$$

Then approximately, the requirement is $\rho^r m \leq E$, or

$$r \geq \frac{\ln(m/E)}{-\ln \rho} \quad (\text{since } \rho < 1 \text{ and } -\ln \rho > 0). \tag{1.4.18}$$

Thus r, the number of iterations required to reduce to E the error in each component of $\mathbf{x}^r$ is inversely proportional to $-\ln \rho$.

How do we know when to terminate the iteration?

Realistically, our only measure is to test

$$\mathbf{\Delta}^r = \mathbf{x}^{r+1} - \mathbf{x}^r.$$

Now

$$\mathbf{x} \simeq \mathbf{x}^r + \mathbf{\Delta}^r + \mathbf{\Delta}^{r+1} + \mathbf{\Delta}^{r+2} + \cdots$$

and

$$\mathbf{e}^r \simeq \lambda_1 \mathbf{e}^{r-1} \Rightarrow \mathbf{e}^{r+1} - \mathbf{e}^r \simeq \lambda_1 (\mathbf{e}^r - \mathbf{e}^{r-1})$$

or

$$\mathbf{x}^{r+1} - \mathbf{x}^r \simeq \lambda_1 (\mathbf{x}^r - \mathbf{x}^{r-1}) \Rightarrow \mathbf{\Delta}^r \simeq \mathbf{\Delta}^{r-1}.$$

Thus, for sufficiently large r,

$$\mathbf{x} \simeq \mathbf{x}^r + \mathbf{\Delta}^r (1 + \lambda_1 + \lambda_1^2 + \cdots) = \mathbf{x}^r + \frac{\mathbf{\Delta}^r}{1 - \lambda_1}. \tag{1.4.19}$$

It follows that if we are to expect errors no greater than E in the components of $\mathbf{x}^r$ we must continue our iterations until

$$\max_{1 \leq i \leq n} \left| \frac{\Delta_i^r}{1 - \lambda_1} \right| < E. \tag{1.4.20}$$

This result tells us that the current correction $\boldsymbol{\Delta}^r$ should really be multiplied by $(1 - \lambda_1)^{-1}$. This is an important result because if we want an approximation to $\mathbf{x}$ with an error no greater than E we might be tempted to terminate the iteration at the first r for which $\|\boldsymbol{\Delta}^r\| \leq E$, where $\| \, . \, \|$ denotes the infinity norm $\|\boldsymbol{\Delta}\| = \max_{1 \leq i \leq n} |\Delta_i|$.

If $\lambda_1 = 0.99$ (which is quite possible), such a termination would give a very poor result, the iteration should be continued until $\|\boldsymbol{\Delta}^r\| \simeq 0.01E$.

For most problems λ_1 will not be known analytically, in which case its value must be estimated. One straightforward way of doing this is as follows.

For sufficiently large r

$$\boldsymbol{\Delta}^r \simeq \lambda_1 \boldsymbol{\Delta}^{r-1}. \tag{1.4.21}$$

Hence

$$\|\boldsymbol{\Delta}^r\| \simeq |\lambda_1| \, \|\boldsymbol{\Delta}^{r-1}\|,$$

so

$$|\lambda_1| = \rho \simeq \frac{\|\boldsymbol{\Delta}^r\|}{\|\boldsymbol{\Delta}^{r-1}\|}$$

where $\|\boldsymbol{\Delta}^r\|$ can be defined as

$$\|\boldsymbol{\Delta}^r\| = \max_i |x_i^{r+1} - x_i^r|$$

or

$$\|\boldsymbol{\Delta}^r\| = |x_1^{r+1} - x_1^r| + |x_2^{r+1} - x_2^r| + \cdots + |x_n^{r+1} - x_n^r|$$

or

$$\|\boldsymbol{\Delta}^r\| = \left[(x_1^{r+1} - x_1^r)^2 + (x_2^{r+1} - x_2^r)^2 + \cdots + (x_n^{r+1} - x_n^r)^2 \right]^{\frac{1}{2}}.$$

Equation (1.4.21) justifies the basis of the SOR iterative method because it proves that when λ_1 is positive the corresponding components of successive correction or displacement vectors are of the same sign. The following set of exercises may now be attempted on iterative methods of solution.

EXERCISES

1.10 Use both Jacobi's iterative method and that of Gauss–Seidel to find iteratively the solution of the linear equations

$$\begin{bmatrix} 2 & 1 & 0.5 \\ -1 & 3 & 1 \\ 0.5 & -1 & 4 \end{bmatrix} \mathbf{x} = \begin{bmatrix} 1 \\ 2 \\ 1 \end{bmatrix}.$$

1.11 Find the spectral radii of the G matrices of equation (1.4.12) for the matrix in Exercise 1.10 and hence find the theoretical rates of convergence. How do these rates compare with the actual rates obtained in the first exercise?

1.12 Now attempt the same problem with the SOR method with a range of ω from $1 < \omega < 1.4$. For this you will probably need to program the algorithm. Draw a graph of the rate of convergence against the relaxation parameter ω.

1.13 Investigate the rates of convergence of Jacobi's method and the Gauss–Seidel method on the matrix

$$A = \begin{bmatrix} 1 & 0 & 1 \\ -1 & 1 & 0 \\ 1 & 2 & -3 \end{bmatrix}.$$

This is a pathological example: normally the Gauss–Seidel method is more rapidly convergent than Jacobi.

1.14 The SOR method for tridiagonal matrices has an optimum ω given by

$$\omega = \frac{2}{1 + \sqrt{1 - \rho(C_J)}}$$

where

$$C_J = D^{-1}(L + U)$$

is the iteration matrix for Jacobi's method. Apply this optimised method to the set of equations:

$$\begin{bmatrix} 2 & -2 & 0 & 0 \\ -1 & 3 & -1 & 0 \\ 0 & -1 & 6 & -1 \\ 0 & 0 & -1 & 11 \end{bmatrix} \begin{bmatrix} x_1 \\ x_2 \\ x_3 \\ x_4 \end{bmatrix} = \begin{bmatrix} 1 \\ 1 \\ 1 \\ 1 \end{bmatrix}.$$

Experiment with values of ω slightly away from the optimum to show the sensitivity of the convergence rate to the ω value used.

1.15 Consider the matrix

$$\begin{bmatrix} 2 & -2 & 0 & 1 \\ -1 & 3 & -1 & 0 \\ 0 & -1 & 6 & -1 \\ 0 & 0 & -1 & 11 \end{bmatrix} \begin{bmatrix} x_1 \\ x_2 \\ x_3 \\ x_4 \end{bmatrix} = \begin{bmatrix} 1 \\ 1 \\ 1 \\ 1 \end{bmatrix}.$$

which differs from the one in Exercise 1.14 by just the element (1,4). Apply the SOR iteration to this matrix to see how much the change of one element affects the optimum ω. The new matrix is not tridiagonal so the theorem of Exercise 1.14 does not apply.

1.5 Further Results on Eigenvalues and Eigenvectors

In this section various results and proofs concerning eigenvalues and eigenvectors are collected together. These results are used freely in the following chapters. Let a square matrix A have eigenvector $\mathbf{x}$ and corresponding eigenvalue λ, then $A\mathbf{x} = \lambda\mathbf{x}$.

Hence

$$A(A\mathbf{x}) = A^2\mathbf{x} = \lambda A\mathbf{x} = \lambda^2\mathbf{x} \qquad (1.5.1)$$

resulting in A^2 having eigenvalue λ^2 and eigenvector $\mathbf{x}$. Similarly

$$A^p\mathbf{x} = \lambda^p\mathbf{x}, \quad p = 3, 4, \dots \qquad (1.5.2)$$

and A^p has eigenvalue λ^p and eigenvector $\mathbf{x}$. These results may be generalised by defining

$$f(A) = a_p A^p + a_{p-1} A^{p-1} + \cdots + a_0 I.$$

This is a polynomial in A when $a_p, \dots, a_0$ are scalars. Then,

$$f(A)\mathbf{x} = (a_p\lambda^p + \cdots + a_0)\mathbf{x} = f(\lambda)\mathbf{x} \qquad (1.5.3)$$

and $f(A)$ has eigenvalue $f(\lambda)$ and eigenvector $\mathbf{x}$. More generally we have the following simple theorem.

Theorem 1.4

The eigenvalue of $[f_1(A)]^{-1} f_2(A)$ corresponding to the eigenvector $\mathbf{x}$ is $f_2(\lambda)/f_1(\lambda)$, where $f_1(A)$ and $f_2(A)$ are polynomials in A.

Proof

We have

$$f_1(A)\mathbf{x} = f_1(\lambda)\mathbf{x}, \quad f_2(A)\mathbf{x} = f_2(\lambda)\mathbf{x}$$

Pre-multiply by $[f_1(A)]^{-1}$ to give

$$[f_1(A)]^{-1}[f_1(A)]\mathbf{x} = [f_1(A)]^{-1} f_1(\lambda)\mathbf{x}$$

and hence

$$[f_1(A)]^{-1}\mathbf{x} = [f_1(\lambda)]^{-1}\mathbf{x}$$

and

$$[f_1(A)]^{-1} f_2(A)\mathbf{x} = f_2(\lambda)[f_1(A)]^{-1}\mathbf{x}.$$

Eliminating $[f_1(A)]\mathbf{x}$ gives

$$[f_1(A)]^{-1} f_2(A)\mathbf{x} = \frac{f_2(\lambda)}{f_1(\lambda)}\mathbf{x}.$$

Similarly the eigenvalue of $f_2(A)[f_1(A)]^{-1}$ corresponding to the eigenvector $\mathbf{x}$ is $f_2(\lambda)/f_1(\lambda)$.

The second set of results concerns the eigenvalues of an order n tridiagonal matrix and forms the next theorem.

Theorem 1.5

The eigenvalues of the order n tridiagonal matrix

$$\begin{pmatrix} a & b & & & & \\ c & a & b & & & \\ & c & a & b & & \\ & & \ddots & \ddots & \ddots & \\ & & & c & a & b \\ & & & & c & a \end{pmatrix}$$

are

$$\lambda_s = a + 2\left[\sqrt{bc}\right]\cos\frac{s\pi}{n+1}, \quad s = 1(1)n \qquad (1.5.4)$$

where a, b and c may be real or complex. This class of matrices arises commonly in the study of stability of the finite difference processes, and a knowledge of its eigenvalues leads immediately into useful stability conditions.

Proof

Let λ represent an eigenvalue of A and $\mathbf{v}$ the corresponding eigenvector with components $v_1, v_2, \ldots, v_n$. Then the eigenvalue equation $A\mathbf{v} = \lambda\mathbf{v}$ gives

$$(a - \lambda)v_1 + bv_2 = 0$$
$$cv_1 + (a - \lambda)v_2 + bv_3 = 0$$
$$\vdots$$
$$cv_{j-1} + (a - \lambda)v_j + bv_{j+1} = 0$$
$$\vdots$$
$$cv_{n-1} + (a - \lambda)v_n = 0.$$

Now define $v_0 = v_{n+1} = 0$ and these n equations can be combined into one difference equation

$$cv_{j-1} + (a - \lambda)v_j + bv_{j+1} = 0, \quad j = 1, \ldots, n. \qquad (1.5.5)$$

The solution is of the form $v_j = Bm_1^j + Cm_2^j$ where B and C are arbitrary constants and m_1, m_2 are roots of the equation

$$C + (a - \lambda)m + bm^2 = 0. \qquad (1.5.6)$$

Hence the conditions

$$v_0 = v_{n+1} = 0$$

give

$$0 = B + C, \quad \text{and} \quad 0 = Bm_1^{n+1} + Cm_2^{n+1}$$

which implies

$$\left(\frac{m_1}{m_2}\right)^{n+1} = 1 = e^{i2s\pi}, \quad s = 1, 2, \dots, n,$$

or

$$\frac{m_1}{m_2} = e^{\frac{i2s\pi}{n+1}}.$$

From (1.5.6), $m_1 m_2 = c/b$ and $m_1 + m_2 = -(a - \lambda)/b$. Hence

$$m_1 = \left(\frac{c}{b}\right)^{\frac{1}{2}} e^{\frac{is\pi}{n+1}} \quad \text{and} \quad m_2 = \left(\frac{c}{b}\right)^{\frac{1}{2}} e^{\frac{-is\pi}{n+1}}$$

which gives

$$\lambda = a + b(m_1 + m_2) = a + b\left(\frac{c}{b}\right)^{\frac{1}{2}} \left(e^{\frac{is\pi}{n+1}} + e^{\frac{-is\pi}{n+1}}\right).$$

Hence the n eigenvalues are

$$\lambda_s = a + 2\sqrt{bc}\cos\frac{s\pi}{n+1}, \quad s = 1, 2, \dots, n \qquad (1.5.7)$$

as required.

The jth component of the corresponding eigenvector is

$$
\begin{aligned}
v_j &= Bm_1^j + Cm_2^j \\
&= B\left(\frac{c}{b}\right)^{\frac{j}{2}} \left(e^{\frac{ijs\pi}{n+1}} - e^{\frac{-ijs\pi}{n+1}}\right) \\
&= 2iB\left(\frac{c}{b}\right)^{\frac{j}{2}} \sin\left(\frac{js\pi}{n+1}\right).
\end{aligned}
$$

So the eigenvector $\mathbf{v}_s$ corresponding to λ_s is

$$
\begin{aligned}
\mathbf{v}_s^T = &\left[\left(\frac{c}{b}\right)^{\frac{1}{2}} \sin\frac{s\pi}{n+1}, \left(\frac{c}{b}\right) \sin\frac{2s\pi}{n+1}, \right. \\
&\left. \left(\frac{c}{b}\right)^{\frac{3}{2}} \sin\frac{3s\pi}{n+1}, \dots, \left(\frac{c}{b}\right)^{\frac{n}{2}} \sin\frac{ns\pi}{n+1}\right].
\end{aligned} \qquad (1.5.8)
$$

As an example consider the tridiagonal matrix

$$
\begin{pmatrix}
1 - 2r & r & & & & \\
r & 1 - 2r & r & & & \\
& r & 1 - 2r & r & & \\
& & \ddots & \ddots & \ddots & \\
& & & r & 1 - 2r & r \\
& & & & r & 1 - 2r
\end{pmatrix}
$$

of order $n - 1$ with

$$a = 1 - 2r, \quad b = r, \quad c = r.$$

Then the previous theorem tells us that the eigenvalues are

$$
\begin{aligned}
\lambda_s &= (1 - 2r) + 2r \left(\frac{r}{r}\right)^{\frac{1}{2}} \cos \frac{s\pi}{n} \\
&= 1 - 2r \left[1 - \cos \frac{s\pi}{n}\right] \\
&= 1 - 4r \sin^2 \frac{s\pi}{2n}.
\end{aligned}
$$

Many of the methods which arise in the solution of partial differential equations require the solution of a tridiagonal set of linear equations, and for this special case the usual elimination routine can be simplified. The algorithm which results is called the Thomas algorithm for tridiagonal systems, and is described below.

Suppose that it is required to solve

$$
\begin{pmatrix}
b_1 & -c_1 & & & & \\
-a_2 & b_2 & -c_2 & & & \\
& -a_3 & b_3 & -c_3 & & \\
& & \ddots & \ddots & \ddots & \\
& & & -a_{n-1} & b_{n-1} & -c_{n-1} \\
& & & & -a_n & b_n
\end{pmatrix}
\begin{pmatrix}
x_1 \\ x_2 \\ x_3 \\ \vdots \\ x_{n-1} \\ x_n
\end{pmatrix}
=
\begin{pmatrix}
d_1 \\ d_2 \\ d_3 \\ \vdots \\ d_{n-1} \\ d_n
\end{pmatrix}.
$$

The algorithm is based on Gauss elimination. In each column only one subdiagonal element is to be removed. In each equation b_i and d_i, $i = 2, \ldots, n$, change as a result of the elimination. Denote the quantities that replace b_i and d_i by α_i and s_i respectively. For convenience set $\alpha_1 = b_1$ and $s_1 = d_1$ then

$$
\begin{aligned}
\alpha_2 &= b_2 - \frac{c_1 a_2}{\alpha_1}, \quad s_2 = d_2 + \frac{s_1 a_2}{\alpha_1}, \\
\alpha_3 &= b_3 - \frac{c_2 a_3}{\alpha_2}, \quad s_3 = d_3 + \frac{s_2 a_3}{\alpha_2},
\end{aligned}
$$

etc.

In general

$$\alpha_i = b_i - \frac{c_{i-1} a_i}{\alpha_{i-1}}, \quad s_i = d_i + \frac{s_{i-1} a_i}{\alpha_{i-1}}. \tag{1.5.9}$$

Once the elimination is complete the x_i, $i = 1, \ldots, n$, are found recursively by back substitution.

The complete algorithm may be expressed as:

$$
\begin{aligned}
\alpha_1 &= b_1, \quad s_1 = d_1, \\
\alpha_i &= b_i - \frac{c_{i-1} a_i}{\alpha_{i-1}}, \quad s_i = d_i - \frac{s_{i-1} a_i}{\alpha_{i-1}}, \quad i = 2, \ldots, n, \\
x_n &= \frac{s_n}{\alpha_n}, \quad x_i = \frac{(s_i + c_i x_{i+1})}{\alpha_i}, \quad i = (n-2), \ldots, 1.
\end{aligned}
$$

Conditions for the applicability of the method are considered next.

We have not used partial pivoting and so we need to investigate the conditions for which the multipliers a_i/α_{i-1}, $i = 2,\ldots,n$, have magnitude not exceeding unity for stable forward elimination and c_i/α_i, $i = 2,\ldots,n-1$, have magnitude not exceeding unity for stable back substitution.

Suppose that $a_i > 0$, $b_i > 0$, $c_i > 0$ then,

(i) assuming that $b_i > a_{i+1} + c_{i-1}$, $i = 1,\ldots,n-1$, the forward elimination is stable; and

(ii) assuming that $b_i > a_i + c_i$, $i = 1,\ldots,n-1$, the back-substitution is stable.

The proof can be found in Smith (1978).

Some assorted exercises on these ideas are now presented.

EXERCISES

1.16 Use the characteristic polynomial directly to confirm that the eigen-values given in (1.5.4) are correct for $n = 2$.

1.17 Find the characteristic polynomial and hence the eigenvalues of the matrix

$$\begin{bmatrix} 4 & 1 & 0 \\ 2 & 4 & 1 \\ 0 & 2 & 4 \end{bmatrix}$$

and compare the result with the formula (1.5.4).

1.18 Use the Thomas algorithm to solve the tridiagonal set of equations

$$\begin{bmatrix} 4 & 1 & 0 \\ 2 & 4 & 1 \\ 0 & 2 & 4 \end{bmatrix} \mathbf{x} = \begin{bmatrix} 1 \\ 2 \\ 3 \end{bmatrix}.$$

1.19 By counting operations establish that Gaussian elimination requires the order of $n^3/3$ multiplication and division operations. This is a measure of the work load in the algorithm. The easiest way to establish this result is to code up the algorithm (which will be a useful tool for later anyway) and then use the formulae:

$$\sum_{i=1}^{n} i = \frac{n(n+1)}{2},$$

$$\sum_{i=1}^{n} i^2 = \frac{n(n+1)(2n+1)}{6},$$

$$\sum_{i=1}^{n} i^3 = \left[\frac{n(n+1)}{2}\right]^2.$$

What is the equivalent count for Thomas's algorithm?

1.20 Compare the work load in Thomas' algorithm with that for say m iterations of Gauss–Seidel. Given the convergence rate from the eigenvalues of the G matrix of (1.4.12), construct advice for prospective users on whether to use the Thomas algorithm or Gauss–Seidel.

1.21 Extend the Thomas algorithm to deal with upper Hessenberg matrices with the form

$$\begin{bmatrix} a_{11} & a_{12} & a_{13} & \cdots & \cdots & a_{1n} \\ b_2 & a_{22} & a_{23} & \cdots & \cdots & a_{2n} \\ 0 & b_3 & a_{33} & \cdots & \cdots & a_{3n} \\ \vdots & & & & & \vdots \\ \vdots & & & & & \vdots \\ 0 & 0 & 0 & \cdots & b_n & a_{nn} \end{bmatrix}$$

which is tridiagonal with non-zero elements in the top right-hand part of the matrix.

1.22 Extend Thomas's algorithm to quindiagonal matrices which have in general diagonal elements with two non-zero elements on either side in each row, except in the first two and last two rows which just have two non-zero elements on one side for the first row, and in addition one non-zero element on the opposite side in the second row.

1.6 Classification of Second Order Partial Differential Equations

Consider a general second order quasi-linear equation defined by the equation

$$Rr + Ss + Tt = W \tag{1.6.1}$$

where

$$p = \frac{\partial z}{\partial x}, \quad q = \frac{\partial z}{\partial y}, \quad r = \frac{\partial^2 z}{\partial x^2}, \quad s = \frac{\partial^2 z}{\partial x \partial y} \quad \text{and} \quad t = \frac{\partial^2 z}{\partial y^2} \tag{1.6.2}$$

with

$$R = R(x,y), \quad S = S(x,y), \quad T = T(x,y) \quad \text{and} \quad W = W(x,y,z,p,q). \tag{1.6.3}$$

Then the characteristic curves for this equation are defined as curves along which highest partial derivatives are not uniquely defined. In this case these derivatives are the second order derivatives r, s and t. The set of linear algebraic equations which these derivatives satisfy can be written down in terms of differentials, and the condition for this set of linear equations to have a non-unique solution will yield the equations of the characteristics, whose significance will then become more apparent. Hence the linear equations follow as $dz = pdx + qdy$ and also

$$\begin{aligned} dp &= rdx + sdy \\ dq &= sdx + tdy \end{aligned} \tag{1.6.4}$$

to give the linear equations

$$\left. \begin{aligned} Rr + Ss + Tt &= W \\ rdx + sdy &= dp \\ sdx + tdy &= dq \end{aligned} \right\} \tag{1.6.5}$$

and there will be no unique solution when

$$\begin{vmatrix} R & S & T \\ dx & dy & 0 \\ 0 & dx & dy \end{vmatrix} = 0 \tag{1.6.6}$$

which expands to give the differential equation

$$R \left(\frac{dy}{dx} \right)^2 - S \left(\frac{dy}{dx} \right) + T = 0. \tag{1.6.7}$$

But when the determinant in (1.6.6) is zero, the other determinants in Cramer's rule for the solution of (1.6.5) will also be zero, for we assume that (1.6.5) does not have a unique solution. Hence the condition

$$\begin{vmatrix} R & T & W \\ dx & 0 & dp \\ 0 & dy & dq \end{vmatrix} = 0 \tag{1.6.8}$$

also holds, and gives an equation which holds along a characteristic, namely

$$-Rdy\, dp - T\, dx\, dq + W\, dxdy = 0 \tag{1.6.9}$$

or

$$R \frac{dp}{dx} \frac{dy}{dx} + T \frac{dq}{dx} - W \frac{dy}{dx} = 0. \tag{1.6.10}$$

Returning now to (1.6.6), this equation is a quadratic in dy/dx and there are three possible cases which arise. If the roots are real the characteristics form two families of real curves. A partial differential equation resulting in real characteristics is said to be hyperbolic. The condition is that

$$S^2 - 4RT > 0. \tag{1.6.11}$$

The second case is when the roots are equal to give the parabolic case and the condition

$$S^2 - 4RT = 0, \tag{1.6.12}$$

and when the roots are complex the underlying equation is said to be elliptic with the condition

$$S^2 - 4RT < 0. \tag{1.6.13}$$

The importance of characteristics only becomes apparent at this stage. The first feature is the use of characteristics to classify equations. The methods that will be used subsequently are quite different from type to type. In the case of hyperbolic equations, the characteristics are real and are used directly in the solution. Characteristics also play a role in reducing equations to a standard or canonical form. Consider the operator

$$R\frac{\partial^2}{\partial x^2} + S\frac{\partial^2}{\partial x \partial y} + T\frac{\partial^2}{\partial y^2} \tag{1.6.14}$$

and put $\xi = \xi(x,y)$, $\eta = \eta(x,y)$ and $z = \zeta$ to see what a general change of variable yields. The result is the operator

$$A(\xi_x, \xi_y)\frac{\partial^2 \zeta}{\partial \xi^2} + 2B(\xi_x, \xi_y, \eta_x, \eta_y)\frac{\partial^2 \zeta}{\partial \xi \partial \eta}$$

$$+A(\eta_x, \eta_y)\frac{\partial^2 \zeta}{\partial \eta^2} = F(\xi, \eta, \zeta, \zeta_\zeta, \zeta_\eta) \tag{1.6.15}$$

where

$$A(u, \nu) = Ru^2 + Su\nu + T\nu^2 \tag{1.6.16}$$

and

$$B(u_1, \nu_1, u_2, \nu_2) = Ru_1 u_2 + \frac{1}{2}S(u_1 \nu_2 + u_2 \nu_1) + T\nu_2 \nu_2. \tag{1.6.17}$$

The question is now asked for what ξ and η do we get the simplest form? Certainly if ξ and η can be found to make the coefficients A equal to zero, then a simplified form will result. However the condition that A should be zero is a partial differential equation of first order which can be solved analytically (Sneddon, 1957). Different cases arise in the three classifications. In the hyberbolic case when $S^2 - 4RT > 0$, let $R\alpha^2 + S\alpha + T = 0$ have roots λ_1 and λ_2 then $\xi = f_1(x,y)$ and $\eta = f_2(x,y)$ where $f_1(x,y)$ and $f_2(x,y)$ are the solutions of the two factors in the related ordinary differential equations

$$\left[\frac{dy}{dx} + \lambda_1(x,y)\right]\left[\frac{dy}{dx} + \lambda_2(x,y)\right] = 0. \tag{1.6.18}$$

Hence the required transformations are precisely the defining functions of the characteristic curves. It follows that with this change of variable the partial differential equation becomes

$$\frac{\partial^2 \zeta}{\partial \eta \partial \xi} = \phi(\xi, \eta, \zeta, \zeta_\zeta, \zeta_\eta) \tag{1.6.19}$$

which is the canonical form for the hyperbolic case.

In the parabolic case, $S^2 - 4RT = 0$, there is now only one root, and any independent function is used for the other variable in the transformation. Hence $A(\xi_x, \xi_y) = 0$, but it is easy to show in general that

$$A(\xi_x, \xi_y)A(\eta_x, \eta_y) - B^2(\xi_x, \xi_y, \eta_x, \eta_y) = (4RT - S^2)(\xi_x \eta_y - \xi_y \eta_x)^2$$

and therefore as $S^2 = 4RT$, we must have $B(\xi_x, \xi_y, \eta_x, \eta_y) = 0$ and $A(\eta_x, \eta_y) \neq 0$ as η is an independent function of x and y. Hence when $S^2 = 4RT$, the transformation $\xi = f_1(x, y)$ and $\eta =$ any independent function yields

$$\frac{\partial^2 \zeta}{\partial \eta^2} = \phi_1(\xi, \eta, \zeta, \zeta_\zeta, \zeta_\eta) \tag{1.6.20}$$

which is the canonical form for a parabolic equation.

In the elliptic case there are again two sets of characteristics but they are now complex. Writing $\xi = \alpha + i\beta$ and $\eta = \alpha - i\beta$ gives the real form

$$\frac{\partial^2 \zeta}{\partial \xi \partial \nu} = \frac{1}{4}\left(\frac{\partial^2 \zeta}{\partial \alpha^2} + \frac{\partial^2 \zeta}{\partial \beta^2}\right) \tag{1.6.21}$$

and hence the elliptic canonical form

$$\frac{\partial^2 \zeta}{\partial \alpha^2} + \frac{\partial^2 \zeta}{\partial \beta^2} = \psi(\alpha, \beta, \zeta, \zeta_\alpha, \zeta_\beta). \tag{1.6.22}$$

Note that Laplace's equation is in canonical form as is the heat equation, but the wave equation is not. As an example of reduction to canonical form consider the linear second order partial differential equation

$$\frac{\partial^2 u}{\partial x^2} + 2\frac{\partial^2 u}{\partial x \partial y} + \frac{\partial^2 u}{\partial y^2} + c^2\frac{\partial u}{\partial y} = 0. \tag{1.6.23}$$

Then the equation of the characteristic curves is

$$\left(\frac{dy}{dx}\right)^2 - 2\frac{dy}{dx} + 1 = 0 \tag{1.6.24}$$

or factorising

$$\left(\frac{dy}{dx} - 1\right)^2 = 0. \tag{1.6.25}$$

Therefore the transformation for the canonical form is:

$$\left.\begin{array}{ccc} p & = & x - y \\ q & = & x \end{array}\right\} \tag{1.6.26}$$

and the required partial derivatives are:

$$\frac{\partial^2 u}{\partial x^2} = \frac{\partial^2 u}{\partial p^2}\left(\frac{\partial p}{\partial x}\right)^2 + \frac{\partial^2 u}{\partial q^2}\left(\frac{\partial q}{\partial x}\right)^2 \tag{1.6.27}$$

and

$$\frac{\partial^2 u}{\partial x \partial y} = \frac{\partial^2 u}{\partial p^2}\left(\frac{\partial p}{\partial y}\frac{\partial p}{\partial x}\right) \tag{1.6.28}$$

which yields the reduced form

$$\frac{\partial^2 u}{\partial x^2} + 2\frac{\partial^2 u}{\partial x \partial y} + \frac{\partial^2 u}{\partial y^2} = \frac{\partial^2 u}{\partial p^2} + \frac{\partial^2 u}{\partial q^2} - 2\frac{\partial^2 u}{\partial p^2} + \frac{\partial^2 u}{\partial p^2}$$

$$= \frac{\partial^2 u}{\partial q^2} \tag{1.6.29}$$

with the transformed equation being

$$\frac{1}{c^2}\frac{\partial^2 u}{\partial q^2} = \frac{\partial u}{\partial p}. \tag{1.6.30}$$

From a numerical point of view, the canonical forms reduce the number of different types of equation for which solutions need to be found. Effectively effort can be concentrated on the canonical forms alone, though this is not always the best strategy, and in this spirit the parabolic type will now be considered in detail in the next chapter. Before considering this work the reader may wish to pursue some of the ideas of the previous section in the following exercises.

EXERCISES

1.23 Classify the following partial differential equations as parabolic, elliptic or hyperbolic:

(a) $\dfrac{\partial^2 \phi}{\partial x^2} + \dfrac{\partial^2 \phi}{\partial x \partial y} + \dfrac{\partial^2 \phi}{\partial y^2} = 0$

(b) $\dfrac{\partial^2 \phi}{\partial t^2} - \dfrac{\partial^2 \phi}{\partial x^2} + \dfrac{\partial \phi}{\partial x} = 0$

(c) $\dfrac{\partial \phi}{\partial t} - \dfrac{\partial^2 \phi}{\partial x^2} - \dfrac{\partial \phi}{\partial x} = 0$

(d) $\dfrac{\partial^2 \phi}{\partial x^2} + x\dfrac{\partial^2 \phi}{\partial y^2} = 0.$

1.24 Find the regions of parabolicity, ellipticity and hyperbolicity for the partial differential equation:

$$\frac{\partial^2 u}{\partial x^2} + 3x^2 y^2 \frac{\partial^2 u}{\partial x \partial y} + (x+y)\frac{\partial^2 u}{\partial y^2} = u$$

and sketch the resulting regions in the (x, y) plane.

1.25 Find the analytic form of the characteristic curves for the partial differential equation

$$\frac{\partial^2 u}{\partial x^2} + 2\left(x + \frac{1}{y}\right)\frac{\partial^2 u}{\partial x \partial y} + \frac{4x}{y}\frac{\partial^2 u}{\partial y^2} = xy$$

and hence categorise the equation.

1.26 Reduce the equation

$$\frac{\partial^2 z}{\partial x^2} - 6\frac{\partial^2 z}{\partial x \partial y} + 9\frac{\partial^2 z}{\partial y^2} = \frac{\partial z}{\partial y}$$

to canonical form.

1.27 Reduce the equation

$$\frac{\partial^2 z}{\partial x^2} + 3\frac{\partial^2 z}{\partial x \partial y} + \frac{\partial^2 z}{\partial y^2} = 0$$

to canonical form, and hence find the general analytic solution.

1.28 Reduce the equation

$$\frac{\partial^2 z}{\partial x^2} + 2\frac{\partial^2 z}{\partial x \partial y} + 3\frac{\partial^2 z}{\partial y^2} = z$$

to canonical form. Make a further transformation to obtain a real canonical form.

2

Finite Differences and Parabolic Equations

2.1 Finite Difference Approximations to Derivatives

The first part of the chapter is concerned with finding finite difference approximations to partial derivatives with a view to replacing the full partial differential equation by a difference representation which then allows a numerical approach to be pursued. The result is that the partial derivatives are replaced by relationships between the function values at nodal points of some grid system. Hence the partial differential equations are approximated by a set of algebraic equations, for the function values at the nodal points. In order to perform this discretisation, expressions for such terms as

$$\frac{\partial \phi}{\partial x} \quad \text{and} \quad \frac{\partial^2 \phi}{\partial x^2} \tag{2.1.1}$$

at typical grid points are required in terms of the ϕ values at neighbouring grid points.

Consider part of a typical grid system as shown in Figure 2.1. A mesh with a constant increment h in the x direction and constant increment k in the y direction is chosen. Using Taylor series in one variable, keeping the other

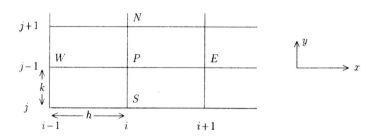

Fig. 2.1.

variable fixed, gives

$$\begin{aligned}
\phi_E = {}& \phi_P + h\left(\frac{\partial\phi}{\partial x}\right)_P + \frac{h^2}{2!}\left(\frac{\partial^2\phi}{\partial x^2}\right)_P \\
& + \frac{h^3}{3!}\left(\frac{\partial^3\phi}{\partial x^3}\right)_P + \frac{h^4}{4!}\left(\frac{\partial^4\phi}{\partial x^4}\right)_P + \cdots
\end{aligned} \tag{2.1.2}$$

and

$$\begin{aligned}
\phi_W = {}& \phi_P - h\left(\frac{\partial\phi}{\partial x}\right)_P + \frac{h^2}{2!}\left(\frac{\partial^2\phi}{\partial x^2}\right)_P \\
& - \frac{h^3}{3!}\left(\frac{\partial^3\phi}{\partial x^3}\right)_P + \frac{h^4}{4!}\left(\frac{\partial^4\phi}{\partial x^4}\right)_P + \cdots
\end{aligned} \tag{2.1.3}$$

These formulae give a range of difference approximations whose names arise from the archaic finite difference operators used in hand calculations (Hildebrand, 1974). Hence (2.1.2) gives the forward difference approximation:

$$\left(\frac{\partial\phi}{\partial x}\right)_P = \frac{\phi_E - \phi_P}{h} + O(h). \tag{2.1.4}$$

The remainder term of $O(h)$ is called the local truncation error and is calculated explicitly in the following exercises. This important concept is discussed fully in Section 2.3. Similarly (2.1.3) gives the backward difference approximation

$$\left(\frac{\partial\phi}{\partial x}\right)_P = \frac{\phi_P - \phi_W}{h} + O(h) \tag{2.1.5}$$

and (2.1.2) and (2.1.3) together give the central difference approximation

$$\phi_E - \phi_W = 2h\left(\frac{\partial\phi}{\partial x}\right)_P + 2\frac{h^3}{3!}\left(\frac{\partial^3\phi}{\partial x^3}\right)_P + O(h^5)$$

which leads to

$$\left(\frac{\partial\phi}{\partial x}\right)_P = \frac{\phi_E - \phi_W}{2h} + O(h^2). \tag{2.1.6}$$

It is equally feasible to generate a central difference approximation to $\left(\partial^2\phi/\partial x^2\right)_P$ to get

$$\phi_E + \phi_W = 2\phi_P + h^2\left(\frac{\partial^2\phi}{\partial x^2}\right)_P + 2\frac{h^4}{4!}\left(\frac{\partial^4\phi}{\partial x^4}\right)_P + O(h^6)$$

which gives

$$\left(\frac{\partial^2\phi}{\partial x^2}\right)_P = \frac{\phi_E + \phi_W - 2\phi_P}{h^2} + O(h^2). \tag{2.1.7}$$

Clearly similar approximations can be obtained for

$$\left(\frac{\partial^2\phi}{\partial y^2}\right)_P \quad \text{and} \quad \left(\frac{\partial\phi}{\partial y}\right)_P.$$

Using the mesh line number notation in Figure 2.1, the central difference approximations may be written as

$$\left(\frac{\partial\phi}{\partial x}\right)_{i,j} = \frac{\phi_{i+1,j} - \phi_{i-1,j}}{2h} + O(h^2) \tag{2.1.8}$$

$$\left(\frac{\partial^2\phi}{\partial x^2}\right)_{i,j} = \frac{\phi_{i+1,j} + \phi_{i-1,j} - 2\phi_{i,j}}{h^2} + O(h^2). \tag{2.1.9}$$

At interior points we can approximate derivatives to $O(h^2)$, but at boundaries, however, this is seldom the case. For Dirichlet boundary conditions, with ϕ specified, there is no problem, but for Neumann boundary conditions, normal derivatives of ϕ are given and the representation of the partial differential equations at the boundary is required.

The finite difference forms arising in dealing with derivative boundary conditions such as at the point P in Figure 2.2 are now considered.

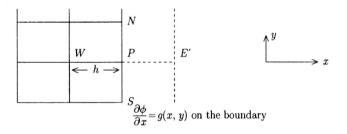

$$\frac{\partial\phi}{\partial x} = g(x, y) \text{ on the boundary}$$

Fig. 2.2.

One way to deal with this is to use a backward difference formula such as

$$\begin{aligned}
\phi_W &= \phi_P - h\left(\frac{\partial\phi}{\partial x}\right)_P + \frac{h^2}{2!}\left(\frac{\partial^2\phi}{\partial x^2}\right)_P - \frac{h^3}{3!}\left(\frac{\partial^3\phi}{\partial x^3}\right)_P + \cdots \\
&= \phi_P - hg_P + \frac{h^2}{2!}\left(\frac{\partial^2\phi}{\partial x^2}\right)_P + O(h^3)
\end{aligned}$$

to yield

$$\left(\frac{\partial^2 \phi}{\partial x^2}\right)_P = \frac{2(\phi_W - \phi_P + hg_P)}{h^2} + O(h). \tag{2.1.10}$$

A technique that is also used is the 'fictitious' point idea. Greater accuracy is sometimes claimed for this method, but in fact the same equation is obtained (see (2.1.14) below). Hence the fictitious point E' is introduced to give

$$\left(\frac{\partial^2 \phi}{\partial x^2}\right)_P = \frac{\phi_{E'} + \phi_W - 2\phi_P}{h^2} + O(h^2) \tag{2.1.11}$$

while the boundary condition gives

$$\frac{\phi_{E'} - \phi_W}{2h} = g_P + O(h^2) \tag{2.1.12}$$

using the central difference, and so

$$\phi_{E'} = 2hg_P + \phi_W + O(h^3). \tag{2.1.13}$$

Then

$$\begin{aligned}
\left(\frac{\partial \phi^2}{\partial x^2}\right)_P &= \frac{2hg_P + \phi_W + O(h^3) + \phi_W - 2\phi_P}{h^2} + O(h^2) \\
&= \frac{2(\phi_W - \phi_P + hg_P)}{h^2} + O(h). \tag{2.1.14}
\end{aligned}$$

Thus, one-sided difference approximations for derivative boundary conditions handled properly give the same result as implementing the ficticious point idea. Note that for the boundary condition where the normal derivative is zero and the boundary is also a plane or axis of symmetry, the second derivative can be approximated to $O(h^2)$.

The point where a mesh line meets the boundary may not always be a natural node of the mesh system. Such a situation is illustrated in Figure 2.3.

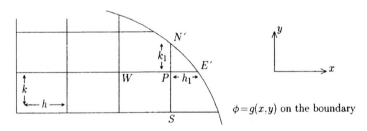

Fig. 2.3.

In this case the resulting finite difference forms are modified according to the following derivation:

$$\phi_W = \phi_P - h\left(\frac{\partial \phi}{\partial x}\right)_P + \frac{h^2}{2!}\left(\frac{\partial^2 \phi}{\partial x^2}\right)_P - \frac{h^3}{3!}\left(\frac{\partial^3 \phi}{\partial x^3}\right)_P + \cdots \tag{2.1.15}$$

and

$$\phi_{E'} = \phi_P + h_1 \left(\frac{\partial \phi}{\partial x}\right)_P + \frac{h_1^2}{2!}\left(\frac{\partial^2 \phi}{\partial x^2}\right)_P + \frac{h_1^3}{3!}\left(\frac{\partial^3 \phi}{\partial x^3}\right)_P + \cdots \qquad (2.1.16)$$

to give

$$h_1 \phi_W + h \phi_{E'} = (h + h_1)\phi_P + \frac{h^2 h_1 + h_1^2 h}{2!}\left(\frac{\partial^2 \phi}{\partial x^2}\right)_P + O(h^4).$$

(Note that at worst h_1 is of order h). Hence the irregular boundary formulae are

$$\left(\frac{\partial^2 \phi}{\partial x^2}\right)_P = \frac{2(h_1 \phi_W + h \phi_{E'} - (h + h_1)\phi_P)}{h_1 h(h_1 + h)} + O(h) \qquad (2.1.17)$$

and similarly

$$\left(\frac{\partial^2 \phi}{\partial y^2}\right)_P = \frac{2(k_1 \phi_S + k \phi_{N'} - (k + k_1)\phi_P)}{k_1 k(k_1 + k)} + O(k). \qquad (2.1.18)$$

Also for the first order derivative:

$$\left(\frac{\partial \phi}{\partial x}\right)_P = \frac{(\phi_{E'} h^2 - \phi_W h_1^2 + (h_1^2 - h^2)\phi_P)}{h_1 h(h_1 + h)} + O(h^2). \qquad (2.1.19)$$

For Neumann conditions on inconvenient boundaries the problem becomes rather more complicated and will be dealt with later.

The following exercises can now be attempted, based on this section.

EXERCISES

2.1 Use the Taylor series approach to find the local truncation error, E, when the approximation:

$$\frac{\partial \phi}{\partial x} = \frac{\phi_{i+1,j} - \phi_{i,j}}{h} + E$$

is made in the form $E = Ah^n$ where A and n should be found.

2.2 Repeat the procedure above for the approximation

$$\frac{\partial^2 \phi}{\partial x^2} = \frac{\phi_{i+1,j} - 2\phi_{i,j} + \phi_{i-1,j}}{h^2} + E.$$

2.3 Use the Taylor series to obtain a set of linear equations for the constants A, B, C, D and E in the expansion:

$$\frac{\partial^2 \phi}{\partial x^2} = A\phi_{i-2,j} + B\phi_{i-1,j} + C\phi_{i+1,j} + D\phi_{i+2,j} + E.$$

2.4 Construct a mesh in the (x, y) plane with x increments of h and y increments of k. At the mesh point (i, j) obtain finite difference expressions for:

(i) $\partial^2 \phi / \partial x \partial y$

(ii) $(\partial^2 \phi / \partial x^2) + x(\partial^2 \phi / \partial y^2)$

correct to $O(\max(h^2, k^2))$.

2.5 Using Taylor series expansions about x_0 as shown in Figure 2.4.

Fig. 2.4.

obtain finite difference expressions for

$$\left(\frac{\partial \phi}{\partial x} \right)_{x_0} \quad \text{and} \quad \left(\frac{\partial \phi^2}{\partial x^2} \right)_{x_0}$$

to $O(\max(h_1^2, h_2^2))$.

(Your formulae may contain terms involving $\partial^3 \phi / \partial x^3$: leave these in your expressions.)

Show that for $h_2 = -h_1$, the usual formulae for a regular mesh result.

2.2 Parabolic Equations

The finite difference forms from the last paragraph will now be employed to generate numerical methods for parabolic equations, characterised by the heat equation. The canonical form for parabolic equations is from Chapter 1

$$\frac{\partial U}{\partial T} = K \frac{\partial^2 U}{\partial X^2} \qquad (2.2.1)$$

with K a constant, and U is required in the interval $0 \le X \le L$. Let U_0 be some particular value of U (such as the minimum or maximum value at $t = 0$), then let

$$x = \frac{X}{L}, \quad u = \frac{U}{U_0}, \quad t = \frac{T}{T_0}$$

$(T_0$ to be chosen), then

$$\frac{\partial U}{\partial T} = \frac{U_0}{T_0}\frac{\partial u}{\partial t}$$

since

$$\frac{\partial U}{\partial T} = \frac{\partial U}{\partial t}\frac{dt}{dT} = \frac{\partial(U_0 u)}{\partial t}\frac{1}{T_0}$$
$$= \frac{U_0}{T_0}\frac{\partial u}{\partial t}$$

and

$$\frac{\partial^2 U}{\partial X^2} = \frac{U_0}{L^2}\frac{\partial^2 u}{\partial x^2}.$$

The equation becomes

$$\frac{U_0}{T_0}\frac{\partial u}{\partial t} = K\frac{U_0}{L^2}\frac{\partial^2 u}{\partial x^2}$$

or

$$\frac{L^2}{KT_0}\frac{\partial u}{\partial t} = \frac{\partial^2 u}{\partial x^2}.$$

Hence if we choose

$$T_0 = \frac{L^2}{K}$$

the equation is

$$\frac{\partial u}{\partial t} = \frac{\partial^2 u}{\partial x^2} \quad \text{where} \quad 0 \le x \le 1.$$

This non-dimensional equation may be used to typify the numerical processes. The first of these processes is the *explicit* method which yields a very simple technique for parabolic equations. The formula expresses one unknown nodal value directly in terms of known nodal values and is illustrated in an example. Consider the problem

$$\frac{\partial \phi}{\partial t} = \frac{\partial^2 \phi}{\partial x^2} \tag{2.2.2}$$

subject to

(i) $\phi = 0$ at $x = 0$ and $x = 1$, $\forall t \in [0,\infty)$

(ii) $\phi = 1 - 4(x - 1/2)^2$ for $t = 0$, $\forall x \in [0,1]$.

This problem and the intended grid are illustrated in Figure 2.5.

A grid time increment k, and space increment h are used. A scheme is required to find a single value at the next time level in terms of known values at an earlier time level. Hence a forward difference approximation for $\partial\phi/\partial t$ and central difference approximation for $\partial^2\phi/\partial x^2$ are used. Because of the different orders for the truncation error, it appears reasonable to set $k = O(h^2)$, and

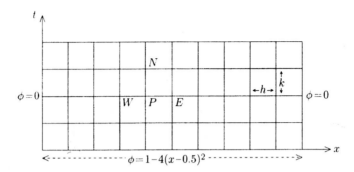

Fig. 2.5.

this condition turns out to be closely related to that for stability in the later analysis. Using these finite difference approximations gives

$$\frac{\phi_N - \phi_P}{k} = \frac{\phi_W - 2\phi_P + \phi_E}{h^2} \qquad (2.2.3)$$

or

$$\phi_N = \phi_P + r(\phi_W - 2\phi_P + \phi_E) \quad \text{where} \quad r = \frac{k}{h^2} \qquad (2.2.4)$$

which gives in mesh line notation

$$\phi_{i,j+1} = \phi_{i,j} + r(\phi_{i-1,j} - 2\phi_{i,j} + \phi_{i+1,j}). \qquad (2.2.5)$$

Hence using (2.2.5) the ϕ values for line $j+1$ can be obtained using the values in line j.

As a specific example, let $h = \frac{1}{10}$ and $k = \frac{1}{1000}$, so that $r = k/h^2 = \frac{1}{10}$ giving

$$\phi_N = \phi_P + \frac{1}{10}[\phi_W - 2\phi_P + \phi_E] = \frac{1}{10}[\phi_W + 8\phi_p + \phi_E].$$

Note that the problem (initial conditions, boundary conditions and finite difference formula) is symmetric and hence we need only calculate the solution for $0 \leq x \leq \frac{1}{2}$. The results are shown in Table 2.1 at a selection of points.

The analytic solution of the partial differential equation using the method of separation of the variables from Evans, Blackledge and Yardley (1999, Chapter 2) is

$$\phi(x,t) = \frac{16}{\pi^3} \sum_{n=1}^{\infty} \frac{1}{n^3}(1 - \cos n\pi) \sin n\pi x \exp(-n^2\pi^2 t) \qquad (2.2.6)$$

which is shown in Table 2.2 for the same points as in Table 2.1, enabling easy comparison.

The accuracy of the computed values follows closely the exact values across the range, despite this being quite a crude computation. In many ways this

Table 2.1.

j	t	$i=0$ $x=0$	$i=1$ 0.1	$i=2$ 0.2	$i=3$ 0.3	$i=4$ 0.4	$i=5$ 0.5
0	0.000	0.0	0.3600	0.6400	0.8400	0.9600	1.0000
1	0.001	0.0	0.3520	0.6320	0.8320	0.9520	0.9920
2	0.002	0.0	0.3448	0.6240	0.8240	0.9440	0.9840
3	0.003	0.0	0.3382	0.6161	0.8160	0.9360	0.9760
4	0.004	0.0	0.3322	0.6083	0.8080	0.9280	0.9680
5	0.005	0.0	0.3266	0.6007	0.8000	0.9200	0.9600
	$\vdots$			$\vdots$			
10	0.01	0.0	0.3030	0.5650	0.7608	0.8801	0.9200

Table 2.2.

j	t	$i=0$ $x=0$	$i=1$ 0.1	$i=2$ 0.2	$i=3$ 0.3	$i=4$ 0.4	$i=5$ 0.5
0	0.000	0.0	0.3600	0.6400	0.8400	0.9600	1.0000
1	0.001	0.0	0.3521	0.6339	0.8320	0.9513	0.9920
2	0.002	0.0	0.3446	0.6252	0.8240	0.9435	0.9840
3	0.003	0.0	0.3379	0.6168	0.8160	0.9357	0.9760
4	0.004	0.0	0.3317	0.6086	0.8080	0.9278	0.9689
5	0.005	0.0	0.3260	0.6007	0.8000	0.9199	0.9600
	$\vdots$			$\vdots$			
10	0.01	0.0	0.3029	0.5646	0.7606	0.8805	0.9200

problem is ideal with continuous boundary conditions. The accuracy will be less good when there are discontinuities in the initial conditions or their derivatives and this will be seen in the exercises following this section. It can be proved analytically that when the boundary values are constant the effect of discontinuities in initial values and initial derivatives upon the solution of a parabolic equation decreases as t increases.

It is of interest now to increase r towards and beyond its limit and to see the effect. Hence if $h = \frac{1}{10}$ and $k = \frac{5}{1000}$ to give $r = k/h^2 = 0.5$ then

$$\phi_N = \frac{1}{2}(\phi_W + \phi_E)$$

with the numerical results shown in Table 2.3 and the corresponding analytic values in Table 2.4.

The value of r is now approaching the upper limit of the stability criterion $r \leq \frac{1}{2}$ which will be derived in Section 2.6, and the accuracy has fallen off. It is to be expected that if the grid size is made finer while keeping r at 0.5, then ultimately convergence will occur. In the extreme case of $h = \frac{1}{10}$ and $k = \frac{1}{100}$

Table 2.3.

j	t	$i=0$ $x=0$	$i=1$ 0.1	$i=2$ 0.2	$i=3$ 0.3	$i=4$ 0.4	$i=5$ 0.5
0	0.000	0.0	0.3600	0.6400	0.8400	0.9600	1.0000
1	0.005	0.0	0.3200	0.6000	0.8000	0.9200	0.9600
2	0.010	0.0	0.3000	0.5600	0.7600	0.8800	0.9200
3	0.015	0.0	0.2800	0.5300	0.7200	0.8400	0.8800
4	0.020	0.0	0.2650	0.5000	0.6850	0.8000	0.8400
5	0.025	0.0	0.2500	0.4750	0.6500	0.7625	0.8000
⋮	⋮			⋮			
10	0.05	0.0	0.1934	0.3672	0.5059	0.5937	0.6250

Table 2.4.

j	t	$i=0$ $x=0$	$i=1$ 0.1	$i=2$ 0.2	$i=3$ 0.3	$i=4$ 0.4	$i=5$ 0.5
0	0.000	0.0	0.3600	0.6400	0.8400	0.9600	1.0000
1	0.005	0.0	0.3260	0.6007	0.8002	0.9199	0.9600
2	0.010	0.0	0.3024	0.5646	0.7606	0.8801	0.9200
3	0.015	0.0	0.2834	0.5327	0.7230	0.8405	0.8802
4	0.020	0.0	0.2671	0.5041	0.6873	0.8019	0.8408
5	0.025	0.0	0.2526	0.4779	0.6536	0.7645	0.8023
⋮	⋮			⋮			
10	0.05	0.0	0.1951	0.3708	0.5099	0.5990	0.6296

so making $r = 1$, then the regime is well in the unstable region and

$$\phi_N = \phi_W - \phi_P + \phi_E$$

to give the expected unstable values in Table 2.5.

Table 2.5.

j	t	$i=0$ $x=0$	$i=1$ 0.1	$i=2$ 0.2	$i=3$ 0.3	$i=4$ 0.4	$i=5$ 0.5
0	0.000	0.0	0.3600	0.6400	0.8400	0.9600	1.0000
1	0.01	0.0	0.2800	0.5600	0.7600	0.8800	0.9200
2	0.02	0.0	0.2800	0.4800	0.6800	0.8000	0.8400
3	0.03	0.0	0.2000	0.4800	0.6000	0.7200	0.7600
4	0.04	0.0	0.2800	0.3200	0.6000	0.6400	0.6800
5	0.05	0.0	0.0400	0.5600	0.3600	0.6400	0.6000
6	0.06	0.0	0.5200	-0.1600	0.8400	0.3200	0.6800

These effects are shown graphically in Figure 2.6, where the results from the worked example are illustrated. The continuous lines are the analytic solutions for $t = 0.01$, $t = 0.04$ and $t = 0.06$. The points marked with a dot which lie on the analytic lines to graphical accuracy are those from the explicit method with $r = 0.1$ and an extension of Table 2.1. To see the effect of instability, the same computations are carried out with $r = 1$ to give Table 2.5 and the points marked by crosses and circles. Crosses are used for the time steps 0.01 and 0.04 and circles for 0.06 where the points are so wild it was necessary to highlight them in the diagram. Using a finer grid with the same r will not improve the situation. This is instability and not a truncation effect which can be reduced by reducing h and k, while keeping r fixed.

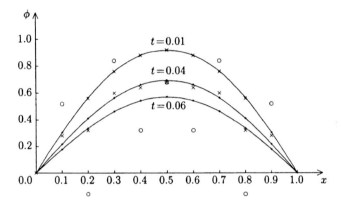

Fig. 2.6.

In the following sections, the theoretical work is developed and the limits on r for stability are established. In the meantime, the following exercises can be attempted using the explicit method.

EXERCISES

2.6 Apply the explicit method to the equation

$$\frac{\partial \phi}{\partial t} = D \frac{\partial^2 \phi}{\partial \phi^2}$$

with $D = 0.1$ and boundary conditions $\phi = 2$ for $0 < x < 20$ for $t = 0$, and for $t \geq 0$, $\phi = 0$ on $x = 0$ and $\phi = 10$ on $x = 20$. Use $h = 4.0$ with $r = 0.25$ and complete four time steps.

2.7 Set up a finite difference approximation for the parabolic problem

$$\frac{\partial \phi}{\partial t} = \frac{\partial}{\partial x}\left((1 + x^2)\frac{\partial \phi}{\partial x}\right), \qquad -1 \leq x \leq 1 \quad \text{and} \quad t \geq 0,$$

where
$$\phi(x,0) = 1000 - |1000x|$$
and
$$\phi(-1,t) = \phi(1,t) = 0 \quad \text{for} \quad t \ge 0$$

Take $h = 0.4$ and $k = 0.04$ and take advantage of symmetry to integrate to $t = 0.12$.

2.8 Consider the partial differential equation
$$\frac{\partial \phi}{\partial t} = \frac{\partial^2 \phi}{\partial x^2}$$

with
$$\phi = 0 \quad \text{at} \quad x = 0 \quad \text{and} \quad x = 1 \quad \text{for} \quad t > 0$$
and
$$\phi = \begin{cases} 2x & 0 \le x \le 1/2 \\ 2(1-x) & 1/2 \le x \le 1 \end{cases} \quad \text{for} \quad t = 0.$$

By taking $h = 0.1$ and $r = 0.5$, find ϕ at the first time step using an explicit method.

2.9 It is required to solve
$$\frac{\partial \phi}{\partial t} = \frac{\partial^2 \phi}{\partial x^2} + \frac{\partial \phi}{\partial x}$$

in the region bounded by
$$\phi = 0 \quad \text{at} \quad x = 0 \quad \text{and} \quad x = 1 \quad \text{for} \quad t \ge 0$$
and
$$\phi = \begin{cases} x & 0 \le x \le 1/2 \\ 1-x & 1/2 \le x \le 1 \end{cases} \quad \text{for} \quad t = 0.$$

Formulate an explicit finite difference scheme, using central differences in the x direction. Hence by taking $h = 0.1$ (and an appropriate k), obtain a solution at $t = 0.02$ using an explicit scheme.

2.10 The function u satisfies the equation
$$\frac{\partial u}{\partial t} = \frac{\partial^2 u}{\partial x^2} - \frac{1}{x}\frac{\partial u}{\partial x}$$

for $0 < x < 1$ with the initial condition $u = 1 - x^2$ when $t = 0$ and $0 \le x \le 1$ and boundary conditions
$$\frac{\partial u}{\partial x} = 0 \quad \text{at} \quad x = 0 \quad \text{for} \quad t > 0$$

and $u = 0$ at $x = 1$ and $t > 0$.

For $h = 0.1$ and $k = 0.001$, calculate a finite difference solution by an explicit method at the points $(0, 0.001)$, $(0.1, 0.001)$ and $(0.9, 0.001)$.

2.11 Formulate an explicit finite difference scheme to solve

$$\frac{\partial \phi}{\partial t} = \frac{\partial^2 \phi}{\partial x^2} - \phi$$

subject to $\phi = x^2$ at $t = 0$ and $\phi = 1$ at $x = 0$ for $t \geq 0$ and

$$\frac{\partial \phi}{\partial x} = 1 - \phi \quad \text{at} \quad x = 1$$

and complete the first few steps of the solution, using $h = 0.1$ and $k = 0.001$.

2.3 Local Truncation Error

In analysing the finite difference approximations used in the last paragraph, the first requirement is to know what errors are involved in using a finite difference approximation. Suppose we approximate the equation

$$\frac{\partial u}{\partial t} - \mathcal{L}u - f = 0 \; (= L(u) \quad \text{say})$$

for $\mathbf{x} \in \mathcal{R}^s$ and where $\mathcal{L}$ is a linear differential operator in the space variables $\mathbf{x}$. In the simple cases considered so far $\mathcal{L} \equiv \partial^2/\partial x^2$. Let $F(\phi) = 0$ represent the difference equation at the (i, j) mesh point in such a way that F approximates $L(u)$, and where ϕ is the exact solution of the difference equation. If ϕ is replaced by Φ at the mesh points of the difference equation where Φ is the exact solution of the partial differential equation, then if $F(\Phi) = O(h^p)$ for every initial condition, the method is said to be of order p. The truncation error $T_{ij}(\psi)$ at the grid point (ih, jk) is defined by $T_{ij}(\psi) = F(\psi_{ij}) - L(\psi_{ij})$. Putting $\psi = \Phi$, and using $L(\Phi) = 0$ gives $T_{ij}(\Phi) = F(\Phi_{ij})$ and the value of $F(\Phi_{ij})$ is called the local truncation error $\tau_{i,j}$ at the (i, j) mesh point.

As an example consider calculating the order of the local truncation error of the above explicit difference approximation to

$$\frac{\partial \Phi}{\partial t} - \frac{\partial^2 \Phi}{\partial x^2} = 0 \tag{2.3.1}$$

at the point (ih, jk). Then

$$F(\phi_{i,j}) = \frac{\phi_{i,j+1} - \phi_{i,j}}{k} - \frac{\phi_{i-1,j} - 2\phi_{i,j} + \phi_{i+1,j}}{h^2} = 0 \tag{2.3.2}$$

and

$$\tau_{i,j} = F(\Phi_{i,j}) = \frac{\Phi_{i,j+1} - \Phi_{i,j}}{k} - \frac{\Phi_{i-1,j} - 2\Phi_{i,j} + \Phi_{i+1,j}}{h^2}. \tag{2.3.3}$$

By Taylor's expansion:

$$\begin{aligned}
\Phi_{i+1,j} &= \Phi[(i+1)h, jk] = \Phi(x_i + h, t_j) \\
&= \Phi_{i,j} + h\left(\frac{\partial \Phi}{\partial x}\right)_{i,j} + \frac{h^2}{2}\left(\frac{\partial^2 \Phi}{\partial x^2}\right)_{i,j} + \frac{h^3}{6}\left(\frac{\partial^3 \Phi}{\partial x^3}\right)_{i,j} + \cdots
\end{aligned}$$

$$\Phi_{i-1,j} = \Phi_{i,j} - h\left(\frac{\partial \Phi}{\partial x}\right)_{i,j} + \frac{h^2}{2}\left(\frac{\partial^2 \Phi}{\partial x^2}\right)_{i,j} - \frac{h^3}{6}\left(\frac{\partial^3 \Phi}{\partial x^3}\right)_{i,j} + \cdots$$

and

$$\Phi_{i,j+1} = \Phi_{i,j} + k\left(\frac{\partial \Phi}{\partial t}\right)_{i,j} + \frac{k^2}{2}\left(\frac{\partial^2 \Phi}{\partial t^2}\right)_{i,j} + \frac{k^3}{6}\left(\frac{\partial^3 \Phi}{\partial t^3}\right)_{i,j} + \cdots .$$

Hence

$$\begin{aligned}
\tau_{i,j} &= \left(\frac{\partial \Phi}{\partial t} - \frac{\partial^2 \Phi}{\partial x^2}\right)_{i,j} + \frac{k}{2}\left(\frac{\partial^2 \Phi}{\partial t^2}\right)_{i,j} - \frac{h^2}{12}\left(\frac{\partial^4 \Phi}{\partial x^4}\right)_{i,j} \\
&\quad + \frac{k^2}{6}\left(\frac{\partial^3 \Phi}{\partial t^3}\right)_{i,j} - \frac{h^4}{360}\left(\frac{\partial^6 \Phi}{\partial x^6}\right)_{i,j} + \cdots .
\end{aligned}$$

But Φ satisfies

$$\left(\frac{\partial \Phi}{\partial t} - \frac{\partial^2 \Phi}{\partial x^2}\right)_{i,j} = 0$$

at the point (i,j), leaving the principal part of the local truncation error as

$$\left(\frac{k}{2}\frac{\partial^2 \Phi}{\partial t^2} - \frac{h^2}{12}\frac{\partial^4 \Phi}{\partial x^4}\right)_{i,j}. \tag{2.3.4}$$

Hence

$$\tau_{i,j} = O(k) + O(h^2). \tag{2.3.5}$$

Note that from the differential equation

$$\frac{\partial}{\partial t} = \frac{\partial^2}{\partial x^2}$$

or

$$\frac{\partial}{\partial t}\left(\frac{\partial \Phi}{\partial t}\right) = \frac{\partial^2}{\partial x^2}\left(\frac{\partial^2 \Phi}{\partial x^2}\right)$$

and so if we put $6k = h^2$ the error becomes

$$\frac{1}{6}k^2\left(\frac{\partial^3 \Phi}{\partial t^3}\right)_{i,j} - \frac{1}{360}h^4\left(\frac{\partial^6 \Phi}{\partial x^6}\right)_{i,j}$$

so that $\tau_{i,j}$ is of order $O(k^2) + O(h^4)$. This is of little use in practice as $k = \frac{1}{6}h^2$ is very small for small h, so needs a huge amount of calculation to advance the solution to large time levels. The advantage of this method is its simplicity and in larger scale problems the disadvantage of small time steps is balanced against the complications of avoiding the issue by using the implicit methods of Section 2.7.

A few exercises on these topics now follow.

EXERCISES

2.12 Find the local truncation error for the Crank–Nicolson method defined by

$$\frac{\phi_{i,j+1} - \phi_{i,j}}{k} = \frac{1}{2}\left[\frac{\phi_{i+1,j+1} - 2\phi_{i,j+1} + \phi_{i-1,j+1}}{h^2}\right.$$
$$\left. + \frac{\phi_{i+1,j} - 2\phi_{i,j} + \phi_{i-1,j}}{h^2}\right]$$

which will be considered in more detail in Section 2.7.

2.13 Find the local truncation error associated with the approximation

$$\frac{\phi_{i+2,j} - \phi_{i+1,j} - \phi_{i-1,j} + \phi_{i-2,j}}{h^2}$$

for the second partial derivative

$$\frac{\partial^2 \phi}{\partial x^2}.$$

2.14 A slightly more tricky problem is to repeat Exercise 2.13. for the approximation

$$\frac{\phi_{i+1,j} - \phi_{i,j+1} - \phi_{i,j-1} + \phi_{i-1,j}}{h^2}.$$

2.15 Find the partial derivative which is approximated by

$$\frac{\phi_{i,j+2} + 2\phi_{i,j+1} - 2\phi_{i,j-1} + \phi_{i,j-2}}{8k}$$

and find the associated truncation error.

2.4 Consistency

It is possible to approximate a parabolic or hyperbolic equation by a finite difference scheme which has a solution that converges to the solution of a different differential equation as the mesh length tends to zero. Such a difference scheme is said to be inconsistent (or incompatible). The relevance of consistency lies in the Lax equivalence theorem, 'If a linear finite difference equation is consistent with a properly posed linear initial-value problem then stability is necessary and sufficient for convergence of ϕ to Φ as the mesh lengths tend to zero.' For further reading on this theorem see Iserles (1996) and for a proof see Richtmyer and Morton (1967).

The general definition of consistency is as follows. Let $L(\Phi) = 0$ be the partial differential equation in independent variables x and t with exact solution Φ. Let $F(\phi) = 0$ be the finite difference equation with exact solution ϕ, and let v be any continuous function of x and t, with a sufficient number of continuous derivatives such that $L(v)$ can be evaluated at (ih, jk). The truncation error is $T_{i,j}(v) = F(v_{i,j}) - L(v_{i,j})$ at the point $v_{ij} = v(ih, jk)$. If $T_{i,j}(v) \to 0$ as $h \to 0$, $k \to 0$ the difference equation is said to be consistent with the partial differential equation in that T_{ij} gives an estimate of the error in replacing $L(v_{i,j})$ by $F(v_{i,j})$.

If we put $v = \Phi$ $(L(\Phi) = 0)$, then $T_{i,j}(\Phi) = F(\Phi_{i,j})$ and the truncation error coincides with the local truncation error. Hence the difference equation is then consistent if the limiting value of the local truncation error is zero as $h \to 0$, $k \to 0$. This modified form of definition will be used here.

It follows that the classical explicit approximation to

$$\frac{\partial \Phi}{\partial t} = \frac{\partial^2 \Phi}{\partial x^2} \tag{2.4.1}$$

is consistent with the differential equation.

By introducing a parameter θ, a more general finite difference approximation to (2.4.1) can be found with the form

$$\frac{\phi_{i,j+1} - \phi_{i,j-1}}{2k} - \frac{[\phi_{i+1,j} - 2[\theta\phi_{i,j+1} + (1 - \theta)\phi_{i,j-1}] + \phi_{i-1,j}]}{h^2} = 0 \tag{2.4.2}$$

to give a local truncation error of

$$\frac{k^2}{6}\frac{\partial^3 \Phi}{\partial t^3} - \frac{h^2}{12}\frac{\partial^4 \Phi}{\partial x^4} + (2\theta - 1)\frac{2k}{h^2}\frac{\partial \Phi}{\partial t} + \frac{k^2}{h^2}\frac{\partial^2 \Phi}{\partial t^2} + O\left(\frac{k^3}{h^2} + h^4 + k^4\right). \tag{2.4.3}$$

Two cases of the consistency of this scheme will be considered, firstly with $k = rh$ and secondly with $k = rh^2$ where r is a positive constant.

(i) With $k = rh$

$$
\begin{aligned}
T_{i,j} = F(\Phi_{i,j}) = \Bigg[&\frac{k^2}{6}\frac{\partial^3 \Phi}{\partial t^3} - \frac{h^2}{12}\frac{\partial^4 \Phi}{\partial x^4} + (2\theta - 1)\frac{2k}{h^2}\frac{\partial \Phi}{\partial t} \\
&+ \frac{k^2}{h^2}\frac{\partial^2 \Phi}{\partial t^2} + O\left(\frac{k^3}{h^2} + h^4 + k^4\right) \Bigg]_{i,j}
\end{aligned}
$$

which gives

$$\lim_{h \to 0} T_{ij} = \left[(2\theta - 1)\frac{2r}{h}\frac{\partial \Phi}{\partial t} + r^2\frac{\partial^2 \Phi}{\partial t^2}\right]_{i,j}.$$

When $\theta \neq \frac{1}{2}$ the first term tends to infinity. However when $\theta = \frac{1}{2}$ the terms tends to $r^2 \partial^2 \Phi / \partial t^2$ and so the truncation error is non-zero. In this case the finite difference equation is consistent with

$$\frac{\partial \Phi}{\partial t} - \frac{\partial^2 \Phi}{\partial x^2} + r^2 \frac{\partial^2 \Phi}{\partial t^2} = 0$$

which is quite a different differential equation from the one that was originally approximated. Hence this difference equation is always inconsistent with

$$\frac{\partial \Phi}{\partial t} - \frac{\partial^2 \Phi}{\partial x^2} = 0$$

when $k = rh$.

(ii) For the other case when $k = rh^2$

$$\lim_{h \to 0} T_{i,j} = 2(2\theta - 1)r \frac{\partial \Phi}{\partial t}.$$

When $\theta \neq \frac{1}{2}$ the difference scheme is consistent with

$$[1 + 2(\theta - 1)r] \frac{\partial \Phi}{\partial t} - \frac{\partial^2 \Phi}{\partial x^2} = 0.$$

It is only when $\theta = \frac{1}{2}$ that the difference scheme is consistent with the given partial differential equation. Clearly, great care must be taken to avoid finite difference schemes which exhibit inconsistency.

The following exercises may now be attempted on this section.

EXERCISES

2.16 Investigate the consistency of the method of Du Fort and Frankel defined by

$$\frac{\phi_{i,j+1} - \phi_{i,j-1}}{2k} = \frac{\phi_{i+1,j} - \phi_{i,j+1} - \phi_{i,j-1} + \phi_{i-1,j}}{h^2}$$

which was originally designed as a method for the heat equation. Consider different ratios of k and h.

2.17 Examine the explicit method used on the equation

$$\frac{\partial \phi}{\partial t} = \frac{\partial^2 \phi}{\partial x^2} + \frac{\partial \phi}{\partial x}$$

from Exercise 2.9 for consistency.

2.5 Convergence

In this section, attention is given to the concept of convergence. To define convergence, consider Φ to represent the exact solution of the partial differential equation with independent variables x and t, and ϕ to represent the exact solution of the finite difference equations used in place of the partial differential equation. Let the vector ϕ_j represent the components of ϕ_{ij} at a particular time level j. Let $\|.\|_k$ be a grid dependent Euclidean norm for the underlying linear space, e.g.

$$\|\mathbf{u}\|_h = \left(h \sum |u_j|^2 \right)^{1/2}$$

where the sum is over all grid points and where the non-standard h in the brackets allows easy passage from vector to function norms in further analysis. Define the discretisation error as $\mathbf{e}_j = \phi_j - \Phi_j$. Then a finite difference equation is said to be convergent in the norm $\|.\|_h$ when

$$\lim_{h \to 0} \left(\max_{j=0,1,\ldots,T/k} \|\mathbf{e}_j\|_h \right) = 0$$

for every initial condition and for every $T > 0$ and for $r = k/h^2$ constant as $h \to 0$. The constant r is called the Courant number.

The convergence of the solution of an approximating set of linear difference equations to the solution of a linear partial differential equation is dealt with most easily using the Lax equivalence theorem. However explicit difference schemes can be investigated by deriving a difference equation for the discretisation error. Consider

$$\frac{\partial \Phi}{\partial t} = \frac{\partial^2 \Phi}{\partial x^2}, \quad 0 < x < 1, \quad t > 0, \qquad (2.5.1)$$

where Φ is known for $0 < x < 1$ when $t = 0$ and at $x = 0$ and $x = 1$ for $t \geq 0$. Denote the discretisation error by $e_{i,j}$ so that

$$\phi_{i,j} = \Phi_{i,j} - e_{i,j}. \qquad (2.5.2)$$

Hence the explicit finite difference approximation

$$\phi_{i,j+1} - \phi_{i,j} - r(\phi_{i-1,j} - 2\phi_{i,j} + \phi_{i+1,j}) = 0 \qquad (2.5.3)$$

with the same boundary conditions as Φ becomes

$$\begin{aligned} \Phi_{i,j+1} - e_{i,j+1} - (\Phi_{i,j} - e_{i,j}) - r(\Phi_{i-1,j} - 2\Phi_{i,j} \\ + \Phi_{i+1,j} - (e_{i-1,j} - 2e_{i,j} + e_{i+1,j})) = 0 \end{aligned} \qquad (2.5.4)$$

and

$$\begin{aligned} e_{i,j+1} = & \ re_{i-1,j} + (1 - 2r)e_{i,j} + re_{i+1,j} + \Phi_{i,j+1} - \Phi_{i,j} \\ & + r(2\Phi_{i,j} - \Phi_{i-1,j} - \Phi_{i+1,j}). \end{aligned} \qquad (2.5.5)$$

Then Taylor's theorem gives

$$
\Phi_{i+1,j} \;=\; \Phi(x_i + h, t_j) = \Phi_{i,j} + h\left(\frac{\partial \Phi}{\partial x}\right)_{i,j} + \frac{h^2}{2!}\left(\frac{\partial^2 \Phi}{\partial x^2}\right)_{(x_i + \theta_1 h, t_j)}
$$

$$
\Phi_{i-1,j} \;=\; \Phi(x_i - h, t_j) = \Phi_{i,j} - h\left(\frac{\partial \Phi}{\partial x}\right)_{i,j} + \frac{h^2}{2!}\left(\frac{\partial^2 \Phi}{\partial x^2}\right)_{(x_i - \theta_2 h, t_j)}
$$

and

$$
\Phi_{i,j+1} = \Phi(x_i, t_j + k) = \Phi_{i,j} + k\left(\frac{\partial \Phi}{\partial t}\right)_{(x_i, t_j + \theta_3 k)}
$$

where $0 < \theta_1, \theta_2, \theta_3 < 1$.

Substituting into (2.5.5) yields

$$
\begin{aligned}
e_{i,j+1} \;=\;& r e_{i-1,j} + (1 - 2r)e_{i,j} + r e_{i+1,j} \\
&+ k\left[\left(\frac{\partial \Phi}{\partial t}\right)_{(x_i, t_j + \theta_3 k)} - \left(\frac{\partial^2 \Phi}{\partial x^2}\right)_{(x_i + \theta_4 h, k)}\right] \qquad (2.5.6) \\
&\text{with} \quad -1 < \theta < 1.
\end{aligned}
$$

Hence the errors, e_{ij} satisfy the same difference equations as the solution, a fact which can be utilised in the analysis of stability in the next section. Now define

$$
E_j = \max_i |e_{ij}| \qquad (2.5.7)
$$

and

$$
M = \max_{i,j}\left|\left(\frac{\partial \Phi}{\partial t}\right)_{(x_j, t_j + \theta_3 k)} - \left(\frac{\partial^2 \Phi}{\partial x^2}\right)_{(x_i + \theta_4 h, t_j)}\right| \qquad (2.5.8)
$$

over all i and j. For $r \leq \frac{1}{2}$ all the coefficients of the e are positive or zero, hence

$$
\begin{aligned}
|e_{i,j+1}| \;\leq\;& r|e_{i-1,j}| + (1 - 2r)|e_{i,j}| + r|e_{i+1,j}| + kM \\
\leq\;& rE_j + (1 - 2r)E_j + rE_j + kM = E_j + kM. \qquad (2.5.9)
\end{aligned}
$$

But (2.5.9) holds for all i and hence for $\max|e_{i,j+1}|$, so

$$
E_{j+1} \leq E_j + kM \leq (E_{j-1} + kM) + kM = E_{j-1} + 2kM
$$

and hence

$$
E_j \leq E_0 + jkM = t_j M \qquad (2.5.10)
$$

Since the initial values of ϕ and Φ are the same, $E_0 = 0$. Further when $h \to 0$, $k = rh^2 \to 0$ and M tends to

$$
\max_{i,j}\left|\left(\frac{\partial \Phi}{\partial t} - \frac{\partial^2 \Phi}{\partial x^2}\right)_{i,j}\right|
$$

then since Φ is a solution to

$$\frac{\partial \Phi}{\partial t} = \frac{\partial^2 \Phi}{\partial x^2}$$

the limiting value of $M = 0$ and so the limiting value of E_j is zero. As

$$|\Phi_{i,j} - \phi_{i,j}| \leq E_j$$

it has been proved that ϕ converges in the max-norm to Φ as $h \to 0$ when $r \leq \frac{1}{2}$ and t is finite. When $r > \frac{1}{2}$ it can be shown that the method is not convergent.

Were exact arithmetic performed on an ideal computer, the errors due to truncation would constitute the whole error. However rounding errors also play a part and depend on the finite word length of the machine employed. This effect is discussed in the next section.

2.6 Stability

The solution of the given partial differential equation is found by solving associated finite difference equations. If no rounding errors were introduced into this process then their exact solution $\phi_{i,j}$ would be obtained at each mesh point (i, j). The essential idea defining stability is that the numerical process should not cause any small perturbations introduced through rounding at any stage to grow and ultimately dominate the solution.

For linear initial-value boundary-value problems the Lax equivalence theorem (Richtmyer and Morton, 1967) relates stability to convergence by defining stability in terms of the boundedness of the solution of the finite difference equations at a fixed time level T as $k \to 0$. This is Lax–Richtmyer stability.

In an actual computation k and h are normally kept constant so the solution moves forward time level by time level and in many books stability is defined in terms of the boundness of this numerical solution as $j \to \infty$, $(t = jk)$ k fixed, which is asymptotic stability.

To make the analysis of stability amenable to mathematical analysis, a definition based on the growth of the exact solution is used. Then, if rounding errors or perturbations are introduced at any stage in time, then these errors will also be bounded if the exact solution is bounded, as the same difference equations (but different initial and boundary conditions) are obeyed by both from (2.5.6). A method is stable if for every $T > 0$, there exists a constant $c(T) > 0$ such that

$$\|\phi_j\|_h \leq c(T), \quad n = 0, 1, \ldots, T/k, \quad \text{and} \quad h \to 0, \quad k \to 0.$$

The first approach to stability analysis is called matrix stability analysis. To establish the criterion for stability, consider the explicit finite difference formula for

$$\frac{\partial \phi}{\partial t} = \frac{\partial^2 \phi}{\partial x^2} \quad \text{on} \quad 0 \leq x \leq 1 \tag{2.6.1}$$

with

$$\phi(0,t) = a, \qquad \phi(1,t) = b, \quad \forall t \in [0,T]$$

and

$$\phi(x,0) = g(x).$$

The difference equation for a general interior point is

$$\phi_{i,j+1} = \phi_{i,j} + r(\phi_{i-1,j} - 2\phi_{i,j} + \phi_{i+1,j}) \qquad (2.6.2)$$

for $i = 0(1)N$ and $Nh = 1$ and $j = 0(1)J$ with $Jk = T$. In addition, on the boundaries,

$$\phi_{1,j+1} = \phi_{1,j} + r(a - 2\phi_{1,j} + \phi_{2,j})$$

and

$$\phi_{N-1,j+1} = \phi_{N-1,j} + r(\phi_{N-2,j} - 2\phi_{N-1,j} + b)$$

where $Nh = 1$.

These can be written in the matrix form

$$\begin{pmatrix} \phi_{1,j+1} \\ \phi_{2,j+1} \\ \vdots \\ \phi_{N-2,j+1} \\ \phi_{N-1,j+1} \end{pmatrix}$$

$$= \begin{pmatrix} 1-2r & r & & & \\ r & 1-2r & r & & \\ & \ddots & \ddots & \ddots & \\ & & r & 1-2r & r \\ & & & r & 1-2r \end{pmatrix} \begin{pmatrix} \phi_{1,j} \\ \phi_{2,j} \\ \vdots \\ \phi_{N-2,j} \\ \phi_{N-1,j} \end{pmatrix} + \begin{pmatrix} ra \\ 0 \\ \vdots \\ 0 \\ rb \end{pmatrix} \qquad (2.6.3)$$

or more compactly as

$$\phi_{j+1} = A\phi_j + \mathbf{d} \qquad (2.6.4)$$

where A and $\mathbf{d}$ are known. Hence in general the difference equations for a parabolic partial differential equation can be written in matrix form and $\mathbf{d}_j$ depends on the boundary conditions and may vary from time step to time step.

The solution domain of the partial differential equation is the finite rectangle $0 \le x \le 1, 0 \le t \le T$, which is subdivided into a uniform mesh with $x_i = ih$, $i = 0(1)N$ $(Nh = 1)$ and $t_j = jk$, $j = 0(1)J$ $(Jk = T)$. Now it will be assumed that h and k are related (for example $k = rh$ or $k = rh^2, r > 0$ and finite) so that $h \to 0$ as $k \to 0$. If the boundary conditions at $i = 0$ and N $(j > 0)$ are known then the equations for $i = 1(1)N - 1$ can be written as

$$\phi_{j+1} = A\phi_j + \mathbf{d}_j. \qquad (2.6.5)$$

Applied recursively (2.6.5) gives

$$
\begin{aligned}
\phi_j &= A\phi_{j-1} + \mathbf{d}_{j-1} = A(A\phi_{j-2} + \mathbf{d}_{j-2}) + \mathbf{d}_{j-1} \\
&= A^2\phi_{j-2} + A\mathbf{d}_{j-2} + \mathbf{d}_{j-1} \\
&\vdots \\
&= A^j\phi_0 + A^{j-1}\mathbf{d}_0 + A^{j-2}\mathbf{d}_1 + \cdots + \mathbf{d}_{j-1}
\end{aligned}
\tag{2.6.6}
$$

where ϕ_0 is the vector of initial values and $\mathbf{d}_0, \ldots, \mathbf{d}_{j-1}$ are vectors of the known boundary values.

The next stage is to consider the propagation of a perturbation, and to this end consider the vector of initial values ϕ_0 perturbed to ϕ_0^* (if we assume no more rounding errors occur), then the exact solution at the jth time step is

$$
\phi_j^* = A^j\phi_0^* + A^{j-1}\phi_0 + A^{j-2}\mathbf{d}_1 + \cdots + \mathbf{d}_{j-1}.
\tag{2.6.7}
$$

Define the perturbation or 'error' vector $\mathbf{e}$ to be $\mathbf{e} = \phi^* - \phi$ then

$$
\mathbf{e}_j = \phi_j^* - \phi_j = A^j(\phi_0^* - \phi_0) = A^j\mathbf{e}_0, \quad j = 1(1)J.
\tag{2.6.8}
$$

(Note that when the finite difference equations are linear we need only consider the propagation of one line of errors because the overall effect of several lines will be given by the addition of the effect produced by each line considered separately.) The finite difference scheme will be stable when $\mathbf{e}_j$ remains bounded as j increases indefinitely. In other words $\mathbf{e}_0$ propagates according to

$$
\mathbf{e}_j = A\mathbf{e}_{j-1} = \cdots = A^j\mathbf{e}_0.
\tag{2.6.9}
$$

Hence, for compatible matrix and vector norms

$$
||\mathbf{e}_j|| \leq ||A^j||\,||\mathbf{e}_0||.
\tag{2.6.10}
$$

Lax defines the difference scheme as stable if there exists $M > 0$ (independent of j, h and k) such that $||A^j|| \leq M$, $j = 1(1)J$. This condition clearly limits the amplification of any initial pertubation and therefore of any rounding errors since $||\mathbf{e}_j|| \leq M||\mathbf{e}_0||$. Since

$$
||A^j|| = ||AA^{j-1}|| \leq ||A||\,||A^{j-1}|| \leq \cdots \leq ||A||^j
\tag{2.6.11}
$$

it follows that the Lax definition of stability is satisfied as long as $||A|| \leq 1$.

When this condition is satisfied it follows immediately that the spectral radius

$$
\rho(A) \leq 1 \quad \text{since} \quad \rho(A) \leq ||A||.
$$

On the other hand, if $\rho(A) \leq 1$, it does not follow that $||A|| \leq 1$, for example the matrix

$$
\begin{pmatrix} -0.8 & 0 \\ 0.4 & 0.7 \end{pmatrix}
$$

has eigenvalues -0.8 and 0.7 to give $\rho(A) = 0.8$, $||A||_1 = 1.2$, $||A||_\infty = 1.1$.
    If however A is real and symmetric then

$$||A||_2 = \sqrt{\rho(A^T A)} = \sqrt{\rho(A^2)} = \sqrt{\rho^2(A)} = \rho(A). \tag{2.6.12}$$

With this theory in place the stability of the classical explicit equations may now be considered as an example. The explicit method has the form:

$$\phi_{i,j+1} = r\phi_{i-1,j} + (1 - 2r)\phi_{i,j} + r\phi_{i+1,j}, \quad i = 1(1)N - 1. \tag{2.6.13}$$

The matrix A is

$$A = \begin{pmatrix} 1 - 2r & r & & & \\ r & 1 - 2r & r & & \\ & \ddots & \ddots & \ddots & \\ & & r & 1 - 2r & r \\ & & & r & 1 - 2r \end{pmatrix} \tag{2.6.14}$$

where $r = k/h^2 > 0$ and it is assumed the boundary values are known for all j.
    There are two ways to proceed: to either find the eigenvalues of A explicitly, or use norms to obtain a bound on the eigenvalues. When $1 - 2r \le 0$ or $0 < r \le \frac{1}{2}$ then

$$||A||_\infty = r + (1 - 2r) + r = 1 \tag{2.6.15}$$

(largest row sum of moduli) and when $1 - 2r < 0$ or $r > \frac{1}{2}$, $|1 - 2r| = 2r - 1$ then

$$||A||_\infty = r + 2r - 1 + r = 4r - 1 > 1.$$

Therefore the scheme is stable for $0 < r \le \frac{1}{2}$ but not for $r > \frac{1}{2}$.
    The alternative is to use the following result that if A is real and symmetric $||A||_2 = \rho(A) = \max|\lambda_s|$. Now $A = I + rS$ where

$$I = \begin{pmatrix} 1 & & 0 \\ & \ddots & \\ 0 & & 1 \end{pmatrix} \quad \text{and} \quad S = \begin{pmatrix} -2 & 1 & & & 0 \\ 1 & -2 & 1 & & \\ & \ddots & \ddots & \ddots & \\ & & 1 & -2 & 1 \\ 0 & & & 1 & -2 \end{pmatrix} \tag{2.6.16}$$

the matrices being of order $(N - 1) \times (N - 1)$. But S has known eigenvalues

$$\lambda_s = -4\sin^2 \frac{s\pi}{2N}, \quad s = 1(1)N - 1 \tag{2.6.17}$$

which may be verified directly, or by solving a three-term difference equation as in Smith (1978, p.113), (see Theorem 1.5). Hence the eigenvalues of A are

$$1 - 4r \sin^2 \frac{s\pi}{2N}$$

and the equations are stable when

$$\|A\|_2 = \max_s \left| 1 - 4r \sin^2 \frac{s\pi}{2N} \right| \leq 1$$

or

$$-1 \leq 1 - kr \sin^2 \frac{s\pi}{2N} \leq 1, \quad s = 1(1)N - 1.$$

The left-hand inequality gives

$$r \leq \frac{1}{2 \sin^2 \left(\frac{(N-1)\pi}{2N} \right)}$$

and as

$$h \to 0, \ N \to \infty \quad \text{and} \quad \sin^2 \frac{(N-1)\pi}{2N} \to 1.$$

Hence $r \leq \frac{1}{2}$ for stability.

There is an obvious problem in applying the above analysis in that in general the eigenvalues of the matrix A are not available in closed form. It may be possible to put bounds on the eigenvalues, or the eigenvalues will need to be found numerically for a range of r and for a range of orders which is not feasible. An alternative is to use Fourier stability analysis. Firstly, consider Fourier stability analysis applied to the explicit method defined in Section 2.2.

It is convenient to represent $\sqrt{(-1)}$ by $\hat{i}$ in the next few lines to avoid notational confusion as i has been used in the definition of the grid. The approach is to introduce errors along $t = 0$ which can be represented by a Fourier series of the form

$$\sum_i A_i \, e^{-\hat{i}\lambda_i x} \qquad (2.6.18)$$

for $x \in (0, L)$ and $\lambda_i = i\pi/Nh$ and $Nh = L$. This form arises from the method of separation of variables (see the companion volume on analytic methods). Because this analysis applies to linear problems only, one term such as $e^{-\hat{i}\lambda x_i}$ may be considered and then superposition used to find the effect of a sum of terms such as in (2.6.18). Let $e^{-\hat{i}\lambda x_i} = e^{-\hat{i}\lambda ih} = \phi_{i,0}$ on $t = 0$, then

$$\phi_{i,j} = e^{\alpha jk}e^{-\hat{i}\lambda ih} \qquad (2.6.19)$$

where in general α is a complex constant, for other t values by direct solution of the difference equations (2.2.5) by separation of the variables. Substituting this solution into (2.2.5) gives

$$e^{\alpha k} = r\left\{ e^{\hat{i}\lambda h} + e^{-\hat{i}\lambda h} \right\} + (1 - 2r)$$

$$= 1 - 4r \sin^2 \frac{1}{2}\lambda h \qquad (2.6.20)$$

where r is non-negative. Hence $e^{\alpha k} < 1$ and if $e^{\alpha k} \geq 0$ then the solution will decay as $j \to \infty$. If $-1 < e^{\alpha k} < 0$ then the solution will be a decaying oscillation. Finally for $e^{\alpha k} < -1$ the solution increases and is unstable.

Hence stability occurs if $-1 \leq 1 - 4r \sin^2 \frac{1}{2}\lambda h$ or

$$r \leq \frac{1}{2\sin^2 \frac{1}{2}h\lambda} = \frac{1}{1 - \cos \lambda h}. \qquad (2.6.21)$$

Inequality (2.6.21) can be eased because $1/(1 - \cos \lambda h)$ has a minimum of $\frac{1}{2}$ giving

$$r \leq \frac{1}{2} \qquad (2.6.22)$$

for stability which is the condition obtained in the matrix approach. Hence for non-oscillatory decay

$$1 - 4r \sin^2 \frac{1}{2}\lambda h \geq 0 \qquad (2.6.23)$$

or

$$r \leq \frac{1}{2(1 - \cos \lambda h)} \qquad (2.6.24)$$

giving $r \leq \frac{1}{4}$.

This theory leaves an outstanding problem. The explicit method has a limited region of stability which restricts the time step to be unduly small. How can this limitation be overcome? Is there a method which is stable for unrestricted r? It is not easy to just invent such a method, even though the analysis of stability is straightforward once a method is set up. Such a method is derived in the next section.

The following exercises on stability may now be considered.

EXERCISES

2.18 As an example of this stability analysis consider the backward difference formula defined by the difference equations:

$$\frac{\phi_{i,j+1} - \phi_{i,j}}{k} = \frac{\phi_{i+1,j+1} - 2\phi_{i,j+1} + \phi_{i-1,j+1}}{h^2}$$

and find the range of r for which this method is stable.

2.19 Consider again Exercise 2.9 in which the partial differential equation

$$\frac{\partial \phi}{\partial t} = \frac{\partial^2 \phi}{\partial x^2} + \frac{\partial \phi}{\partial x}$$

is solved using an explicit method. Use Gerschgorin's theorems to examine the stability of the method.

2.20 Exercise 2.10 considers the solution of the equation

$$\frac{\partial u}{\partial t} = \frac{\partial^2 u}{\partial x^2} - \frac{1}{x}\frac{\partial u}{\partial x}.$$

Find the range of $r = k/h^2$ for which the explicit finite difference equations are stable.

2.21 Investigate the stability of the explicit finite difference scheme used to solve

$$\frac{\partial \phi}{\partial t} = \frac{\partial^2 \phi}{\partial x^2} - \phi$$

in Exercise 2.11.

2.22 Investigate Exercise 2.19, 2.20 and 2.22 using the Fourier stability approach.

2.7 The Crank–Nicolson Implicit Method

The explicit method has the drawback that the step size k has to be very small $\left(0 < k/h^2 \leq \frac{1}{2} \text{ that is } k \leq h^2/2 \right)$ if stability is to be maintained. The explicit scheme is analogous to Euler's method for ordinary differential equations. The Crank–Nicolson method corresponds to the trapezium rule for ordinary differential equations and is derived by replacing $\partial^2 \phi/\partial x^2$ by the mean of its finite difference representations at the jth and $(j+1)$th time levels. This method is stable for all finite values of r.

For

$$\frac{\partial \phi}{\partial t} = \frac{\partial^2 \phi}{\partial x^2}$$

Crank–Nicolson method is

$$\frac{\phi_{i,j+1} - \phi_{i,j}}{k} = \frac{1}{2} \left[\frac{\phi_{i+1,j+1} - 2\phi_{i,j+1} + \phi_{i-1,j+1}}{h^2} \right.$$
$$\left. + \frac{\phi_{i+1,j} - 2\phi_{i,j} + \phi_{i-1,j}}{h^2} \right] \qquad (2.7.1)$$

where an average is taken to represent the second order derivative. This equation can be rewritten as

$$-r\phi_{i-1,j+1} + (2 + 2r)\phi_{i,j+1} - r\phi_{i+1,j+1}$$
$$= r\phi_{i-1,j} + (2 - 2r)\phi_{i,j} + r\phi_{i+1,j} \qquad (2.7.2)$$

with

$$r = \frac{k}{h^2}.$$

The right-hand side is known, but the left-hand side involves three successive unknown values. Applying the equation for $i = 1, 2, \ldots, N - 1$ (internal mesh points) and using the boundary conditions on $x = 0$ and $x = l$ leads to a tridiagonal set of linear algebraic equations which have to be solved at each time level. A method where the calculation of unknown nodal values requires the solution of a set of simultaneous equations is called an *implicit* one. The equations may be solved directly using the Thomas algorithm of Chapter 1.

The method will now be analysed for its stability properties. For the solution of

$$\frac{\partial \phi}{\partial t} = \frac{\partial^2 \phi}{\partial x^2} \tag{2.7.3}$$

subject to the boundary conditions $\phi = 0$ at $x = 0$, $x = 1$, $t > 0$, the finite difference approximation is

$$-r\phi_{i-1,j+1} + (2 + 2r)\phi_{i,j+1} - r\phi_{i+1,j+1}$$
$$= r\phi_{i-1,j} + (2 - 2r)\phi_{i,j} + r\phi_{i+1,j} \tag{2.7.4}$$

where

$$i = 1, 2, \ldots, n - 1, \quad \phi_{0,j} = \phi_{n,j} = 0, \quad \forall j.$$

In matrix form we have

$$\begin{pmatrix} 2+2r & -r & & & \\ -r & 2+2r & -r & & \\ & \ddots & \ddots & \ddots & \\ & & -r & 2+2r & -r \\ & & & -r & 2+2r \end{pmatrix} \begin{pmatrix} \phi_{1,j+1} \\ \phi_{2,j+1} \\ \vdots \\ \phi_{n-2,j+1} \\ \phi_{n-1,j+1} \end{pmatrix}$$

$$= \begin{pmatrix} 2-2r & r & & & \\ r & 2-2r & r & & \\ & \ddots & \ddots & \ddots & \\ & & r & 2-2r & r \\ & & & r & 2-2r \end{pmatrix} \begin{pmatrix} \phi_{1,j} \\ \phi_{2,j} \\ \vdots \\ \phi_{n-2,j} \\ \phi_{n-1,j} \end{pmatrix}, \tag{2.7.5}$$

or in compact form

$$[2I - rT_{n-1}]\phi_{j+1} = [2I + rT_{n-1}]\phi_j \tag{2.7.6}$$

where

$$T_{n-1} = \begin{pmatrix} -2 & 1 & & & \\ 1 & -2 & 1 & & \\ & \ddots & \ddots & \ddots & \\ & & 1 & -2 & 1 \\ & & & 1 & -2 \end{pmatrix}. \tag{2.7.7}$$

This equation becomes

$$\phi_{j+1} = [2I - rT_{n-1}]^{-1}[2I + rT_{n-1}]\phi_j. \tag{2.7.8}$$

Hence let

$$A = [2I - rT_{n-1}]^{-1}[2I + rT_{n-1}]. \tag{2.7.9}$$

Now it can be shown that if the $n \times n$ symmetric matrices B and C commute then $B^{-1}C$, BC^{-1} and $B^{-1}C^{-1}$ are symmetric. Now T_{n-1} is symmetric, so $2I - rT_{n-1}$ and $2I + rT_{n-1}$ are symmetric. By direct evaluation they commute. Hence A is symmetric. Since the eigenvalues of T_{n-1} are

$$-4 \sin^2 \frac{s\pi}{2n}, \quad s = 1, \ldots, n-1,$$

the eigenvalues of A are

$$\frac{2 - 4r \sin^2 \left(\frac{s\pi}{2n}\right)}{2 + 4r \sin^2 \left(\frac{s\pi}{2n}\right)}, \quad s = 1, \ldots, n-1. \tag{2.7.10}$$

Therefore

$$\|A\|_2 = \rho(A) = \max_s \left| \frac{2 - 4r \sin^2 \left(\frac{s\pi}{2n}\right)}{2 + 4r \sin^2 \left(\frac{s\pi}{2n}\right)} \right| < 1, \quad \forall r > 0$$

proving that the Crank–Nicolson method is stable for all r. Such a method is said to be unconditionally stable. The method is also consistent, and therefore convergent. (Note that it can be shown that r must be restricted in order to avoid the possibility of finite oscillations near points of discontinuity.)

The Crank–Nicolson method can be demonstrated by calculating a numerical solution to

$$\frac{\partial u}{\partial t} = \frac{\partial^2 u}{\partial x^2} \quad (0 < x < 1) \tag{2.7.11}$$

with boundary conditions

(i) $u = 0$ at $x = 0$, $x = 1$, $t \geq 0$

(ii) $u = 1 - 4(x - 1/2)^2$ for $t = 0$ and $0 \leq x \leq 1$

whose solution may then be compared with the earlier solution to the same problem using the explicit method, taking $h = \frac{1}{10}$; although the method is valid for all finite $t = k/h^2$, a large value will yield inaccurate approximations for $\partial u/\partial t$. Hence the choice $r = 1$ is made giving $k = 1/100$. The finite difference approximation is then

$$-u_{i-1,j+1} + 4u_{i,j+1} - u_{i+1,j+1} = u_{i-1,j} + u_{i+1,j}. \tag{2.7.12}$$

From the symmetry of the problem $u_{0,j} = u_{4,j}$, etc. giving for the first time step the equations

$$\begin{aligned} -0 + 4u_{11} - u_{21} &= 0.0 + 0.64 \\ -u_{11} + 4u_{21} - u_{31} &= 0.36 + 0.84 \\ -u_{21} + 4u_{31} - u_{41} &= 0.64 + 0.96 \\ -u_{31} + 4u_{41} - u_{51} &= 0.84 + 1.0 \\ -2u_{41} + 4u_{51} &= 0.96 + 0.96 \end{aligned} \tag{2.7.13}$$

which have the solution

$$u_{11} = 0.3014, u_{21} = 0.5657, u_{31} = 0.7615,$$
$$u_{41} = 0.8804, u_{51} = 0.9202.$$

The process is then repeated to obtain u_{21}, u_{22}, etc. The numerical values are shown in Table 2.6.

Table 2.6.

t	$x = 0$	0.1	0.2	0.3	0.4	0.5
0.00	0.0	0.3600	0.6400	0.8400	0.9600	1.0
0.01	0.0	0.3014	0.5657	0.7615	0.8804	0.9202
0.02	0.0	0.2676	0.5048	0.6885	0.8030	0.8417
0.03	0.0	0.2399	0.4548	0.6231	0.7300	0.7665
0.04	0.0	0.2164	0.4110	0.5646	0.6626	0.6963
0.05	0.0	0.1958	0.3721	0.5117	0.6011	0.6319
$\vdots$				$\vdots$		
0.10	0.0	0.1197	0.2277	0.3135	0.3685	0.3875

Table 2.7 shows the corresponding true values for the same range of values of t.

Table 2.7.

t	$x = 0$	0.1	0.2	0.3	0.4	0.5
0.00	0.0	0.3600	0.6400	0.8400	0.9600	1.0
0.01	0.0	0.3024	0.5646	0.7606	0.8800	0.9200
0.02	0.0	0.2671	0.5041	0.6873	0.8019	0.8408
0.03	0.0	0.2394	0.4537	0.6218	0.7284	0.7649
0.04	0.0	0.2157	0.4098	0.5630	0.6607	0.6943
0.05	0.0	0.1950	0.3708	0.5099	0.5990	0.6296
$\vdots$				$\vdots$		
0.10	0.0	0.1189	0.2261	0.3112	0.3658	0.3847

In this example, the accuracy of the Crank–Nicolson method over the time range taken is about the same as for the explicit method which uses ten times as many time steps.

A sequence of exercises now follows to illustrate the use of implicit methods. Readers will find that having a linear equation solver available will be helpful in these exercises.

EXERCISES

2.23 Apply the implicit method to the equation

$$\frac{\partial \phi}{\partial t} = D \frac{\partial^2 \phi}{\partial \phi^2}$$

with $D = 0.1$ and boundary conditions $\phi = 2$ for $0 < x < 20$ for $t = 0$, and for $t \geq 0$, $\phi = 0$ on $x = 0$ and $\phi = 10$ on $x = 20$. Use $h = 4.0$ with $r = 1.00$ and complete four time steps. How does this solution compare with the explicit approach in Exercise 2.6?

2.24 Consider the partial differential equation

$$\frac{\partial \phi}{\partial t} = \frac{\partial^2 \phi}{\partial x^2}$$

with

$$\phi = 0 \quad \text{at} \quad x = 0 \quad \text{and} \quad x = 1 \quad \text{for} \quad t > 0$$

and

$$\phi = \begin{cases} 2x & 0 \leq x \leq 1/2 \\ 2(1-x) & 1/2 \leq x \leq 1 \end{cases} \quad \text{for} \quad t = 0.$$

By taking $h = 0.1$ and $r = 5$, find ϕ at the first time step using an implicit method.

2.25 Solve

$$\frac{\partial \phi}{\partial t} = \frac{\partial^2 \phi}{\partial x^2} + \frac{\partial \phi}{\partial x}$$

in the region bounded by

$$\phi = 0 \quad \text{at} \quad x = 0 \quad \text{and} \quad x = 1 \quad \text{for} \quad t \geq 0$$

and

$$\phi = \begin{cases} x & 0 \leq x \leq 1/2 \\ 1-x & 1/2 \leq x \leq 1 \end{cases} \quad \text{for} \quad t = 0$$

using an implicit method with $h = 0.1$ and an appropriate k to obtain a solution at $t = 0.02$.

2.26 Derive the Crank–Nicolson approximation for the equation

$$\frac{\partial \phi}{\partial t} = \frac{\partial^2 \phi}{\partial x^2} - \phi$$

subject to the conditions

$$\begin{aligned} \phi &= f(x) \quad \text{at} \quad t = 0 \quad \text{for} \quad 0 < x < 1, \\ \phi &= 0 \quad \text{at} \quad x = 0, \\ \phi &= 0 \quad \text{at} \quad x = 1. \end{aligned}$$

Express the set of equations needed to advance the solution by one time step in the form

$$A_1 \phi_{j+1} = A_2 \phi_j$$

giving explicit expressions for A_1 and A_2. Hence show that the method is unconditionally stable.

2.8 Parabolic Equations in Cylindrical and Spherical Polar Coordinates

A common method of handling non-rectangular geometries is to transform the relevant partial differential equation into a different coordinate system in which the new boundaries form constant surfaces. The general forms of the vector operators div, grad and curl from which the majority of physical partial differential equations arise are discussed in Evans, Blackledge and Yardley (1999, Chapter 1). These operators can be written in terms of a general orthogonal coordinate system which makes the process of obtaining a specific transformation very easy. The non-dimensional form of the heat conduction equation in cylindrical polar coordinates (r, θ, z) is

$$\frac{\partial u}{\partial t} = \frac{\partial^2 u}{\partial r^2} + \frac{1}{r}\frac{\partial u}{\partial r} + \frac{1}{r^2}\frac{\partial^2 u}{\partial \theta^2} + \frac{\partial^2 u}{\partial z^2} \qquad (2.8.1)$$

which is obtained from the Cartesian form using $x = r\cos\theta$ and $y = r\sin\theta$. For simplicity, assume that u is independent of z, to give

$$\frac{\partial u}{\partial t} = \frac{\partial^2 u}{\partial r^2} + \frac{1}{r}\frac{\partial u}{\partial r} + \frac{1}{r^2}\frac{\partial^2 u}{\partial \theta^2}. \qquad (2.8.2)$$

For non-zero values of r, and for an increment δr in r and $\delta \theta$ in θ

$$\frac{\partial u}{\partial t}(r_i, \theta_j, t^l) \approx \frac{u_{i,j}^{l+1} - u_{i,j}^l}{\delta t}, \quad \frac{\partial u}{\partial r}(r_i, \theta_j, t^l) \approx \frac{u_{i+1,j}^l - u_{i-1,j}^l}{2\delta t}$$

$$\frac{\partial^2 u}{\partial r^2}(r_i, \theta_j, t^l) \approx \frac{u_{i+1,j}^l - 2u_{i,j}^l + u_{i-1,j}^l}{(\delta r)^2}$$

$$\frac{\partial^2 u}{\partial \theta^2}(r_i, \theta_j, t^l) \approx \frac{u_{i,j+1}^l - 2u_{i,j}^l + u_{i,j-1}^l}{(\delta \theta)^2} \quad \text{and} \quad r = i\delta r \qquad (2.8.3)$$

where superscripts have been introduced for the time index. At $r = 0$ the right-hand side appears to contain singularities, which can be approximated in the following manner. Construct a circle of radius δr centre the origin and denote the four points in which $0x, 0y$ meet this circle by $1, 2, 3, 4$ and the corresponding function values by u_1, u_2, u_3, u_4 and the value of the origin by u_0. Then

$$\nabla^2 u = \frac{u_1 + u_2 + u_3 + u_4 - 4u_0}{(\delta r)^2} + O(\delta r^2). \qquad (2.8.4)$$

Rotation of axes through small angles leads to a similar equation. Repetition of this rotation and the addition of all such equations gives

$$\nabla^2 u = \frac{4(u_m - u_0)}{\delta r^2} + O(\delta r^2) \tag{2.8.5}$$

where u_m is a mean value of u round the circle. (The best mean value available is given by adding all values and dividing by their number.)

When a two-dimensional problem in cylindrical coordinates possesses circular symmetry

$$\frac{\partial^2 u}{\partial \theta^2} = 0 \tag{2.8.6}$$

to leave

$$\frac{\partial u}{\partial t} = \frac{\partial^2 u}{\partial r^2} + \frac{1}{r} \frac{\partial u}{\partial r}. \tag{2.8.7}$$

If the problem is symmetrical with respect to the origin then $\partial u/\partial r = 0$ and $(1/r)\partial u/\partial r$ assumes the indeterminate form at this point.

By Maclaurin's expansion

$$u'(r) = u'(0) + ru''(0) + \frac{1}{2}r^2 u'''(0) + \cdots \tag{2.8.8}$$

but $u'(0) = 0$ so the limiting value of $(1/r)\partial u/\partial r$ as $r \to 0$ is the value of $\partial^2 u/\partial r^2$ at $r = 0$ (or use L'Hospital's rule

$$\lim_{r \to 0} \frac{\partial u}{\partial r} / r = \lim_{r \to 0} \left(\frac{\partial^2 u}{\partial r^2} \right) / 1 \right).$$

Hence the equation at $r = 0$ can be replaced by

$$\frac{\partial u}{\partial t} = 2 \frac{\partial^2 u}{\partial r^2}. \tag{2.8.9}$$

(This result can also be deduced from

$$\frac{\partial u}{\partial t} = \frac{\partial^2 u}{\partial x^2} + \frac{\partial^2 u}{\partial y^2}$$

since

$$\frac{\partial^2 u}{\partial x^2} = \frac{\partial^2 u}{\partial y^2}$$

by circular symmetry and we can make the x-axis coincide with the direction of r.)

The finite difference representation of (2.8.9) is further simplified by the condition $\partial u/\partial r$ at $r = 0$ as this gives $u_{-1,j} = u_{1,j}$. The explicit approximation is then

$$\frac{u_{0,j+1} - u_{0,j}}{\delta t} = \frac{2(u_{1,j} - 2u_{0,j} + u_{-1,j})}{(\delta r)^2} \tag{2.8.10}$$

$$\frac{u_{0,j+1} - u_{0,j}}{\delta t} = \frac{4(u_{1,j} - u_{0,j})}{\delta r^2}. \tag{2.8.11}$$

An identical complication arises at $r = 0$ with spherical polars for which the Laplacian operator assumes the form:

$$\nabla^2 u = \frac{\partial^2 u}{\partial r^2} + \frac{2}{r}\frac{\partial u}{\partial r} + \frac{\cot\theta}{r}\frac{\partial u}{\partial \theta}$$
$$+ \frac{1}{r^2}\frac{\partial^2 u}{\partial \theta^2} + \frac{1}{r^2\sin^2\theta}\frac{\partial^2 u}{\partial \phi^2} \tag{2.8.12}$$

where the transformation from Cartesian form is now $x = r\sin\theta\cos\phi$, $y = r\sin\theta\sin\phi$ and $z = r\cos\theta$. By the same argument (2.8.12) can be replaced at $r = 0$ by

$$\frac{\partial^2 u}{\partial x^2} + \frac{\partial^2 u}{\partial u^2} + \frac{\partial^2 u}{\partial z^2} \tag{2.8.13}$$

and approximated by

$$\frac{6(u_m - u_0)}{\delta r^2} \tag{2.8.14}$$

where u_m is the mean of u over the sphere of radius δr.

If the problem is symmetrical with respect to the origin, that is independent of θ and ϕ, $\nabla^2 u$ reduces to

$$\frac{\partial^2 u}{\partial r^2} + \left(\frac{2}{r}\right)\frac{\partial u}{\partial r} \tag{2.8.15}$$

with $\partial u/\partial r = 0$ at $r = 0$. By either of the above arguments, the equation reduces to

$$\frac{\partial u}{\partial t} = 3\frac{\partial^2 u}{\partial r^2} \quad \text{at} \quad r = 0. \tag{2.8.16}$$

In the case of symmetrical heat flow problems for hollow cylinders and spheres that exclude $r = 0$ simpler equations than the above may be employed by suitable changes of variable:

(i) The change of independent variable $R = \log_e r$ transforms the cylindrical equation

$$\frac{\partial u}{\partial t} = \frac{\partial^2 u}{\partial r^2} + \frac{1}{r}\frac{\partial u}{\partial r} \tag{2.8.17}$$

to

$$e^{2r}\frac{\partial u}{\partial t} = \frac{\partial^2 u}{\partial R^2}. \tag{2.8.18}$$

(ii) The change of dependent variable given by $u = w/r$ transforms the spherical equation

$$\frac{\partial u}{\partial t} = \frac{\partial^2 u}{\partial r^2} + \frac{2}{r}\frac{\partial u}{\partial r} \tag{2.8.19}$$

to

$$\frac{\partial w}{\partial t} = \frac{\partial^2 w}{\partial r^2}. \tag{2.8.20}$$

As a worked example of the above methods, consider the equation

$$\frac{\partial u}{\partial t} = \frac{\partial^2 u}{\partial x^2} + \frac{2}{x}\frac{\partial u}{\partial x}, \quad 0 < x < 1,$$

with the initial condition $u = 1 - x^2$ when $t = 0$, $0 \le x \le 1$, and the boundary conditions $\partial u/\partial x = 0$ at $x = 0$, $t > 0$, $u = 0$ at $x = 1$, $t > 0$.

(i) Find the range of $r = k/h^2$ for which the explicit finite difference equations will be stable.

(ii) For $h = 0.1$, $k = 0.001$ calculate a finite difference solution to four significant digits by an explicit method at the points $(0, 0.001)$, $(0.1, 0.001)$ and $(0.9, 0.001)$ in the xt-plane.

At $x = 0$, $(2/x)\partial u/\partial x$ is indeterminate. As

$$\lim_{x \to 0} \frac{2}{x}\frac{\partial u}{\partial x} = \lim_{x \to 0} 2\frac{\partial^2 u}{\partial x^2}$$

the equation is replaced at $x = 0$ by

$$\frac{\partial u}{\partial t} = 3\frac{\partial^2 u}{\partial x^2}$$

which may be approximated by

$$\frac{u_{0,j+1} - u_{0,j}}{k} = \frac{3(u_{-1,j} - 2u_{0,j} + u_{1,j})}{h^2}.$$

If $(\partial u/\partial x)_{i,j}$ is approximated by $(u_{i+1,j} - u_{i-1,j})/2h$ we have

$$u_{-1,j} = u_{i,j} \quad \text{as} \quad \left(\frac{\partial u}{\partial x}\right)_{0,j} = 0.$$

Hence

$$u_{0,j+1} = u_{0,j} + 3r(2u_{1,j} - 2u_{0,j}).$$

At $x \ne 0$ the partial differential equation can be approximated by

$$\frac{1}{k}(u_{i,j+1} - u_{i,j}) = \frac{1}{h^2}(u_{i-1,j} - 2u_{i,j} + u_{i+1,j}) + \frac{2}{2i(h)^2}(u_{i+1,j} - u_{i-1,j})$$

or

$$u_{i,j+1} = r\left(1 - \frac{1}{i}\right)u_{i-1,j} + (1 - 2r)u_{i,j} + 2\left(1 + \frac{1}{i}u_{i+1,j}\right)$$

for $i = 0(1)n$ where $nh = 1$. Now $u = 0$ at $x = 1$, $u_{n,j} = 0$, to give

$$u_{n-1,j+1} = r\left(1 - \frac{1}{n-1}\right)u_{n-2,j} + (1 - 2r)u_{n-1,j}.$$

Hence, the matrix equation is

$$
\mathbf{u}_{j+1} = \begin{pmatrix}
1-6r & 6r & & & & \\
0 & 1-2r & 2r & & & \\
\ddots & \ddots & \ddots & & & \\
& r\left(1-\frac{1}{i}\right) & 1-2r & r\left(1+\frac{1}{i}\right) & & \\
& & \ddots & \ddots & \ddots & \\
& & & r\left(1-\frac{1}{n-1}\right) & 1-2r &
\end{pmatrix} \mathbf{u}_j i.
$$

(Since all the off-diagonal elements are one-signed all its eigenvalues are real.)

An application of Gerschgorin's theorem shows that each eigenvalue of A satisfies at least one of

(i) $|\lambda - (1-6r)| \le 6r$;

(ii) $|\lambda - (1-2r)| \le |r(1-1/i)| + |r(1+1/i)|, \quad i = 1, \ldots, n-2$,

(iii) $|\lambda - (1-2r)| \le |r(1-1/(n-1))|$.

Consider (i), where the condition $|\lambda - (1-6r)| \le 6r$ is shown graphically in the top diagram of Figure 2.7. If $|\lambda| \le 1$ then the non-trivial case gives $-1 \le 1-12r$ or $r \le \frac{1}{6}$.

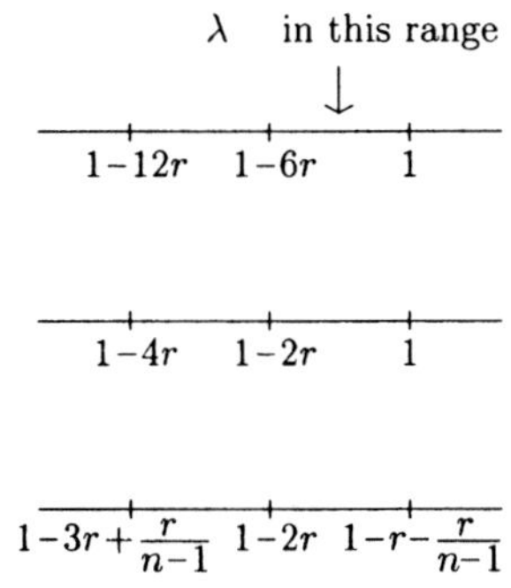

Fig. 2.7.

Consider (ii) in which $|\lambda - (1-2r)| \le 2r$ as illustrated in the central diagram of Figure 2.7. Now $|\lambda| \le 1$ to give the useful case $-1 \le 1 - 4r$ or $r \le \frac{1}{2}$.

Finally, consider (iii) with the inequality $|\lambda - (1-2r)| \le r(1 - 1/(n-1))$ with the relationship shown in the bottom diagram of Figure 2.7. If $|\lambda| \le 1$ then $-1 \le (1-3r) + r/(n-1)$ or $(3 - 1/(n-1))r \le 2$. Hence

$$
r \le \frac{2(n-1)}{3n-4}.
$$

It is also a requirement that

$$
1 - r - \frac{r}{n-1} \le 1
$$

to give $r \geq 0$. Hence $0 < r \leq \frac{1}{6}$ for overall stability.

Hence for the suggested grid steps in this problem, $\delta x = 0.1$ and $\delta t = 0.001$, then $r = 0.1$, which satisfies the stability conditions. The difference equations for the numerical solution are then

$$u_{0,j+1} = \frac{1}{5}(2u_{0,j} + 3u_{1,j})$$

and

$$u_{i,j+1} = 0.1\left(1 - \frac{1}{i}\right)u_{i-1,j} + (0.8)u_{i,j} + 0.1\left(1 + \frac{1}{i}\right)u_{i+1,j}$$

and the results are shown in Table 2.8.

Table 2.8.

$t\backslash x$	0.0	0.1	0.2	0.3	0.4
	0.5	0.6	0.7	0.8	0.9
0.001	0.994	0.984	0.954	0.904	0.834
	0.744	0.634	0.504	0.354	0.184
0.002	0.988	0.978	0.948	0.898	0.828
	0.738	0.628	0.498	0.348	0.1787
0.003	0.982	0.972	0.942	0.892	0.822
	0.732	0.622	0.492	0.3421	0.1739
0.004	0.976	0.9668	0.936	0.886	0.816
	0.726	0.616	0.486	0.3363	0.1695

The work in this chapter has acted as an introduction to the problems associated with the solution of partial differential equations with parabolic equations being used as a medium of instruction. The finite difference concept will now be carried forward to the other equation types, namely hyperbolic and elliptic. The chapter is concluded with some exercises on the use of spherical and cylindrical co-ordinates for the heat equation.

EXERCISES

2.27 A hollow infinite cylinder has internal radius 0.5 and exterior radius 2.0. The external surface is maintained at 0°C and the internal surface at 100°C. Initially the cylinder has a uniform temperature of 15°C and it is required to compute the distribution of temperature across the radius as time progresses. Use an explicit method with a suitable time step to compute the temperature for $r = 0.5(0.25)2.0$ for the first few time steps.

2.28 Repeat Exercise 2.27 using an implicit method.

2.29 The boundary condition on the exterior radius of the above cylinder is a radiation condition that

$$\frac{\partial u}{\partial r} = 1.$$

The internal surface is still maintained at 100°C and the initial uniform temperature at 15°C. Use both explicit and implicit methods to compute the temperature distribution as in Exercise 2.27.

2.30 A hollow sphere has internal radius 0.1 and exterior radius 1.0. The external surface is maintained at 0°C and the internal surface at 100°C. Initially the cylinder has a uniform temperature of 15°C and it is required to compute the distribution of temperature across the radius as time progresses. Use an explicit method with a suitable time step to compute the temperature for $r = 0.1(0.1)1.0$ for the first few time steps.

2.31 Repeat Exercise 2.30 using an implicit method.

2.32 The boundary condition on the exterior radius of the sphere of Exercise 2.30 is a radiation condition that

$$\frac{\partial u}{\partial r} = 1.$$

The internal surface is still maintained at 100°C and the initial uniform temperature at 15°C. Use both explicit and implicit methods to compute the temperature distribution as in Exercise 2.30.

3
Hyperbolic Equations and Characteristics

3.1 First Order Quasi-linear Equations

In Section 1.6, general second order equations were classified using characteristics, and this subject is revisited here. In the first chapter, the characteristics
were used to classify the equations and to form a transformation to allow
reduction to canonical form. In the process an ordinary differential equation
was also obtained which held along a characteristic (1.6.10) and in the case of
real characteristics this equation can form the basis of a method of solution.
The method is based on using finite differences along the characteristic curves
which form a natural grid. This will be covered in Section 3.3 and a slightly
different derivation to that in Chapter 1 will be given, as this approach is
informative in its own right.

Firstly however, first order partial differential equations will be treated using
characteristics, as there is nothing special about the second order cases so far
considered. These concepts are then easily applied to second order hyperbolic
equations. Hence consider

$$a\frac{\partial u}{\partial x} + b\frac{\partial u}{\partial y} = c \tag{3.1.1}$$

where a, b and c are in general functions of x, y and u. For economy of writing
define

$$\frac{\partial u}{\partial x} = p \quad \text{and} \quad \frac{\partial u}{\partial y} = q \tag{3.1.2}$$

as in the first chapter, the equation then becoming

$$ap + bq = c. \tag{3.1.3}$$

To find a solution to this equation, a direction at each point of the xy-plane is sought, along which the integration of the equation transforms to the integration of an ordinary differential equation. In this direction the expression to be integrated will be independent of partial derivatives in other directions such as p and q.

Assume that the problem is solved and the solution values at every point on a curve C are known, where C does not coincide with the curve Γ on which initial values of u are given. Then consider the situation in Figure 3.1 and let $D(x, y)$ and $E(x + \delta x, y + \delta y)$ be adjacent points on C, then $u(x, y)$ at D and the partial derivative values at D satisfy the differential equation $ap + bq = c$. But along the direction of the tangent at D to the curve C

$$du = \frac{\partial u}{\partial x} dx + \frac{\partial u}{\partial y} dy = p\, dx + q\, dy \qquad (3.1.4)$$

where dy/dx is the slope of C at D. This same equation was used in Section 1.6.

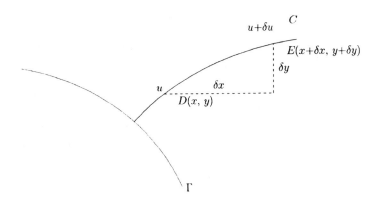

Fig. 3.1.

Hence $ap + bq = c$ and $dx\, p + dy\, q = du$ form two simultaneous equations which on eliminating p give

$$bq\, dx - aq\, dy = c\, dx - a\, du$$

or

$$q(a\, dy - b\, dx) + (c\, dx - a\, du) = 0. \qquad (3.1.5)$$

This equation is independent of p (a and b depend only on x, y and u). Equation (3.1.5) can also be made independent of q by choosing the curve C so that its slope satisfies the equation

$$\frac{dy}{dx} = \frac{b}{a}. \qquad (3.1.6)$$

On this curve, the differential equation reduces to

$$c\,dx - a\,du = 0. \tag{3.1.7}$$

Equation (3.1.6) is the differential equation for curve C (a characteristic) and (3.1.7) is a differential equation for the solution values of u along C. Equations (3.1.6) and (3.1.7) may be written as

$$\frac{dx}{a} = \frac{dy}{b} = \frac{du}{c}. \tag{3.1.8}$$

The same result is obtained by eliminating q in the step to (3.1.5). Indeed (3.1.8) is the curve along which the highest derivative of the partial differential equation (now the first) is not uniquely defined, in accordance with the earlier definition of Chapter 1.

The equation of the characteristic curve and the ordinary differential equation along this curve embodied in equation (3.1.8) can be used to set up a numerical solution to first order equations of the type in (3.1.1). Let u be specified on the initial line Γ which must not be a characteristic and let $R(x_R, y_R)$ be a point on Γ and $P(x_P, y_P)$ be a point on the characteristic C through R such that $x_P - x_R$ is small.

Then the differential equation for the characteristic is $a\,dy = b\,dx$ which gives either dy or dx when the other is known. The differential equation for the solution along a characteristic is either $a\,du = c\,dx$ or $b\,du = c\,dy$. Denoting a first approximation to u at P by $u_P^{(1)}$, and a second approximation by $u_P^{(2)}$, then to compute the first approximation: assume x_P is known, then from $a\,dy = b\,dx$ we have

$$a_R[y_P^{(1)} - y_R] = b_R[x_P - x_R] \tag{3.1.9}$$

which gives a first approximation $y_P^{(1)}$ to y_P and then $a\,du = c\,dx$ gives

$$a_R[u_P^{(1)} - u_R] = c_R[x_P - x_R] \tag{3.1.10}$$

giving $u_P^{(1)}$.

The second and subsequent approximations are found by replacing the coefficients a, b and c by known mean values over the arc RP. Then

$$\frac{1}{2}(a_R + a_P^{(1)})(y_P^{(2)} - y_R) = \frac{1}{2}(b_R + b_P^{(1)})(x_P - x_R) \tag{3.1.11}$$

gives $y_P^{(2)}$ and

$$\frac{1}{2}(a_R + a_P^{(1)})(u_P^{(2)} - u_R) = \frac{1}{2}(c_R + c_P^{(1)})(x_P - x_R) \tag{3.1.12}$$

gives $u_P^{(2)}$. This second procedure can be repeated iteratively until successive iterates agree to a specified number of decimal places.

Consider now an example in which the function u satisfies the equation

$$\frac{1}{x}\frac{\partial u}{\partial x} + u^2\frac{\partial u}{\partial y} = u \qquad (3.1.13)$$

with the condition $u = x$ on $y = 0$ for $0 < x < \infty$. The above finite difference method is used to calculate a first approximation to the solution and to the value of y at the point $P(1.1, y)$, $y > 0$, on the characteristic through the point $R(1, 0)$. A second approximation to these values is then found using the iterative technique. The Cartesian equation of the characteristic through $(1, 0)$ is found and hence the values given by the analytical formulae for y and u.

R is the point $(1, 0)$ which implies that

$$x_R = 1, \quad y_R = 0, \quad u_R = x_R = 1, \quad x_P = 1.1.$$

From equation (3.1.13), the equation of the characteristic and the equation along the characteristic are

$$x\,dx = \frac{dy}{u^2} = \frac{du}{u}. \qquad (3.1.14)$$

Hence the left-hand equation gives the characteristic equation in finite difference form as

$$x_R(x_P - x_R) = \frac{1}{u_R^2}(y_P^{(1)} - y_R) \qquad (3.1.15)$$

which gives

$$y_P^{(1)} = 0.1. \qquad (3.1.16)$$

For the right-hand equation of (3.1.14) which gives the equation holding along the characteristic, the finite difference form is

$$y_P^{(1)} - y_R = u_R(u_P^{(1)} - u_R)$$

to give

$$u_P^{(1)} = 1.1. \qquad (3.1.17)$$

These results can now be improved upon by using average approximations based on the first estimates of y and u at the point P. Hence $y_P^{(2)}$ satisfies

$$\frac{1}{2}(x_P + x_R)(x_P - x_R) = \frac{1}{2}\left(\frac{1}{u_P^{(1)\,2}} + \frac{1}{u_R^2}\right)(y_P^{(2)} - y_R). \qquad (3.1.18)$$

In this particular example, the use of averages for x_P and x_R could have been employed at the first stage as in this special case the coefficient a is just a function of x. However the method has been implemented here as it would be were this specialisation not available. The corresponding averaged equation for the equation along the characteristic is

$$y_P^{(2)} - y_R = \frac{1}{2}(u_P^{(1)} + u_R)(u_P^{(2)} - u_R). \tag{3.1.19}$$

These formulae yield $y_P^{(2)} = 0.11$ and $u_P^{(2)} = 1.11524$. Equation (3.1.14) will yield the analytic solution for comparison. The two equations split easily to give

$$x \, dx = \frac{du}{u}$$

and

$$dy = u \, du$$

with solutions

$$x^2 = 2 \ln u + A \quad \text{and} \quad y = (u^2 + B)/2. \tag{3.1.20}$$

To force the characteristic through (1,0) with $u = 1$ gives $A = 1$ and $B = -1$. Hence at $x = 1.1$,

$$u = e^{\frac{x^2-1}{2}} \tag{3.1.21}$$

which gives $u = 1.11071$ and using the second equation of (3.1.20) gives $y = 0.1168$ which shows good agreement and demonstrates the gain in accuracy of using the averaged formulae.

This first order case will be used as a model in the consideration of second order hyperbolic equations in Section 3.4, but first two alternative first order methods are outlined. Before attempting the next section the reader is encouraged to try the following illustrative examples.

EXERCISES

3.1 The function u satisfies the equation

$$x^2 u \frac{\partial u}{\partial x} + e^{-y} \frac{\partial u}{\partial y} = -u^2$$

and the condition $u = 1$ on $y = 0$ for $0 < x < \infty$. Calculate the Cartesian equation of the characteristic through the point $R(x_R, 0)$ with $x_R > 0$ and the solution along this characteristic. Use a finite difference method to calculate the first approximations to the solution and to the value of y at the point $P(1.1, y)$ with $y > 0$ on the characteristic through the point $R(1, 0)$.

Calculate the second approximation to the above values by an iterative method. Compare your final results with those given by the analytic solution.

3.2 The function u satisfies the equation

$$\frac{\partial u}{\partial x} + \frac{x}{\sqrt{u}}\frac{\partial u}{\partial y} = 2x$$

and the conditions $u = 0$ on $x = 0$ for $y \geq 0$ and $u = 0$ on $y = 0$ for $x > 0$. Calculate the analytic solutions at the points $(2, 5)$ and $(5, 4)$. Sketch the characteristics through these two points. If the initial condition along $y = 0$ is replaced by $u = x$, calculate the approximate solution and the value of y at the point $P(4.05, y)$ on the characteristic through the point $R(4, 0)$. Compare these results with the analytic values.

3.3 Calculate the analytic solution of the equation

$$\frac{\partial u}{\partial x} + 3x^2\frac{\partial u}{\partial y} = x + y$$

at the point $(3, 19)$ given that $u(x, 0) = x^2$.

Now solve the same problem numerically using the characteristic method. This is not the usual type of problem. It is normal to give conditions at a point on an initial line and then compute along the characteristic through that point. Hence the point at which the solution will be found in not known a *priori*. Hence the point of intersection of the characteristic through $(3, 19)$ with the initial line $y = 0$ needs to be found first. Modify the characteristic method to achieve this variation. What is special about this problem which allows such an approach?

3.4 Calculate by an iterative method, the solution of the equation

$$(x - y)\frac{\partial u}{\partial x} + u\frac{\partial u}{\partial y} = x + y$$

at the point $P(1.1, y)$ on the characteristic through $R(1, 0)$ given that $u(x, 0) = 1$.

3.2 Lax–Wendroff and Wendroff Methods

The first method to be considered in this section is the Lax–Wendroff explicit method, which will be applied to the equation:

$$\frac{\partial u}{\partial t} + a\frac{\partial u}{\partial x} = 0, \quad a > 0 \tag{3.2.1}$$

as an alternative to the above characteristic approach. By Taylor's expansion

$$
\begin{aligned}
u_{i,j+1} &= u(x_i, t_j + k) \\
&= u_{i,j} + k\left(\frac{\partial u}{\partial t}\right)_{i,j} + \frac{1}{2}k^2\left(\frac{\partial^2 u}{\partial t^2}\right)_{i,j} + \cdots
\end{aligned}
\tag{3.2.2}
$$

where

$$
x_i = ih, \quad t_j = jR, \quad i = 0, \pm 1, \pm 2, \ldots, \quad j = 0, 1, 2, \ldots
$$

The differential equation is used to eliminate the t derivatives as it gives

$$
\frac{\partial}{\partial t} = -a\frac{\partial}{\partial x}
\tag{3.2.3}
$$

or

$$
u_{i,j+1} = u_{i,j} - ka\left(\frac{\partial u}{\partial x}\right)_{i,j} + \frac{1}{2}k^2 a^2\left(\frac{\partial^2 u}{\partial x^2}\right)_{i,j} + \cdots
\tag{3.2.4}
$$

Finally replace the x derivatives by central difference approximations to give

$$
\begin{aligned}
u_{i,j+1} &= u_{i,j} - \frac{ka}{2h}(u_{i+1,j} - u_{i-1,j}) + \frac{k^2 a^2}{2h^2}(u_{i-1,j} - 2u_{i,j} + u_{i+1,j}) + \cdots \\
&= \frac{1}{2}(ap)(1 + ap)u_{i-1,j} + (1 - a^2 p^2)u_{i,j} \\
&\quad - \frac{1}{2}ap(1 - ap)u_{i+1,j} + \cdots
\end{aligned}
\tag{3.2.5}
$$

where $p = k/h$. This is the Lax–Wendroff scheme. Equation (3.2.5) may be used for both initial value and boundary value problems. The method can be shown to be stable for $0 < ap \leq 1$ and its local truncation error is

$$
\frac{1}{6}k^2\frac{\partial^3 u}{\partial t^3} + \frac{1}{6}ah^2\frac{\partial^3 u}{\partial x^3}.
\tag{3.2.6}
$$

As an example of the Lax–Wendroff method, consider the simplest problem with $a = 1$, namely

$$
\frac{\partial u}{\partial t} + \frac{\partial u}{\partial x} = 0, \quad 0 < x < \infty, \quad t > 0,
$$

subject to the boundary condition

$$
u(0, t) = t, \quad t > 0,
$$

and initial condition

$$
u(x, 0) = \begin{cases} x^2 & 0 \leq x \leq 2 \\ x & 2 \leq x. \end{cases}
$$

The problem will be solved:

(i) analytically, and

(ii) numerically using the explicit Lax–Wendroff equation, with $h = 1/4$, $k = 1/8$.

Hence for (i)

$$\frac{\partial u}{\partial t} + \frac{\partial u}{\partial x} = 0 \quad \text{gives} \quad \frac{dx}{1} = \frac{dt}{1} = \frac{du}{0}$$

which implies that u is constant along the straight line characteristic through $(x_R, 0)$ which is $t = x - x_R$. If $u(x,0)$ is $\phi(x)$ (say) then the solution along the characteristic is

$$u(x,t) = \phi(x_R) = \phi(x - t).$$

Similarly if $u(0,t) = \Psi(t)$ the solution along the characteristic $t - t_s = x$ from $(0, t_s)$ is $u(x,t) = \Psi(t-x)$. The solution is shown graphically in Figure 3.2.

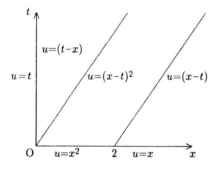

Fig. 3.2.

The numerical solution of the problem is shown in Table 3.1 for a range of values of t and x, which cross the three regions shown in Figure 3.2.

To achieve the range of values in the x direction at $t = 0.75$, the initial range of x at $t = 0$ was from $x = 0.0$ to $x = 3.75$, and then at each time step there is one less x point on the right-hand side. These extra x values are not shown in the table.

For comparison, the analytic values are evaluated from the formulae in Figure 3.2 at the same points and are shown in Table 3.2.

The change of boundary condition at $x = 2$ will transmit the discontinuity into the solution along the characteristic $t = x - 2$ as will the change at the origin which will transmit along $t = x$. Along these lines where the solution is in fact 1.0 and 0.0 respectively, the computed solution will be expected to be least accurate as in effect there are no characteristics built into the calculation. These points nevertheless still follow the true solution quite well. Away from these regions of difficulty the computed solution maintains good accuracy.

Table 3.1.

$t\backslash x$	0.25	0.5	0.75	1.0	1.25
	1.5	1.75	2.0	2.25	2.5
0.125	0.01563	0.1406	0.3906	0.8047	1.125
	1.375	1.625	1.875	2.125	2.375
0.25	0.04102	0.0625	0.2451	0.6094	0.9736
	1.25	1.5	1.75	2.00	2.25
0.375	0.1167	0.0316	0.1311	0.4273	0.8025
	1.115	1.375	1.625	1.875	2.125
0.5	0.2242	0.0511	0.0568	0.2693	0.6227
	0.9654	1.246	1.5	1.75	2.0
0.625	0.3493	0.1153	0.0281	0.1454	0.4473
	0.8018	1.109	1.374	1.625	1.875
0.75	0.4819	0.2139	0.0461	0.0637	0.2898
	0.6304	0.9609	1.243	1.500	1.75

Table 3.2.

$t\backslash x$	0.25	0.5	0.75	1.0	1.25
	1.5	1.75	2.0	2.25	2.5
0.125	0.01563	0.1406	0.3906	0.7656	1.125
	1.375	1.625	1.875	2.125	2.375
0.25	0.00	0.0625	0.2500	0.5625	1.000
	1.25	1.5	1.75	2.00	2.25
0.375	0.125	0.0156	0.1406	0.3906	0.7656
	1.125	1.375	1.625	1.875	2.125
0.5	0.2500	0.0000	0.0625	0.2500	0.5625
	1.0000	1.250	1.5	1.75	2.0
0.625	0.3750	0.1250	0.0156	0.1406	0.3906
	0.7656	1.125	1.375	1.625	1.875
0.75	0.5000	0.2500	0.0000	0.0625	0.2500
	0.5625	1.0000	1.250	1.500	1.75

Before moving on to more general first order problems and a further finite difference method, the condition for convergence is now derived.

This condition is the Courant–Friedrich–Lewy (CFL) condition for first order equations. A typical section of a characteristic curve on a rectangular grid is illustrated in Figure 3.3.

Assume that the first order hyperbolic differential equation (3.1.1) has been

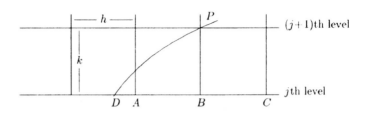

Fig. 3.3.

approximated by a difference equation of the form

$$u_{i,j+1} = au_{i-1,j} + bu_{i,j} + cu_{i+1,j}. \tag{3.2.7}$$

Then u_P depends on u_A, u_B and u_C. Assume that the characteristic through P of the hyperbolic equation meets the line AC at D and consider AC as an initial line segment. If the initial values along AC are altered then the solution value of P of the finite difference equation will change, but these alterations will not affect the solution value at P of the partial differential equation which depends on the initial value of D. In this case the solution of the difference equation cannot converge to the solution of the partial differential equation as $h \to 0, k \to 0$. For convergence D must lie between A and C, and this is the CFL condition.

Consider the Lax–Wendroff scheme for

$$\frac{\partial u}{\partial t} + a\frac{\partial u}{\partial x} = 0,$$

namely

$$u_{i,j+1} = \frac{1}{2}ap(1 + ap)u_{i-1,j} + (1 - a^2p^2)u_{i,j} - \frac{1}{2}ap(1 - ap)u_{i+1,j}. \tag{3.2.8}$$

The characteristic equation of the partial differential equation is

$$\frac{dt}{1} = \frac{dx}{a}. \tag{3.2.9}$$

For convergence of the difference equation, the slope of $PD = dt/dx \geq$ the slope of PA to give

$$\frac{1}{a} \geq \frac{k}{h} \quad \text{or} \quad ap \leq 1 \tag{3.2.10}$$

which coincides with the condition for stability, namely

$$0 < ap \leq 1 \quad \text{since} \quad a > 0, p > 0.$$

There are some shortcomings with the Lax–Wendroff method which are apparent from the previous example. These arise because the method involves computing $u_{i,j+1}$ from $u_{i-1,j}$, $u_{i,j}$ and in particular $u_{i+1,j}$. If at the first step

u is known to say $i = N$, then at the second step the right-most value of u which can be found is at $i = N - 1$, unless a numerical boundary condition is supplied at $i = N$. At each step the end value will move back one grid point. Hence to get a specific range in the x direction at some later time t will require carrying the requisite number of extra x points at $t = 0$, as implemented for Table 3.1. If there were no conditions at $x = 0$ for a range of t as in the previous example the same would occur at the opposite end. Also the method relies on the analytic substitution of $\partial/\partial t$ by $a\partial/\partial x$. Hence, if for example a were a function of x and t, then the replacement of the second derivative would then involve partial derivatives of a. Equally if there were a right-hand side function the same would be true, and this case is considered later.

An alternative to both Lax–Wendroff and the characteristic method is to use finite differences on the original equation. This results in the Wendroff method. Hence to place

$$a\frac{\partial u}{\partial x} + b\frac{\partial u}{\partial y} = c \qquad (3.2.11)$$

in finite difference form use the average of the central difference in the x direction at y and $y + k$ and the equivalent in the y direction to yield:

$$\frac{a}{2}\left[\frac{u_{i+1,j} - u_{i,j}}{h} + \frac{u_{i+1,j+1} - u_{i,j+1}}{h}\right]$$
$$+ \frac{b}{2}\left[\frac{u_{i,j+1} - u_{i,j}}{k} + \frac{u_{i+1,j+1} - u_{i+1,j}}{k}\right] = c. \qquad (3.2.12)$$

This equation can only be used when there are known initial values on $t = 0$ and known boundary values on $x = 0$. The formula can then be used explicitly in the form:

$$u_{i+1,j+1} = u_{i,j} + \frac{b - ap}{b + ap}(u_{i+1,j} - u_{i,j+1}) + \frac{2kc}{b + ap} \qquad (3.2.13)$$

with $p = k/h$. This method is unconditionally stable.

The application of the Wendroff method to re-compute the previous Lax–Wendroff example is shown in Table 3.3. Again the same step sizes are employed, and now there is no requirement to compute extra x values because of the rectangular nature of the computational molecule in (3.2.13).

As expected from the previous discussion, it is around the two characteristics through the discontinuities as shown in Figure 3.2 where the maximum error occurs. These results are less accurate than the Lax–Wendroff approach, but there the analytic removal of the partial t derivative is replaced by a relatively crude finite difference. The results reflect the table of true values very well despite this loss of accuracy in the approximation.

It is quite feasible to generalise the Lax–Wendroff method to equations of the form:

$$\frac{\partial u}{\partial t} + a(x,t)\frac{\partial u}{\partial x} = c(x,t). \qquad (3.2.14)$$

Table 3.3.

| $t \backslash x$ | 0.25 | 0.5 | 0.75 | 1.0 | 1.25 |
	1.5	1.75	2.0	2.25	2.5
0.125	−0.0208	0.1528	0.3866	0.7670	1.161
	1.363	1.629	1.874	2.125	2.375
0.25	0.0347	0.01852	0.2755	0.5504	0.9705
	1.292	1.475	1.762	1.995	2.252
0.375	0.1366	−0.0046	0.1119	0.4216	0.7334
	1.157	1.398	1.597	1.895	2.114
0.5	0.2539	0.0504	0.0159	0.2471	0.5837
	0.9244	1.315	1.492	1.731	2.022
0.625	0.3763	0.1452	0.0073	0.0958	0.4098
	0.7552	1.111	1.442	1.589	1.875
0.75	0.5004	0.2579	0.0617	0.0187	0.2262
	0.5861	0.9301	1.281	1.544	1.699

With this equation

$$\frac{\partial u}{\partial t} = -a\frac{\partial u}{\partial x} + c \qquad (3.2.15)$$

and hence

$$
\begin{aligned}
\frac{\partial^2 u}{\partial t^2} &= \frac{\partial}{\partial t}\left[-a\frac{\partial u}{\partial x} + c\right] \\
&= -a\frac{\partial^2 u}{\partial t \partial x} - \frac{\partial a}{\partial t}\frac{\partial u}{\partial x} + \frac{\partial c}{\partial t} \\
&= a^2\frac{\partial^2 u}{\partial x^2} + a\frac{\partial a}{\partial x}\frac{\partial u}{\partial x} - a\frac{\partial c}{\partial x} - \frac{\partial a}{\partial t}\frac{\partial u}{\partial x} + \frac{\partial c}{\partial t}. \qquad (3.2.16)
\end{aligned}
$$

Expansion (3.2.2) can now be used with the above derivatives of t to give the method

$$
\begin{aligned}
u_{i,j+1} = \; & u_{i-1,j}\left(\frac{ap}{2} + \frac{a^2 p^2}{2} - \frac{ak^2}{4h}\frac{\partial a}{\partial x} + \frac{k^2}{4h}\frac{\partial a}{\partial t}\right) \\
& + u_{i,j}(1 - a^2 p^2) \\
& + u_{i+1,j}\left(-\frac{ap}{2} + \frac{a^2 p^2}{2} + \frac{ak^2}{4h}\frac{\partial a}{\partial x} - \frac{k^2}{4h}\frac{\partial a}{\partial t}\right) \\
& + ck + \frac{k^2}{2}\frac{\partial c}{\partial t} - \frac{ak^2}{2}\frac{\partial c}{\partial x}. \qquad (3.2.17)
\end{aligned}
$$

This formula now requires knowledge of $\partial a/\partial x$, $\partial c/\partial x$, $\partial a/\partial t$ and $\partial c/\partial t$, and is required for some of the exercises at the end of the section.

If a, b and c are functions of x and y in the Wendroff method, then in (3.2.12), a, b and c will vary from point to point. The obvious value consistent with the approach used to finite difference the derivatives around the central point $u_{i+1/2,j+1/2}$ is to take a as

$$(a(x_i, y_j) + a(x_{i+1}, y_j) + a(x_i, y_{j+1}) + a(x_{i+1}, y_{j+1}))/4.0 \qquad (3.2.18)$$

so averaging the a values at the four computational points. A similar formula is used for both b and c. Such a generalisation of the Wendroff formula is used in the exercises.

In summary, all the alternatives to using characteristics have some limitations in their use, but when employed in a suitable environment, these alternatives can provide rapid and simple methods.

This section is rounded off with a selection of exercises, and then attention is turned to second order hyperbolic equations.

EXERCISES

3.5 Use the Lax–Wendroff method to solve numerically the problem

$$\frac{\partial u}{\partial t} + \frac{\partial u}{\partial x} = 0$$

with the boundary conditions $u = t$ on $x = 0$ for $t > 0$, and $u = x(x-1)$ for $t = 0$ and $0 \le x \le 1$ and $u = 1 - x$ for $t = 0$ and $1 \le x$. Use $h = 0.25$ and compute the solution for four time steps of 0.125, beginning with the x range from 0 to 4 at $t = 0$.

3.6 Use the Lax–Wendroff method in extended form to solve the equation

$$\frac{\partial u}{\partial t} + 3t^2 \frac{\partial u}{\partial x} = t + x$$

with $u(x, 0) = x^2$. Take an initial range of x from 0 to 4 in steps of 0.25 and use time steps of 0.125 to complete four steps.

With these steps, what range of x would be needed at $t = 0$ to compute u at $t = 4$ and $x = 1$?

3.7 Use Lax–Wendroff in extended form on the problem

$$\frac{\partial u}{\partial t} + (x - t) \frac{\partial u}{\partial x} = x + t$$

with $u(x, 0) = 1$. Using a step size of 0.1 in x and 0.05 in t, find the value at the point P with $x = 0.5$ and $t = 0.5$. Take a suitable range of x at $t = 0$ to achieve this end.

3.8 Satisfy yourself that the problems in equations (3.1.1) and (3.1.2) would defeat Lax–Wendroff in its general form. Is there any further generalisation involving a, b or c being functions of u which could be attempted?

3.9 Repeat Exercise 3.5 with the Wendroff method. Use a suitable range of x to arrive at the same final range as the Lax–Wendroff method.

3.10 Apply the Wendroff method to Exercise 3.6. To do this the condition along $x = 0$ is required. Obtain this analytically to get the solution started. Here is a clear shortcoming of this method.

3.11 Consider the Wendroff method when there is u dependence in one or other of a, b or c.

3.3 Second Order Quasi-linear Hyperbolic Equations

An alternative approach to the derivation of the characteristics is given here. The method is completely equivalent to the derivation in Chapter 1, and allows all the relevant equations to be marshalled in this section.

Consider the second order quasi-linear partial differential equation

$$a\frac{\partial^2 u}{\partial x^2} + b\frac{\partial^2 u}{\partial x \partial y} + c\frac{\partial^2 u}{\partial y^2} + e = 0 \qquad (3.3.1)$$

where a, b, c and d may be functions of x, y, u, $\partial u/\partial x$ or $\partial u/\partial y$ but not second derivatives.

We will show that for hyperbolic equations, there are, at every point of the xy-plane, two directions in which the integration of the partial differential equations reduces to the integration of an equation involving total differentials only. In these directions the equation to be integrated is not complicated by the presence of partial derivatives in other directions.

As in Chapter 1, denote first and second derivatives by

$$\frac{\partial u}{\partial x} = p, \ \frac{\partial u}{\partial y} = q, \ \frac{\partial^2 u}{\partial x^2} = r, \ \frac{\partial^2 u}{\partial x \partial y} = s, \ \frac{\partial^2 u}{\partial y^2} = t. \qquad (3.3.2)$$

Let C be a curve in the xy-plane on which values of u, p and q are such that they and the second order derivatives r, s and t derivable from them satisfy equation (3.3.1). (C is not a curve on which initial values of u, p and q are given.) Then the differentials of p and q in directions tangential to C satisfy the equations

$$dp = \frac{\partial p}{\partial x}dx + \frac{\partial p}{\partial y}dy = r\,dx + s\,dy \qquad (3.3.3)$$

and

$$dq = \partial q/\partial x\,dx + \frac{\partial q}{\partial y}dy = s\,dx + t\,dy \qquad (3.3.4)$$

where

$$ar + bs + ct + e = 0 \qquad (3.3.5)$$

and dy/dx is the slope of the tangent to C. Eliminating r and t from (3.3.5) using (3.3.3) and (3.3.4) gives

$$a\left(\frac{dp}{dx} - s\frac{dy}{dx}\right) + bs + c\left(\frac{dq}{dy} - s\frac{dx}{dy}\right) + e = 0 \qquad (3.3.6)$$

or

$$s\left[a\left(\frac{dy}{dx}\right)^2 - b\left(\frac{dy}{dx}\right) + c\right] - \left[a\frac{dp}{dx}\frac{dy}{dx} + c\frac{dq}{dx} + e\frac{dy}{dx}\right] = 0. \qquad (3.3.7)$$

Now choose C so that the slope of the tangent at every point satisfies

$$a\left(\frac{dy}{dx}\right)^2 - b\left(\frac{dy}{dx}\right) + c = 0 \qquad (3.3.8)$$

so that s is also eliminated. This is the usual equation of the characteristics, and in this direction the equation reduces to

$$a\frac{dp}{dx}\frac{dy}{dx} + c\frac{dq}{dx} + e\frac{dy}{dx} = 0. \qquad (3.3.9)$$

Thus if

$$a\left(\frac{dy}{dx}\right)^2 - b\left(\frac{dy}{dx}\right) + c = 0$$

has two real roots for dy/dx that is $b^2 - 4ac > 0$ (which is the condition for the equation to be hyperbolic) then at every point $P(x, y)$ of the solution domain, there are two directions, given by these roots along which there is a relationship between the total differentials dp and dq. This relation can be used to evaluate the solution by a step-by-step method, which is called the characteristic method.

Let the real roots of

$$a\left(\frac{dy}{dx}\right)^2 - b\left(\frac{dy}{dx}\right) + c = 0 \qquad (3.3.10)$$

be

$$\frac{dy}{dx} = f, \quad \frac{dy}{dx} = g \qquad (3.3.11)$$

Let Γ be a non-characteristic curve along which initial values for u, p and q are known. Let P and Q be two points on Γ that are close together and let the characteristic with slope f through P intersect the characteristic with slope g through Q at the point R as shown in Figure 3.4.

As a first approximation, the arcs PR and QR may be considered to be straight lines of slope f_P and g_Q respectively. Then we have

$$y_R - y_P = f_P(x_R - x_P) \quad \text{and} \quad y_R - y_Q = g_Q(x_R - x_Q) \qquad (3.3.12)$$

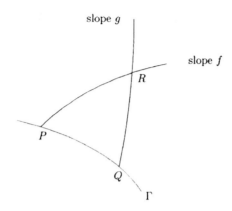

Fig. 3.4.

giving two equations for the two unknowns y_R and x_R.

On PR, the relationship $af\,dp + c\,dq + e\,dy = 0$ holds, and on QR, the equivalent relationship is $ag\,dp + c\,dq + e\,dy = 0$. These relationships can be approximated by:

on PR

$$a_P f_P(p_R - p_P) + c_P(q_R - q_P) + e_P(y_R - y_P) = 0 \qquad (3.3.13)$$

on QR

$$a_Q g_Q(p_R - p_Q) + c_Q(q_R - q_Q) + e_Q(y_R - y_Q) = 0. \qquad (3.3.14)$$

Equations (3.3.13) and (3.3.14) can be solved for p_R and q_R once x_R and y_R are known from (3.3.12). The value of u at R can then be obtained from

$$du = \frac{\partial u}{\partial x}\,dx + \frac{\partial u}{\partial y}\,dy = p\,dx + q\,dy \qquad (3.3.15)$$

by replacing the values of p and q along PR by their average values and approximating the last equation by

$$u_R - u_P = \frac{1}{2}(p_P + p_R)(x_R - x_P) + \frac{1}{2}(q_P + q_R)(y_R - y_P) \qquad (3.3.16)$$

(or we could use the values along QR).

This first approximation for u_R can now be used to improve the results by replacing the coefficients by average values:

$$y_R - y_P = \frac{1}{2}(f_P + f_R)(x_R - x_P) \qquad (3.3.17)$$

$$y_R - y_Q = \frac{1}{2}(g_P + g_R)(x_R - x_Q) \qquad (3.3.18)$$

and

$$\frac{1}{2}(a_P + a_R)\frac{1}{2}(f_P + f_R)(p_R - p_P) \tag{3.3.19}$$

$$+\frac{1}{2}(c_P + c_R)(q_R - q_P) + \frac{1}{2}(e_P + e_R)(y_R - y_P) = 0$$

$$\frac{1}{2}(a_Q + a_R)\frac{1}{2}(g_Q + g_R)(p_R - p_Q) \tag{3.3.20}$$

$$+\frac{1}{2}(c_Q + c_R)(q_R - q_Q) + \frac{1}{2}(e_Q + e_R)(y_R - y_Q) = 0.$$

An improved value of u_R can then be found from (3.3.16), and repetition of this last cycle of operations will eventually yield u_R to the accuracy warranted by these finite difference approximations. The process is then continued as shown in Figure 3.5 so generating a natural grid. From the points A, B, C and D the values at P, Q and R are found, followed by those at S and T.

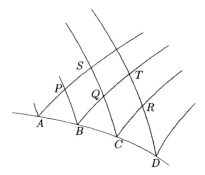

Fig. 3.5.

The method of characteristics is now used to derive a solution of the quasi-linear equation

$$\frac{\partial^2 u}{\partial x^2} - (x + y)^2 \frac{\partial^2 u}{\partial y^2} = 0 \tag{3.3.21}$$

at the first characteristic grid point between $x = 0.1$ and $x = 0.2$, for $y > 0$. Along the initial line $y = 0$ for $0 \le x \le 1$, $u = 1 + x^2/2$ and $q = 2x$.

Since u is given as a continuous function of x on $[0, 1]$ the initial value of $p = \partial u / \partial x = x$. The equation for the slope of the characteristics is

$$\left(\frac{dy}{dx}\right)^2 - (x + y)^2 = 0$$

and the slopes are

$$\frac{dy}{dx} = x + y, \quad \frac{dy}{dx} = -(x + y)$$

giving $f = x + y$ and $g = -(x + y)$. Now initially on $y = 0$

$$u = 1.0 + x^2/2, \quad f = x + y, \quad g = -(x + y)$$

and

$$p = x, \quad q = 2x.$$

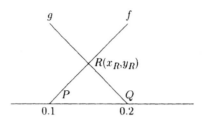

Fig. 3.6.

Let P be $x = 0.1$ and Q be $x = 0.2$ giving the situation shown in Figure 3.6. Now $a = 1$, $b = e = 0$, $c = -(x + y)^2$ and

$$
\begin{aligned}
f_P &= 0.1, \quad p_P = \left.\frac{\partial u}{\partial x}\right|_P = 0.1, \quad q_P = \left.\frac{\partial u}{\partial y}\right|_P = 0.2 \\
a_P &= 1, b_P = e_P = 0, c_P = -0.01 \\
g_Q &= -0.2, p_Q = 0.2, q_Q = 0.4 \\
a_Q &= 1, b_Q = e_Q = 0, c_Q = -0.04
\end{aligned}
$$

are the values of the relevant variables. On PR

$$dy = f dx$$

to give

$$y_R - y_P = f_P(x_R - x_P)$$

or

$$y_R = 0.1(x_R - 0.1)$$

and on QR

$$dy = g dx$$

to yield

$$y_R - y_Q = g_Q(x_R - x_Q)$$

or

$$y_R = -0.2(x_R - 0.2).$$

These equations give the first approximation

$$x_R = 0.16667, \quad y_R = 0.006667.$$

On PR

$$af\,dp + c\,dq + e\,dy = 0$$

which gives

$$a_P f_P(p_R - p_P) + c_P(q_R - q_P) + e_P(y_R - y_P) = 0$$

and hence

$$0.1(p_R - 0.1) - 0.01(q_R - 0.2) = 0.$$

Whereas on QR

$$ag\,dp + c\,dq + e\,dy = 0$$

to yield

$$a_Q g_Q(p_R - p_Q) + c_Q(q_R - q_Q) + e_Q(y_R - y_Q) = 0$$

and the second equation

$$0.2(p_R - 0.2) - 0.04(q_R - 0.4) = 0$$

which together give a first approximation

$$p_R = 0.14667, \quad q_R = 0.6667.$$

Then $du = p\,dx + q\,dy$ on PR gives

$$
\begin{aligned}
u_R &= u_P + \frac{1}{2}(p_P + p_R)(x_R - x_P) + \frac{1}{2}(q_P + q_R)(y_R - y_P) \\
&= 1.005 + \frac{1}{2}(0.1 + 0.14667)(0.0667) + \frac{1}{2}(0.6667 + 0.2)(0.006667) \\
&= 1.0161
\end{aligned}
$$

for a first approximation.

For the second approximation

$$y_R - y_P = \frac{1}{2}(f_P + f_R)(x_R - x_P)$$

or

$$y_R = \frac{1}{2}(0.1 + 0.17334)(x_R - 0.1)$$

and

$$y_R - y_Q = \frac{1}{2}(g_Q + g_R)(x_R - x_Q)$$

or

$$y_R = -\frac{1}{2}(0.2 + 0.17334)(x_R - 0.2)$$

which give
$$x_R = 0.15773, \quad y_R = 0.007890.$$

Equation (3.3.19) gives

$$\frac{1}{2}(1+1)\frac{1}{2}(0.1+0.17334)(p_R - 0.1) - \frac{1}{2}(0.01+0.03005)(q_R - 0.2) = 0$$

and equation (3.3.20) gives

$$\frac{1}{2}(1+1)\frac{1}{2}(-0.2 - 0.17334)(p_R - 0.2) - \frac{1}{2}(0.04 + 0.03005)(q_R - 0.4) = 0$$

to give the second approximation to p_R and q_R:

$$p_R = 0.16030, \quad q_R = 0.6116$$

and hence the second approximation to u_R is

$$\begin{aligned}
u_R &= 1.005 + \frac{1}{2}(0.1 + 0.16030)(0.05773) + \frac{1}{2}(0.2 + 0.6116)(0.00789) \\
&= 1.0157.
\end{aligned}$$

Further iterations using averages will result in little gain of accuracy in these finite difference values, and of course there is a much larger truncation error associated with the finite difference approximations so that there is no significance in getting over-accurate results to the approximate problem.

A number of general issues will now be pursued, beginning with the possibility of a characteristic as an initial curve. If the curve on which the initial values are given is itself a characteristic then the equation can have no solution unless the initial conditions themselves satisfy the necessary differential relationship for this characteristic. If they do, the solution will be unique along the initial curve but nowhere else – it is impossible to use the method of characteristics to extend the solution from points on the initial curve to points off it, so that the locus of all points is the initial curve itself. As an example of this situation consider the problem

$$\frac{\partial^2 u}{\partial x^2} - \frac{\partial^2 u}{\partial x \partial y} - 6\frac{\partial^2 u}{\partial y^2} = 0. \tag{3.3.22}$$

The characteristic directions are given by

$$\left(\frac{dy}{dx}\right)^2 + \left(\frac{dy}{dx}\right) - 6 = 0$$

or

$$\frac{dy}{dx} = -3, \quad \frac{dy}{dx} = 2.$$

Hence the characteristics are the straight lines

$$y + 3x = \text{const} \quad \text{and} \quad y - 2x = \text{const}. \tag{3.3.23}$$

Let the initial curve be the characteristic $y - 2x = 0$ on which

$$u = 2, \quad \frac{\partial u}{\partial x} = -2, \quad \frac{\partial u}{\partial y} = 1.$$

The differential relationship along this line from

$$a\frac{dy}{dx}\, dp + c\, dq + e\, dy = 0$$

is

$$2\, dp - 6\, dq = 0$$

or

$$p - 3q = \text{const.}$$

This is obviously satisfied by the initial conditions. It is easily verified that one solution satisfying these conditions is

$$u = 2 + (y - 2x) + A(y - 2x)^2$$

where A is an arbitrary constant. This is unique along $y - 2x = 0$ but nowhere else in the xy-plane.

As a second general issue consider the propagation of discontinuities in the solution of hyperbolic partial differential equations. The solutions of elliptic and parabolic equations can be proved to be analytic even when the boundary or initial conditions are discontinuous. However with hyperbolic equations discontinuities in initial conditions are propagated as discontinuities into the solution domain along the characteristics.

Consider the situation in Figure 3.7. Let Γ be a non-characteristic curve along which initial values for u, p and q are known. Let P and Q be distinct points on Γ and let the f characteristic (at P) cut the g characteristic (at Q) at R. The solution at R is determined by the values at P and Q. Assuming no two characteristics of the same family cut then the solution for every point S in PQR is determined by the values between P and Q. In $PRVT$, the value of the solution at U is determined by the initial condition at a point on TP and a point on PQ. When the initial conditions along TP are analytically different from those on PQ then the solution in $PRVT$ will be analytically different from the solution inside PQR. As T tends to P the strip tends to the characteristic PR proving that the discontinuity in the initial condition at P is propagated along a characteristic. Hence a characteristic can separate two different solutions. An alternative method of solution for hyperbolic equations based on the ideas of Chapter 2 and finite difference grids is now presented, but first however, here are a selection of exercises on the last section.

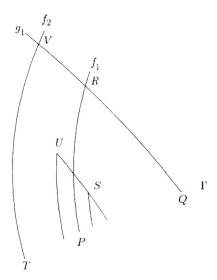

Fig. 3.7.

EXERCISES

3.12 The function u satisfies

$$\frac{\partial^2 u}{\partial x^2} - 4x^2 \frac{\partial^2 u}{\partial y^2} = 0$$

and the initial conditions $u = x^2$ and $\partial u/\partial y = 0$ on $y = 0$ for $-\infty < x < \infty$. Show from first principles with the usual notation that $dp - 2x\,dq = 0$ along the characteristic of slope $2x$ and that $dp + 2x\,dq = 0$ along the characteristic of slope $(-2x)$. The characteristic with positive slope through $A(0.3, 0)$ intersects the characteristic with negative slope through $B(0.4, 0)$ at R. Calculate an approximation to the solution at R, and improve the solution iteratively using averaged values.

3.13 The equation

$$\frac{\partial^2 u}{\partial x^2} + (1 - 2x)\frac{\partial^2 u}{\partial x \partial y} + (x^2 - x - 2)\frac{\partial^2 u}{\partial y^2} = 0$$

satisfies the initial conditions $u = x$ and $\partial u/\partial y = 1$ on $y = 0$ for $0 \le x \le 1$. Use the method of characteristics to calculate the solution at the points R, S and T for $y > 0$, the coordinates P, Q and W being $(0.4, 0)$, $(0.5, 0)$ and $(0.6, 0)$ as shown in Figure 3.8. Use one step of iterative refinement on each point.

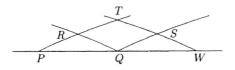

Fig. 3.8.

3.14 The function u is the solution of

$$\frac{\partial^2 u}{\partial x^2} + \left(\frac{\partial u}{\partial x} - u\right)\frac{\partial^2 u}{\partial x \partial y} - u\frac{\partial u}{\partial x}\frac{\partial^2 u}{\partial y^2} + x = 0$$

and satisfies the initial conditions $u = 1+x^2$ and $\partial u/\partial y = 1$ on $y = 0$ for $-\infty < x < 0$. Show from first principles that $p\,dp - up\,dq + x\,dy = 0$ along the characteristic of slope p, and that $-u\,dp - up\,dq + x\,dy = 0$ along the characteristic of slope $(-u)$ where $p = \partial u/\partial x$ and $q = \partial u/\partial y$.

The characteristic with positive slope through $A(0.5, 0)$ intersects the characteristic with negative slope through $B(0.6, 0)$ at the point $R(x_R, y_R)$. Calculate the first approximate values for the coordinates of R. Write down the equations giving the first approximate values for p and q at R and find their values. Calculate a first approximate value for u at R.

Use iterative refinement to find a second approximation to the solution.

3.15 Conventionally the characteristic curves are approximated by piecewise straight lines in the previous section. Derive a quadratic approximation $y = \alpha x^2 + \beta x + \gamma$ for the characteristic curves, finding α, β and γ explicitly in terms of known values at some initial point P. Demonstrate any problems which arise when such an approximation is used to set up a piecewise quadratic characteristic grid.

3.16 The partial differential equation

$$2\frac{\partial^2 z}{\partial x^2} - e^{-x^2}\frac{\partial^2 z}{\partial x \partial y} - e^{-2x^2}\frac{\partial^2 z}{\partial y^2} = x$$

has boundary conditions $z = 4x^2$ and $\partial z/\partial y = 0$ on $y = 0$.

Apply the characteristic method to this problem using a quadratic approximation for the characteristics to find a solution to z in the region $y > 0$ by considering the characteristics through the initial points (x, y) given by $(0.0, 0.0)$ and $(0.5, 0.0)$.

3.17 Considerable insight into why the explicit method for parabolic equations has stability limitations and why the method of Crank–Nicolson has not can be obtained by the following investigation.

Consider the following hyperbolic partial differential equation

$$\frac{\partial u}{\partial t} = \frac{\partial^2 u}{\partial x^2} - \epsilon^2 \frac{\partial^2 u}{\partial t^2}.$$

Find the equations of the characteristics for small ϵ. Sketch these on a typical heat equation grid system with known values at $t = 0$ and at $x = 0$ and $x = L$ for some finite L. Now consider what happens as $\epsilon \to 0$. Remember the boundary information effectively transmits itself along the characteristics, and hence where two intersect there are two equations for two unknowns to give the solution. Compare this limiting situation when the hyperbolic equation has reduced to the parabolic heat equation, to where information is moving in the explicit and implicit cases to see why problems can be expected in the explicit case. You should see that to get sensible compatibility with the limiting case any implicit method will do for the heat equation.

3.4 Rectangular Nets and Finite Difference Methods for Second Order Hyperbolic Equations

Normally the method of characteristics provides the most accurate process for solving hyperbolic equations; it is probably the most convenient method when the initial data is discontinuous, which is difficult to deal with on any grid other than a grid of characteristics. Problems involving no discontinuities can be solved satisfactorily using appropriate finite difference methods on rectangular grids. Consider the wave equation

$$\frac{\partial^2 u}{\partial x^2} = \frac{\partial^2 u}{\partial t^2} \quad (t > 0) \tag{3.4.1}$$

where initially

$$u(x,0) = f(x), \quad \frac{\partial u}{\partial t}(x,0) = g(x).$$

The slopes of the characteristics are given by

$$\left(\frac{dt}{dx}\right)^2 - 1 = 0 \tag{3.4.2}$$

so the characteristics through $P(x_P, t_P)$ are the straight lines $t - t_P = \pm(x - x_P)$ meeting the x-axis at $D(x_P - t_P, 0)$ and $E(x_P + t_P, 0)$. D'Alembert's solution to the wave equation is

$$u(x,t) = \frac{1}{2}\left[f(x+t) + f(x-t) + \int_{x-t}^{x+t} g(s)\,ds \right]. \tag{3.4.3}$$

Hence the solution at $P(x_P, t_P)$ is

$$u(x_P, t_P) = \frac{1}{2}\left[f(x_P + t_P) + f(x_P - t_P) + \int_{x_P - t_P}^{x_P + t_P} g(s)\, ds \right]. \qquad (3.4.4)$$

Hence the solution at P depends on the value of $f(x)$ at D and E and upon the value of $g(x)$ between D and E, as shown in Figure 3.9. DE is the *interval of dependence*, the area PDE is the *domain of dependence* for P.

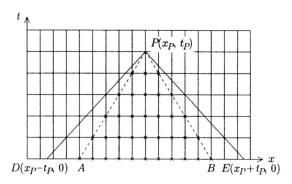

Fig. 3.9.

A central difference approximation to

$$\frac{\partial^2 u}{\partial x^2} = \frac{\partial^2 u}{\partial t^2} \qquad (3.4.5)$$

at mesh point

$$(x_i, t_j) = (ih, jk)$$

is

$$\frac{1}{h^2}(u_{i+1,j} - 2u_{i,j} + u_{i-1,j}) = \frac{1}{k^2}(u_{i,j+1} - 2u_{i,j} + u_{i,j-1})$$

or

$$u_{i,j+1} = r^2 u_{i-1,j} + 2(1 - r^2)u_{i,j} + r^2 u_{i+1,j} - u_{i,j-1} \qquad (3.4.6)$$

where $r = k/h$. To start the scheme, $u_{i,-1}$ is required since

$$u_{i,1} = r^2 u_{i-1,0} + 2(1 - r^2)u_{i,0} + r^2 u_{i+1,0} - u_{i,-1}$$

the central difference approximation to the initial derivative condition gives

$$\frac{1}{2k}(u_{i,1} - u_{i,-1}) = g_i$$

which implies that

$$u_{i,1} = \frac{1}{2}[r^2 f_{i-1} + 2(1 - r^2)f_i + r^2 f_{i+1} + 2kg_i]. \qquad (3.4.7)$$

The recurrence relations show that u_P at point P depends on the values of $u_{i,j}$ at the mesh points marked by crosses. This set of points is called the numerical domain of dependence of P and lines PA and PB are called the numerical characteristics. A problem can then arise. For if the initial conditions on DA and BE change, these alter the analytical solution of the partial differential equation at P but not the numerical one (as given by the equations above). In this case the numerical solution cannot possibly converge to the analytical solution for all such arbitrary changes as $h \to 0, k \to 0$ (r remaining constant). When, however, the numerical characteristics lie outside the domain of dependence, PDE, Courant, Friedrich and Lewy have shown that the effect of the initial data on DA and BE on the solution at P of the finite difference equations tends to zero as $h \to 0, k \to 0$, for r constant, and P remaining fixed.

This CFL condition for convergence is expressed as: "for convergence the numerical domain of dependence of the difference equation must include the domain of dependence of the differential equation." For the case considered we have $0 < r \leq 1$.

It is both amusing and instructive to be able to try physical examples whose behaviour is easy to anticipate intuitively, and the motion of a string is such a case. Consider the problem of a string stretched between two points with $x = 0$ and $x = 1$. Initially let the string be displaced so that $u = x(1-x)/2$, to give a maximum displacement of $1/8$ at the midpoint. Track the behaviour as time progresses once the string is released from rest.

A step of 0.1 was taken in the x direction and 0.05 in t. For the first step formula (3.4.7) was used with $f = x(1-x)/2$ and $g = 0$, and from then on the scheme (3.4.6) was employed. The results are shown in Table 3.4 for the range $0 < x < 1/2$. The other half of the range is symmetrical. Only every other time step is recorded.

The table shows that the string takes about 0.5 units to reach zero

Table 3.4.

$t\backslash x$	0.1	0.2	0.3	0.4	0.5
0.1	0.0403	0.0750	0.1000	0.1150	0.1200
0.2	0.0301	0.0605	0.0850	0.1000	0.1050
0.3	0.0198	0.0403	0.0607	0.0750	0.0800
0.4	0.0101	0.0196	0.0305	0.0409	0.0452
0.5	0.0000	0.0000	−0.0004	0.0008	0.0021
0.6	−0.0101	−0.0199	−0.0298	−0.0392	−0.0432
0.7	−0.0199	−0.0399	−0.0585	−0.0740	−0.0809
0.8	−0.0298	−0.0585	−0.0839	−0.1002	−0.1055
0.9	−0.0385	−0.0735	−0.1004	−0.1155	−0.1193
1.0	−0.0437	−0.0805	−0.1054	−0.1198	−0.1250
1.1	−0.0421	−0.0760	−0.0998	−0.1148	−0.1204
1.2	−0.0327	−0.0616	−0.0852	−0.1003	−0.1048

displacement along its length with a small amount of error creeping in. It is expected that it will take the same time to reach the maximum negative displacement and this is observed at around $t = 1.0$. There is no damping term so this behaviour will be repeated. Rounding will slowly distort the solution.

Implicit methods cannot be used without simplifying assumptions to solve pure initial value problems as they lead to an infinite number of simultaneous equations. They can be used for initial boundary value problems, that is initial conditions on $0 \leq x \leq 1$ for $t = 0$ and boundary conditions on $x = 0, 1, t > 0$. One satisfactory scheme approximating the wave equation at (ih, jk) is

$$\frac{1}{k^2}\delta_t^2 u_{i,j} = \frac{1}{h^2}\left(\frac{1}{4}\delta_x^2 u_{i,j+1} + \frac{1}{2}\delta_x^2 u_{i,j} + \frac{1}{4}\delta_x^2 u_{i,j-1}\right) \tag{3.4.8}$$

where

$$\delta_t^2 u_{i,j} = u_{i,j+1} - 2u_{i,j} + u_{i,j-1} \quad \text{etc.}$$

This leads to a tridiagonal system of equations that can be solved as for parabolic equations. This method is unconditionally stable for all $r = k/h > 0$.

This section is concluded with some illustrative exercises.

EXERCISES

3.18 The transverse displacement U of a point at a distance x from one end of a vibrating string of length L at a time T satisfies the equation

$$\frac{\partial^2 U}{\partial T^2} = c^2\frac{\partial^2 U}{\partial X^2}.$$

Show that this equation can be reduced to the non-dimensional form

$$\frac{\partial^2 u}{\partial t^2} = \frac{\partial^2 u}{\partial x^2}$$

for $0 \leq x \leq 1$ by putting $x = X/L$, $u = U/L$ and $t = cT/L$. A solution of the latter eqation satisfies the boundary conditions $u = 0$ at $x = 0$ and $x = 1$ for $t \geq 0$ and the initial conditions $u = x/4$ for $0 \leq x \leq 1/2$ and $u = (1 - x)/4$ for $1/2 \leq x \leq 1$. The initial velocity of the string, $\partial u/\partial t = 0$ for $0 \leq x \leq 1$ when $t = 0$.

Use the finite difference method to calculate the non-dimensional velocities and displacements to time $t = 0.5$ in steps of 0.05 of the points on the string defined by $x = 0(0.1)0.5$.

3.19 The function u satisfies the equation

$$\frac{\partial^2 u}{\partial x^2} = \frac{\partial^2 u}{\partial t^2}$$

with the boundary conditions $u = 0$ at $x = 0$ and $x = 1$ for $t \geq 0$ and the initial conditions $u = \frac{1}{8} \sin \pi x$ and $\partial u / \partial t = 0$ when $t = 0$ for $0 \leq x \leq 1$. Use the explicit finite difference formula for the equation and a central difference approximation for the derivative condition to calculate a solution for $x = 0(0.1)1$ and $t = 0(0.1)0.5$. Derive the analytic solution $u = \frac{1}{8} \sin \pi x \cos \pi t$ and compare it with the numerical solution at several points.

3.20 Modify the finite difference method of (3.4.6) and (3.4.7) to include a damping term proportional to the string velocity and hence repeat the calculations of the worked example for the equation

$$\frac{\partial^2 u}{\partial t^2} = \frac{\partial^2 u}{\partial x^2} - \kappa \frac{\partial u}{\partial t}.$$

Take $\kappa = 2.5$.

4
Elliptic Equations

4.1 Laplace's Equation

In the case of elliptic equations the canonical form is Laplace's equation which is therefore the basis of the work in this chapter. With elliptic equations in the plane, the boundary conditions are specified round a closed curve, and the finite difference schemes then lead to a large set of linear algebraic equations for the complete set of unknowns. Elliptic equations are like boundary value problems in ordinary differential equations in which there is no step-by-step procedure such as those employed with parabolic equations in Chapter 2 and hyperbolic equations in Chapter 3. Hence most of the difficulty in the solution of elliptic equations lies in the solution of large sets of algebraic equations, and in the representation of curved boundaries. Iterative methods for the solution of linear algebraic equations have been considered in Chapter 1 and further examples of their application will arise here.

Consider Laplace's equation in the unit square subject to Dirichlet boundary conditions. Let the region be described using a uniform mesh with spacing $h = M^{-1}$ for some integer M. Then the conventional finite difference approximation to the second order derivative is

$$\frac{\partial^2 \phi}{\partial x^2} = \frac{\phi_{i-1,j} - 2\phi_{i,j} + \phi_{i+1,j}}{h^2}$$

with an equivalent form in y with the i subscript held constant. Hence the discrete analogue of

$$\frac{\partial^2 \phi}{\partial x^2} + \frac{\partial^2 \phi}{\partial y^2} = 0 \qquad (4.1.1)$$

95

relative to the computational molecule shown in Figure 4.1 correct to $O(h^2)$ is

$$4\phi_P - \phi_E - \phi_W - \phi_N - \phi_S = 0. \tag{4.1.2}$$

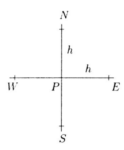

Fig. 4.1.

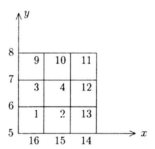

Fig. 4.2.

In order to see how the solution develops consider the simple case with $M = 3$ which is illustrated in Figure 4.2. Applying the difference formula to each interior node gives

$$
\begin{aligned}
4\phi_1 - \phi_2 - \phi_3 - \phi_6 - \phi_{16} &= 0 \\
4\phi_2 - \phi_{13} - \phi_4 - \phi_1 - \phi_{15} &= 0 \\
4\phi_3 - \phi_4 - \phi_9 - \phi_7 - \phi_1 &= 0 \\
4\phi_4 - \phi_{12} - \phi_{10} - \phi_3 - \phi_2 &= 0
\end{aligned}
$$

which in matrix form is

$$
\begin{pmatrix}
4 & -1 & -1 & 0 \\
-1 & 4 & 0 & -1 \\
-1 & 0 & 4 & -1 \\
-1 & -1 & 0 & 4
\end{pmatrix}
\begin{pmatrix}
\phi_1 \\
\phi_2 \\
\phi_3 \\
\phi_4
\end{pmatrix}
=
\begin{pmatrix}
\phi_6 + \phi_{16} \\
\phi_{13} + \phi_{15} \\
\phi_9 + \phi_7 \\
\phi_{12} + \phi_{10}
\end{pmatrix}.
$$

The solution methods of Chapter 1 are then employed, as the matrices will always be sparse in that they can only have 5 elements in each row at most. For a small 4×4 example the sparsity is clearly less marked.

Suppose the boundary conditions are that $\phi = y$ on $x = 0$, $\phi = x$ on $y = 0$, $\phi = 1.0$ on $x = 1$ and $\phi = 1.0$ on $y = 1$. Then the right-hand side vector becomes $\{2/3, 5/3, 5/3, 2\}^T$ and the equations can be solved by Jacobi's method to give $\phi_1 = 0.54667$, $\phi_2 = \phi_3 = 0.76000$ and $\phi_4 = 0.82667$ after 20 iterations, which are shown in Table 4.1.

Table 4.1.

n	ϕ_1	ϕ_2	ϕ_3	ϕ_4
0	0.0	0.0	0.0	0.0
1	0.91667	1.9167	1.9167	3.0000
2	1.1250	1.3958	1.3958	1.2083
3	0.86458	1.0000	1.0000	1.1302
4	0.66667	0.912536	0.91536	0.96615
5	0.62435	0.82487	0.82487	0.89551
6	0.57910	0.79663	0.79663	0.86230
7	0.56498	0.77702	0.77702	0.84393
8	0.55518	0.76890	0.76890	0.83550
9	0.55111	0.76434	0.76434	0.83101
10	0.54883	0.76220	0.76220	0.82886
11	0.54777	0.76109	0.76109	0.82776
12	0.54721	0.76054	0.76054	0.82721
13	0.54694	0.76027	0.76027	0.82694
14	0.54680	0.76014	0.76014	0.82680
15	0.54674	0.76007	0.76007	0.82674
16	0.54670	0.76003	0.76003	0.82670
17	0.54668	0.76002	0.76002	0.82668
18	0.54667	0.76000	0.76000	0.82668
19	0.54667	0.76000	0.76000	0.82667

The symmetry in the original problem is reflected in the numerical solution. As a comparison, the method of Gauss–Seidel was also employed and obtained the same values after just 10 iterations. The approximate halving of the number of iterations from the Jacobi to the Gauss–Seidel method can be established in theory, and is shown to occur in this example. These iterates are shown in Table 4.2.

It must be remembered that the accuracy of the solutions will be limited by the truncation error in the original finite difference approximation to the partial differential equation, so there is no great advantage in iterating the linear equations to machine accuracy. This was carried out purely as an exercise in solving linear equations, and the 42nd Jacobi iterate yielded 12 figure accuracy against the 23rd iterate for Gauss–Seidel. With $\omega = 1.1$, the SOR method

Table 4.2.

n	ϕ_1	ϕ_2	ϕ_3	ϕ_4
0	0.0	0.0	0.0	0.0
1	0.91667	0.64583	0.64583	0.89063
2	0.48958	0.76172	0.76172	0.81283
3	0.54743	0.75675	0.75675	0.82607
4	0.54504	0.75945	0.75945	0.82612
5	0.54639	0.75979	0.75979	0.82655
6	0.54656	0.75994	0.75994	0.82663
7	0.54664	0.75998	0.75998	0.82665
8	0.54666	0.76000	0.76000	0.82666
9	0.54666	0.76000	0.76000	0.82667

achieved the same accuracy at 21 iterations, and this rose to 28 iterations for $\omega = 1.2$. The number of equations is very small here and the SOR gain is not very marked.

In order to see what level of error is arising from the finite difference truncation, the whole problem was repeated with a grid size reduced from $1/3$ to $1/6$. There are now 25 interior unknowns and 25 tedious equations to be set up. For the 9 equations based on points with no boundary conditions, namely the central block of (3×3) grid points, the matrix can be programmed. Some hand setting is required along the boundary edges. The grid is shown in Figure 4.3.

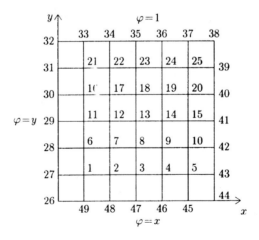

Fig. 4.3.

This setting up problem illustrates the first of the issues in solving in practice a reasonably sized problem. To solve the linear equations, Gauss–Seidel was employed and converged to 12 figures in 96 iterations. The SOR method was then applied with $\omega = 1.1$ and the iteration count reduced to 86, and with $\omega = 1.2$ there was a further reduction to 78. The ϕ values corresponding to the earlier four unknowns are now ϕ_7, ϕ_9, ϕ_{17} and ϕ_{19} and these have the values $\phi_7 = 0.555556$, $\phi_9 = \phi_{17} = 0.777778$ and $\phi_{19} = 0.888889$. These demonstrate that the original values are correct to almost 2 figures which is remarkable considering the crudity of the first approximation.

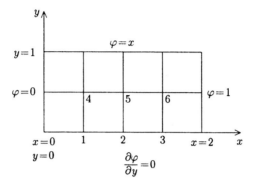

Fig. 4.4.

As a second example, consider the following region shown in Figure 4.4 in which $\nabla^2 \phi = 0$, subject to the boundary conditions shown in the figure. The algebraic problem which results from using the standard five-point formula with the above mesh is obtained as follows. The equation

$$\nabla^2 \phi = 0$$

gives the finite difference form

$$4\phi_P - \phi_N - \phi_W - \phi_S - \phi_E = 0$$

at each point as shown in Figure 4.1. Applying this computational molecule at each point gives

$$4: \quad 4\phi_4 - \frac{1}{2} - \phi_1 - \phi_5 = 0$$
$$5: \quad 4\phi_5 - 1 - \phi_2 - \phi_4 - \phi_6 = 0$$
$$6: \quad 4\phi_6 - \frac{3}{2} - \phi_3 - \phi_5 - 1 = 0.$$

For points 1, 2 and 3 we need a discretisation technique for $\partial\phi/\partial y = 0$ at a point P on the bottom of the boundary, which is simply achieved using

$$\frac{\partial \phi}{\partial y} = \frac{\phi_N - \phi'_S}{2h}$$

where the point S is a fictitious mesh point. This implies

$$\phi'_S = \phi_N \quad \text{as} \quad \frac{\partial \phi}{\partial y} = 0$$

to give the specific edge representation of

$$\nabla^2 \phi = 0$$

as

$$4\phi_P - 2\phi_N - \phi_E - \phi_W = 0.$$

Note that by considering a Taylor expansion, this is accurate to $O(h)$ unless $\partial \phi / \partial y = 0$ actually implies symmetry about the line $y = 0$, in which case $\partial^3 \phi / \partial y^3 = 0$ as well and the formula is accurate to $O(h^3)$. Applying the edge formula gives the additional equations

$$1 \quad 4\phi_1 - 2\phi_4 - \phi_2 = 0$$
$$2 \quad 4\phi_2 - 2\phi_5 - \phi_1 - \phi_3 = 0$$
$$3 \quad 4\phi_3 - 2\phi_6 - \phi_2 - 1 = 0.$$

Hence the six difference equations give the matrix equation

$$\begin{pmatrix} 4 & -1 & 0 & -2 & 0 & 0 \\ -1 & 4 & -1 & 0 & -2 & 0 \\ 0 & -1 & 4 & 0 & 0 & -2 \\ -1 & 0 & 0 & 4 & -1 & 0 \\ 0 & -1 & 0 & -1 & 4 & -1 \\ 0 & 0 & -1 & 0 & -1 & 4 \end{pmatrix} \begin{pmatrix} \phi_1 \\ \phi_2 \\ \phi_3 \\ \phi_4 \\ \phi_5 \\ \phi_6 \end{pmatrix} = \begin{pmatrix} 0 \\ 0 \\ 1 \\ \frac{1}{2} \\ 1 \\ \frac{5}{2} \end{pmatrix}$$

which is noted for its high degree of sparsity. In fact it is not usually necessary to formally construct this matrix equation. Usually, the separate equations are written as

$$\phi_1 = \frac{1}{4}(2\phi_4 + \phi_2)$$

$$\phi_2 = \frac{1}{4}(2\phi_5 + \phi_1 + \phi_3)$$

$$\phi_3 = \frac{1}{4}(2\phi_6 + \phi_2 + 1)$$

$$\phi_4 = \frac{1}{4}\left(\phi_1 + \phi_5 + \frac{1}{2}\right)$$

$$\phi_5 = \frac{1}{4}(\phi_2 + \phi_4 + \phi_6 + 1)$$

$$\phi_6 = \frac{1}{4}\left(\phi_3 + \phi_5 + \frac{5}{2}\right)$$

where the diagonal term is isolated on the left-hand side. It is then trivial to set up the Jacobi iteration in the form:

$$\phi_1^{n+1} = \frac{1}{4}(2\phi_4^n + \phi_2^n)$$

$$\phi_2^{n+1} = \frac{1}{4}(2\phi_5^n + \phi_1^n + \phi_3^n)$$

$$\phi_3^{n+1} = \frac{1}{4}(2\phi_6^n + \phi_2^n + 1)$$

$$\phi_4^{n+1} = \frac{1}{4}\left(\phi_1^n + \phi_5^n + \frac{1}{2}\right)$$

$$\phi_5^{n+1} = \frac{1}{4}(\phi_2^n + \phi_4^n + \phi_6^n + 1)$$

$$\phi_6^{n+1} = \frac{1}{4}\left(\phi_3^n + \phi_5^n + \frac{5}{2}\right).$$

The simple change to using current iterates yields the Gauss–Seidel method with the form:

$$\phi_1^{n+1} = \frac{1}{4}(2\phi_4^n + \phi_2^n)$$

$$\phi_2^{n+1} = \frac{1}{4}(2\phi_5^n + \phi_1^{n+1} + \phi_3^n)$$

$$\phi_3^{n+1} = \frac{1}{4}(2\phi_6^n + \phi_2^{n+1} + 1)$$

$$\phi_4^{n+1} = \frac{1}{4}\left(\phi_1^{n+1} + \phi_5^n + \frac{1}{2}\right)$$

$$\phi_5^{n+1} = \frac{1}{4}(\phi_2^{n+1} + \phi_4^{n+1} + \phi_6^n + 1)$$

$$\phi_6^{n+1} = \frac{1}{4}\left(\phi_3^{n+1} + \phi_5^{n+1} + \frac{5}{2}\right).$$

Equivalently the SOR method could be used with the iterations:

$$\phi_1^{n+1} = \frac{\omega}{4}(2\phi_4^n + \phi_2^n) - (\omega - 1)\phi_1^n$$

$$\phi_2^{n+1} = \frac{\omega}{4}(2\phi_5^n + \phi_1^{n+1} + \phi_3^n) - (\omega - 1)\phi_2^n$$

$$\phi_3^{n+1} = \frac{\omega}{4}(2\phi_6^n + \phi_2^{n+1} + 1) - (\omega - 1)\phi_3^n$$

$$\phi_4^{n+1} = \frac{\omega}{4}\left(\phi_1^{n+1} + \phi_5^n + \frac{1}{2}\right) - (\omega - 1)\phi_4^n$$

$$\phi_5^{n+1} = \frac{\omega}{4}(\phi_2^{n+1} + \phi_4^{n+1} + \phi_6^n + 1) - (\omega - 1)\phi_5^n$$

$$\phi_6^{n+1} = \frac{\omega}{4}\left(\phi_3^{n+1} + \phi_5^{n+1} + \frac{5}{2}\right) - (\omega - 1)\phi_6^n$$

where it is common to take $\omega \sim 1.1$ or 1.2 for small numbers of equations.

For the SOR iterations we note that in this case we can write

$$\phi_j^{n+1} = \omega(\text{RHS of the GS iteration equation}) - (\omega - 1)\phi_i^n$$

where GS is the Gauss–Seidel iteration. All three iterations converge in the manner expected from the earlier examples to the values:

$$
\begin{aligned}
\phi_1 &= 0.401786 \\
\phi_2 &= 0.750000 \\
\phi_3 &= 0.973214 \\
\phi_4 &= 0.428571 \\
\phi_5 &= 0.812500 \\
\phi_6 &= 1.071429
\end{aligned}
$$

and for 12-figure accuracy Jacobi requires 79 iterations, Gauss–Seidel uses 43, and with $\omega = 1.1$ the SOR method takes 37, and with $\omega = 1.2$ this count reduces to 33. Again the high accuracy obtained here is purely to illustrate the performance of the iterative methods, and is well beyond the finite difference errors in the original approximation. Some exercises on solutions of Laplace's equation over rectangular regions are now presented.

EXERCISES

4.1 The function ϕ satisfies the equation

$$\frac{\partial^2 \phi}{\partial x^2} + \frac{\partial^2 \phi}{\partial y^2} + 2 = 0$$

at every point inside the square bounded by the straight lines $x = \pm 1$ and $y = \pm 1$ and is zero on the boundary. Calculate a finite difference solution using a square mesh of side $1/2$.

(Hint: use symmetry.)

4.2 The function ϕ satisfies Laplace's equation in the unit square $0 < x < 1$ and $0 < y < 1$ and is subject to the following Dirichlet boundary conditions:

(i) $\phi = 0$ on $y = 0$

(ii) $\phi = y^2$ on $x = 0$

(iii) $\phi = 1 + x^2$ on $y = 1$

(iv) ϕ is continuous and linear on $x = 1$.

Take a square mesh of side $1/5$ and determine the matrix equation for the 16 internal points. Solve the matrix problem.

4.3 The function ϕ satisfies Laplace's equation in the unit square $0 < x < 1$ and $0 < y < 1$ and is subject to the following boundary conditions:

(i) $\phi = 0$ on $y = 0$

(ii) $\partial\phi/\partial x = 1$ on $x = 0$

(iii) $\phi = 1 + x^2$ on $y = 1$

(iv) $\partial\phi/\partial x = 0$ on $x = 1$.

Take a square mesh of side $1/5$ and determine the matrix equation for the 16 internal points. Solve the matrix problem.

4.4 Some irregular shapes can be handled as long as the grid nodes fall on the boundary. As an example solve Laplace's equation in the triangular region defined by the boundaries $x = 0$, $y = 0$ and $x + y = 1$, with $0 < x < 1$ and $0 < y < 1$. Take the grid nodes with spacings $h = k = 0.25$, so that grid nodes occur on the diagonal $x + y = 1$. Take the boundary conditions to be $\phi = 0$ on $y = 0$, $\phi = y$ on $x = 0$ and $\phi = y$ on $x + y = 1$.

4.5 A second irregular shape to be considered is the L-shaped region with boundaries $y = 0$ with $1/2 < x < 1$, $x = 1$ with $0 < y < 1$, $y = 1$ with $0 < x < 1$, $x = 0$ with $1/2 < y < 1$ and then the indentation $y = 1/2$ with $0 < x < 1/2$, and $x = 1/2$ with $0 < y < 1/2$. Take a grid size of $h = k = 1/6$ to give 16 interior nodes, with boundary conditions $\phi = 0$ in the indentation, $\phi = 1$ on both $x = 1$ and $y = 1$, and $\phi = 2(y - 1/2)$ on $x = 0$ for $1/2 < y < 1$, and symmetrically $\phi = 2(x - 1/2)$ on $y = 0$ for $1/2 < x < 1$. Use the symmetry to reduce the number of equations which are to be solved to just nine.

4.6 Modify the above problem by placing a diagonal across the indention which will now run from $(0, 1/2)$ to $(1/6, 1/2)$ (as before), and then diagonally to $(1/2, 1/6)$, and then to the lower edge at $(1/2, 0)$. Keep the boundary condition at $\phi = 0$ in the new indentation, and the same conditions as Exercise 4.5 otherwise. How much change does this alteration make to the solution?

4.7 To introduce some of the practical difficulties in solving elliptic problems, reconsider Exercise 4.5 with the grid size halved. With this still quite small number of equations, attempt to automate the setting of the coefficients in your program. The irregular shape of the area means that the coding will require some thought. You will need to match the grid coordinates with the equation numbers. If you are using a language such as C++ or Pascal then the use of a data structure will help. Compare the more accurate solution with that of Exercise 4.5, especially near to the corner $(1/2, 1/2)$.

4.8 Introduce the same finer grid into Exercise 4.6. and again compare the results with the earlier values.

4.9 A different complication is introduced in this problem, namely using grids of varying mesh to gain accuracy in regions of maximum variation of the solution. Consider this time the square region $0 < x < 1$ and $0 < y < 1$ with boundary conditions $\phi = 0$ on $x = 0$, $y = 1$ and $x = 1$. On $y = 0$, $\phi = 1$, so suggesting that a finer grid is used near to $y = 0$. Take the basic grid to be of size $h = k = 1/4$, but near $y = 0$ add the grid line $y = 1/8$ and $x = 1/8$, $x = 3/8$, $x = 5/8$ and $x = 7/8$, with each of these ranging from $y = 0$ to $y = 1/4$. Hence there are the basic nine interior points, and seven extra points along $y = 1/8$. In addition there are four points $(1/8, 1/4)$, $(3/8, 1/4)$, $(5/8, 1/4)$ and $(7/8, 1/4)$ which are not proper nodes on which the computational molecule can be applied, as the vertical grid lines terminate at these points to effect the change of grid size. Clearly the conventional computational molecule can be applied at the other interior grid points, and this will leave four required equations from these 'loose-end points'. Use simple interpolation to obtain these equations, to give for example

$$\phi(3/8, 1/4) = (\phi(1/4, 1/4) + \phi(1/2, 1/4))/2$$

at the point $(3/8, 1/4)$. Hence find the grid values for this double gauge grid.

4.2 Curved Boundaries

There are two major approaches to deal with curved boundaries. The first leans on the analytic approach in which a coordinate system is selected which allows the boundary to be a constant curve. Hence for a cheese-shaped region the use of polar coordinates allows the boundary condition on the circular segment to become a condition on $r = $ const. Hence in r, θ space the curvilinear grid transforms to a conventional rectangular form. The limitation of this approach is that most practical curves (with the exception of structures like power station cooling towers) are not geometric shapes which lend themselves to a special coordinate system. Cylindrical and spherical polars deal with circular sections in some circumstances (not if two different circular curves are involved). Toroidal coordinates can handle doughnut shapes (such as in toroidal containment vessels in fusion machines), and elliptic coordinates can be used for some limited elliptic and hyperbolic shapes.

For a general curve, some numerical approach is required, and it is straight-forward to find finite difference relations in which the computational molecule has unequal length arms, so allowing the intersection with a boundary to be handled easily. There is a computer studies problem which arises here. How is the boundary represented computationally? This book is primarily about the mathematics of partial differential equations, but practical solution of complex problems requires these ideas to be converted into efficient code. It used to be considered that the finite difference methods were hard to handle in a curvilinear situation, but the common use of data structures in programming languages has meant that grid systems can now be represented in a very natural way. Languages such as C++, Fortran 90 and Pascal all allow the user to define structures which are effectively arrays whose elements are not of the same type. If this data structure approach is allied to the use of pointers or addresses then the linkage in a grid can be set up with pointers to the relevant neighbours of a computational molecule. The structure can additionally contain the node value and the required distances to neighbouring nodes so that irregular grids can be handled as part of the general set up. A typical structure in C++ would have a form such as

```
struct node{double node_value;
            int node_number;
            node *north, *south, *east, *west;
            double h1,k1,h2,k2;}
```

where *north is a pointer to the address of the node to the north of the current node for example and the lengths of the arms of the computational molecule are h1, h2, k1 and k2. The grid is set up initially and the pointers allow trivial movement around the grid. The information about the specific node is then immediately available for use in the finite difference formulae. Boundary values can also be imposed on the structure so that the data structure in effect defines the curvilinear boundary, the boundary values and the grid system, which can itself be irregular to reflect fast varying parts of the solution. More recent developments involve software in which light pens are used to define the boundary curves with remarkable ease.

To return now to the mathematics, consider an example with curved boundaries which can be solved by transforming to polar coordinates. Suppose that Laplace's equation is satisfied within a cylinder of circular cross-section. In addition, let the condition $\partial \phi / \partial \theta = 0$ for axial symmetry hold. The boundary conditions are shown in Figure 4.5.

The problem will be solved in three stages: it will first be shown that for $\delta r = \delta z = h$ that for an interior point, P, to order $O(h^2)$

$$4\phi_P - \phi_W - \phi_E - \phi_N \left(1 + \frac{h}{2r_P}\right) - \phi_S \left(1 - \frac{h}{2r_P}\right) = 0.$$

In the second stage it is shown that for a point on the axis of symmetry, to

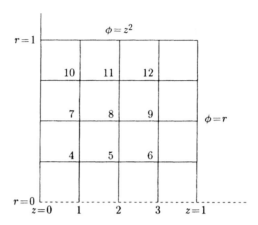

Fig. 4.5.

$O(h^2)$

$$6\phi_P - 4\phi_N - \phi_W - \phi_E = 0.$$

Finally, the finite difference equation for the case $\delta r = \delta z = \frac{1}{4}$ is set up.

Laplace's equation in cylindrical polars from Chapter 1 is

$$\frac{\partial^2 \phi}{\partial r^2} + \frac{\partial^2 \phi}{\partial z^2} + \frac{1}{r}\frac{\partial \phi}{\partial r} + \frac{1}{r^2}\frac{\partial^2 \phi}{\partial \theta^2} = 0 \qquad (4.2.1)$$

which reduces to

$$\frac{\partial^2 \phi}{\partial r^2} + \frac{\partial^2 \phi}{\partial z^2} + \frac{1}{r}\frac{\partial \phi}{\partial r} = 0 \qquad (4.2.2)$$

with axial symmetry.

(a) For a point not on the axis of symmetry

$$\left(\frac{\partial^2 \phi}{\partial r^2}\right)_P = \frac{(\phi_N + \phi_S - 2\phi_P)}{h^2} + O(h^2)$$

with the notation in Figure 4.1 as before.

$$\left(\frac{\partial^2 \phi}{\partial z^2}\right)_P = \frac{(\phi_W + \phi_E - 2\phi_P)}{h^2} + O(h^2)$$

$$\left(\frac{1}{r}\frac{\partial \phi}{\partial r}\right)_P = \frac{1}{r_P}\frac{(\phi_N - \phi_S)}{2h} + O(h^2)$$

to give the finite difference forms

$$\phi_N + \phi_S - 2\phi_P + \phi_W + \phi_E - 2\phi_P + \frac{h}{2r_P}(\phi_N - \phi_S) + O(h^2) = 0$$

and

$$4\phi_P - \phi_W - \phi_E - \left(1 + \frac{h}{2r_P}\right)\phi_N - \left(1 - \frac{h}{2r_P}\right)\phi_S = 0$$

to order h^2.

(b) On the axis $r = 0$,

$$\frac{\partial\phi}{\partial r} = 0 \quad \text{and} \quad \lim_{r\to\infty}\frac{1}{r}\frac{\partial\phi}{\partial r} = \lim_{r\to 0}\frac{\partial^2\phi}{\partial r^2}.$$

As axial symmetry holds at $r = 0$ then $\partial^3\phi/\partial r^3 = 0$, and indeed all other odd order derivatives are zero which means that

$$\phi_N = \phi_P + \frac{h^2}{2!}\left(\frac{\partial^2\phi}{\partial r^2}\right)_P + O(h^4)$$

which gives

$$\left(\frac{\partial^2\phi}{\partial r^2}\right)_P = 2\frac{(\phi_N - \phi_P)}{h^2} + O(h^2).$$

Hence Laplace's equation on the axis reduces to

$$2\frac{\partial^2\phi}{\partial r^2} + \frac{\partial^2\phi}{\partial z^2} = 0$$

which in finite difference form is

$$2\frac{2(\phi_N - \phi_P)}{h^2} + \frac{\phi_W - 2\phi_P + \phi_E}{h^2} + O(h^2) = 0$$

or

$$6\phi_P - 4\phi_N - \phi_W - \phi_E = 0$$

to order h^2.

(c) Hence the three axial equations are:

$$
\begin{aligned}
6\phi_1 - 4\phi_4 - \phi_2 &= 1 \\
6\phi_2 - 4\phi_5 - \phi_1 - \phi_3 &= 0 \\
6\phi_3 - 4\phi_6 - \phi_2 &= 0.
\end{aligned}
$$

The other nine equations follow by the direct application of the finite difference forms, and hence for the computational molecule centred on point 4 the equation is:

$$4\phi_4 - \phi_5 - \left(1 + \frac{\frac{1}{4}}{2\frac{1}{4}}\right)\phi_7 - \left(1 - \frac{\frac{1}{4}}{2\frac{1}{4}}\right)\phi_1 = \frac{3}{4}$$

which simplifies to

$$4\phi_4 - \phi_5 - \frac{3}{2}\phi_7 - \frac{1}{2}\phi_1 = \frac{3}{4}.$$

In a similar and tedious manner the final eight equations are:

$$4\phi_5 - \phi_6 - \phi_4 - \frac{3}{2}\phi_8 - \frac{1}{2}\phi_2 = 0$$

$$4\phi_6 - \phi_5 - \frac{3}{2}\phi_9 - \frac{1}{2}\phi_3 = \frac{1}{4}$$

$$4\phi_7 - \phi_8 - \frac{5}{4}\phi_{10} - \frac{3}{4}\phi_4 = \frac{1}{2}$$

$$4\phi_8 - \phi_9 - \phi_7 - \frac{5}{4}\phi_{11} - \frac{3}{4}\phi_5 = 0$$

$$4\phi_9 - \phi_8 - \frac{5}{4}\phi_{12} - \frac{3}{4}\phi_6 = \frac{1}{2}$$

$$4\phi_{10} - \phi_{11} - \frac{5}{6}\phi_7 = \frac{79}{96}$$

$$4\phi_{11} - \phi_{12} - \phi_{10} - \frac{5}{6}\phi_8 = \frac{7}{24}$$

$$4\phi_{12} - \phi_{11} - \frac{5}{6}\phi_9 = \frac{45}{32}.$$

The SOR method was applied to these equations with $\omega = 1.2$ and the resulting values after 36 iterations were correct to 12 figures. Quoting these to just 5 figures gives:

$$\phi_1 = 0.48368, \qquad \phi_2 = -0.12179$$
$$\phi_3 = 0.21596, \qquad \phi_4 = 0.50597$$
$$\phi_5 = 0.35759, \qquad \phi_6 = 0.35438$$
$$\phi_7 = 0.44963, \qquad \phi_8 = 0.42061$$
$$\phi_9 = 0.46797, \qquad \phi_{10} = 0.39873$$
$$\phi_{11} = 0.39732, \qquad \phi_{12} = 0.548387.$$

In the more general situation the following curved boundary form of the finite difference approximations is required. Consider a typical curved boundary intersecting part of a rectangular grid system as shown in Figure 4.6.

A Taylor expansion can be used to find u_A and u_3 in terms of the values of u at the grid point u_0 to yield

$$u_A = u_0 + h\theta_1 \left(\frac{\partial u}{\partial x}\right)_0 + \frac{1}{2}h^2\theta_1^2 \left(\frac{\partial^2 u}{\partial x^2}\right)_0 + O(h^3) \qquad (4.2.3)$$

$$u_3 = u_0 - h\left(\frac{\partial u}{\partial x}\right)_0 + \frac{1}{2}h^2 \left(\frac{\partial^2 u}{\partial x^2}\right)_0 + O(h^3). \qquad (4.2.4)$$

Eliminating $\left(\partial^2 u/\partial x^2\right)_0$ gives

$$\left(\frac{\partial u}{\partial x}\right)_0 = \frac{1}{h}\left\{\frac{1}{\theta_1(1+\theta_1)}u_A - \frac{(1-\theta_1)}{\theta_1}u_0 - \frac{\theta_1}{(1+\theta_1)}u_3\right\} + O(h^2) \qquad (4.2.5)$$

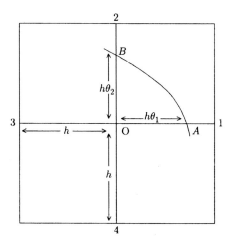

Fig. 4.6.

and eliminating $(\partial u/\partial x)_0$ gives

$$\left(\frac{\partial^2 u}{\partial x^2}\right)_0 = \frac{1}{h^2}\left\{\frac{2}{\theta_1(1+\theta_1)}u_A + \frac{2}{(1+\theta_1)}u_3 - \frac{2}{\theta_1}u_0\right\} + O(h). \qquad (4.2.6)$$

Hence approximate derivatives at points near to the boundary in terms of values within the boundary are available. The result is that amongst the defining equations these special ones for points near to the boundary are included. The boundary itself can be represented in the computer program by the values of θ_1 and θ_2 in the form of a data structure or Pascal RECORD, as shown in the introduction to this section. It is then possible to generate quite elegant code which takes account of the curved boundary.

To illustrate the process on a simple example consider the solution of Laplace's equation with the grid and boundary conditions shown in Figure 4.7.

Conversion of the grid to polar coordinates gives again the transformed equation

$$\frac{\partial^2 \phi}{\partial r^2} + \frac{1}{r}\frac{\partial \phi}{\partial r} + \frac{1}{r^2}\frac{\partial^2 \phi}{\partial \theta^2} = 0 \qquad (4.2.7)$$

with the new grid shown in Figure 4.8.

Now only grid point 3 will require the curved approximations of (4.2.5), and four equations for four unknowns are easily found. The point A is defined by

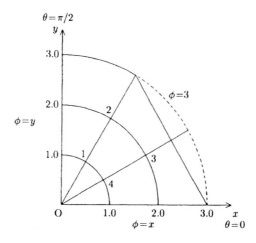

Fig. 4.7.

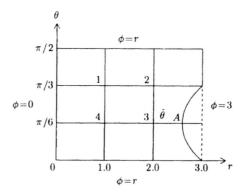

Fig. 4.8.

$\theta = \pi/6$ and $r = 3\sqrt{3}/2$, and hence the linear equations are

$$\phi_2 - 2\phi_1 + \frac{\phi_2}{2} + \frac{(1 - 2\phi_1 + \phi_4)}{\left(\frac{\pi}{6}\right)^2} = 0$$

$$\phi_3 - 2\phi_4 + \frac{\phi_3}{2} + \frac{(\phi_1 - 2\phi_4 + 1)}{\left(\frac{\pi}{6}\right)^2} = 0$$

$$3 - 2\phi_2 + \phi_1 + \frac{(3 - \phi_1)}{4} + \frac{(2 - 2\phi_2 + \phi_3)}{4\left(\frac{\pi}{6}\right)^2} = 0$$

$$\frac{2(3 - (1 + \hat{\theta})\phi_3 + \hat{\theta}\phi_4)}{\hat{\theta}(1 + \hat{\theta})} + \frac{(3 - (1 - \hat{\theta}^2)\phi_3 - \hat{\theta}^2\phi_4)}{\hat{\theta}(1 + \hat{\theta})} + \frac{(\phi_2 - 2\phi_3 + 2)}{4\left(\frac{\pi}{6}\right)^2} = 0$$

$$(4.2.8)$$

where $\hat{\theta} = r_A - 3$. These equations reduce to

$$\begin{bmatrix} -9.295 & 1.5 & 0 & 3.647 \\ 3.647 & 0 & 1.5 & -9.295 \\ 0.75 & -3.824 & 0.9119 & 0 \\ 0 & 0.9119 & -1.978 & -0.1662 \end{bmatrix} \begin{bmatrix} \phi_1 \\ \phi_2 \\ \phi_3 \\ \phi_4 \end{bmatrix} = \begin{bmatrix} -3.647 \\ -3.647 \\ -5.5738 \\ -2.7865 \end{bmatrix} \qquad (4.2.9)$$

to yield by Gauss elimination the results

$$\begin{aligned} \phi_4 &= 1.2625 \\ \phi_3 &= 2.3458 \\ \phi_2 &= 2.2627 \\ \phi_1 &= 1.2529 \end{aligned} \qquad (4.2.10)$$

with ϕ_2 and ϕ_3 being nearest to the $\phi = 3$ boundary giving the largest values.

This section is again concluded with some exercises to illustrate the work covered.

EXERCISES

4.10 Take $h = 1/4$ and consider

$$\frac{\partial^2 \phi}{\partial x^2} + \frac{\partial^2 \phi}{\partial y^2} = 4xy^2$$

in the region illustrated in Figure 4.9 with the given boundary conditions.

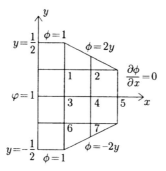

Fig. 4.9.

Using the usual five-point difference formula and methods for dealing with derivative boundary conditions and irregular boundary shapes, obtain a set of finite difference equations for ϕ at the numbered mesh points, using symmetry to reduce the number of unknowns. Use the

SOR method with $\omega = 1.1$ to obtain an approximate solution to these equations.

4.11 Solve Laplace's equation on the semicircular region shown in Figure 4.10. Take $\phi = 1$ on $y = 0$ and on $r = 1$ let $\phi = \cos^2 \theta$. Take a uniform grid spacing with $h = k = 1/3$, using curved boundary conditions on the circular edge.

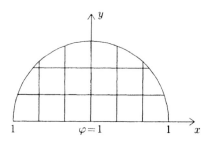

Fig. 4.10.

4.12 Exercise 4.11 might be more easily solved using a transformation to polar coordinates (r, θ). Consider the solution in this light, now with a regular grid of commensurate gauge. The resulting nodes will not of course coincide with the rectangular grid, hence any comparison will need interpolation.

4.13 Solve Laplace's equation in the region $0 < r < 1$ with $0 < \theta < \alpha$, but with an irregular boundary because of a straight line cord replacing the curved boundary from $r = 1$ and for $0 < \theta < \beta$. The suggested method is to transform the region to a near rectangular region using a polar transformation and then used curved boundary conditions on the short length of irregular boundary. Take $\alpha = \pi/3$, and $\beta = \pi/6$, with a grid size in the r direction equal to $1/4$ and in the θ direction as $\pi/12$. Take $\phi = 0$ on $\theta = 0$, $\phi = r$ on $\theta = \pi/3$ and $\phi = 3\theta/\pi$ on $r = 1$.

4.14 Use curved boundary forms to solve Laplace's equation

$$\frac{\partial^2 \phi}{\partial x^2} + \frac{\partial^2 \phi}{\partial y^2} = 0$$

with the boundary conditions $\phi = y$ on $x = 0$ for $0 \le y \le 1$, $\phi = x/2$ on $y = 0$ for $0 \le x \le 2$ and $\phi = 1$ on the positive quadrant of the ellipse

$$\frac{x^2}{4} + y^2 = 1.$$

Take a grid size of $1/2$ for both the x and the y directions to leave just three ϕ values to be found.

4.15 Repeat the solution of Exercise 4.14, with a step size of 1/4 to leave 18 unknowns to be found.

4.16 In this problem a combination of curved boundary and changing grid size is employed. The object is to solve Laplace's equation on a semicircular region $r = 1$, $0 < \theta < \pi$ with boundary conditions $\phi = 0$ on all the boundary except at $r = 1$, $\theta = \pi/2$ where $\phi = 1$. This is typical of a contact-type problem where one point has a load applied to it. The discontinuity in the boundary condition gives large variations in the solution nearby. Take a basic grid size with $h = k = 1/3$, but set up a finer grid with step size 1/6 near the discontinuity as in Figure 4.11. Use symmetry to reduce the number of equations.

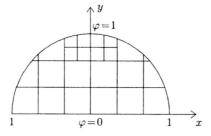

Fig. 4.11.

4.17 Put in a further fine grid of $h = k = 1/12$ in Exercise 4.16 near the discontinuity, and compare the results with the previous values.

4.18 Solve Laplace's equation in the region bounded by the dish-shaped area shown in Figure 4.12. Use the grid shown with curved boundary formulae employed where necessary. In this example there is little to be gained by a polar transformation.

4.3 Solution of Sparse Systems of Linear Equations

It is clear from the previous paragraphs that the main problem which arises in the solution of elliptic problems is the solution of large sparse sets of linear algebraic equations. The principle weapon in the solution of these equations is the iterative solution, and for problems with diagonal dominance the methods discussed so far converge quite well. However, diagonal dominance may be easily destroyed in a practical problem. The use of irregular-shaped grids on

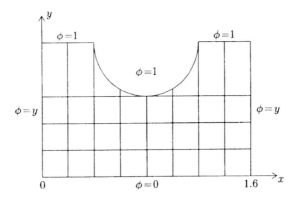

Fig. 4.12.

the curved boundary formulae can clearly move the weighting at the central grid
point towards one of the neighbours so reducing the dominance of the diagonal
element. Applying finite differences to equations in non-Cartesian coordinates
will involve terms such as

$$\frac{1}{r}\frac{\partial u}{\partial r}$$

which will yield large values when r is near the axis. The result of these effects
is that the matrix will have regions in which the diagonal element has modulus
greater than the sum of the other row or column elements, but in other regions
this dominance is lost.

There are two possible ways forward. One is to abandon iterative methods
all together and revert to elimination methods. A straightforward Gaussian
elimination with pivoting will inevitably cause fill-in of the non-zero elements
on two counts. Firstly, the elimination process will cause multiples of one row
to be added to another, and as the non-zero elements do not align from one row
to another, the process will start to fill the matrix in. Secondly, pivoting will
quite seriously fill the non-zero elements in as rows are interchanged. Research
into methods of reducing the fill-in by careful ordering of the matrix and by
reducing the pivoting has generated powerful methods for large problems (Duff,
Reid and Erisman, 1986).

Another approach is to take a hybrid line. The difficult parts of the
matrix from an iterative point of view can usually be isolated into contiguous
submatrices. This suggests the use of a block iterative method in which the
iteration is carried out on blocks of the matrix in a single go. Suppose the first
block consists of the top left-hand m rows and columns of the matrix. In the
usual way the result vector, say $\mathbf{x}$, is set to an initial value. Gaussian elimination
is now applied to the $m \times m$ block with just the relevant m unknowns. All
other terms are taken onto the right hand side using the current values of
$x_{m+1}, \ldots, x_n$. The result is a new set of values for $x_1, \ldots, x_m$ which are used
to update the corresponding old values. The iteration then proceeds onto the

next block and so on to completion. If the blocks have just one element then the method reduces to Gauss–Seidel.

To illustrate block iteration, any worked example will perforce need to be of a fairly large order to enable suitable blocks to be considered. Hence the following 10×10 matrix will be taken:

$$
\begin{bmatrix}
1 & 1 & 2 & 3 & 0 & 0 & 0 & 0 & 0 & 0 \\
1 & 2 & 1 & 2 & 3 & 0 & 0 & 0 & 0 & 0 \\
3 & 2 & 3 & 1 & 2 & 3 & 0 & 0 & 0 & 0 \\
0 & 3 & 2 & 4 & 1 & 2 & 3 & 0 & 0 & 0 \\
0 & 0 & 3 & 2 & 5 & 1 & 2 & 3 & 0 & 0 \\
0 & 0 & 0 & 3 & 2 & 6 & 2 & 0 & 0 & 0 \\
0 & 0 & 0 & 0 & 0 & 3 & 7 & 2 & 0 & 0 \\
0 & 0 & 0 & 0 & 0 & 0 & 3 & 8 & 2 & 0 \\
0 & 0 & 0 & 0 & 0 & 0 & 0 & 3 & 9 & 3 \\
0 & 0 & 0 & 0 & 0 & 0 & 0 & 0 & 3 & 10
\end{bmatrix}
\mathbf{x} =
\begin{bmatrix}
1 \\ 1 \\ 1 \\ 1 \\ 1 \\ 1 \\ 1 \\ 1 \\ 1 \\ 1
\end{bmatrix}.
$$

It is seen that in this matrix, the bottom right-hand corner has marked diagonal dominance, whereas the top left-hand corner has not. This suggests that the blocking that would be suitable might be a 5×5 block in the top left, then a 3×3 block and finally two single-element blocks which just restore the normal Gauss–Seidel process. The Gauss–Seidel philosophy will be employed in that as soon as new x values are found, they will be used in the very next steps.

Hence the first step will involve solving the 5×5 problem

$$
\begin{bmatrix}
1 & 1 & 2 & 3 & 0 \\
1 & 2 & 1 & 2 & 3 \\
3 & 2 & 3 & 1 & 2 \\
0 & 3 & 2 & 4 & 1 \\
0 & 0 & 3 & 2 & 5
\end{bmatrix}
\hat{\mathbf{x}} =
\begin{bmatrix}
1 \\
1 \\
1 - 3x_6 \\
1 - 2x_6 - 3x_7 \\
1 - x_6 - 2x_7 - 3x_8
\end{bmatrix}
$$

where all the elements not in the block have been taken to the right-hand side where current x values are used. The result of conventional Gaussian elimination on these five equations is the new x values:

$$
\begin{aligned}
\hat{x}_1 &= 0.263736, & \hat{x}_2 &= -0.0769231 \\
\hat{x}_3 &= -0.0549451, & \hat{x}_4 &= 0.307692 \\
\hat{x}_5 &= 0.109890,
\end{aligned}
$$

which are used to replace the first five x values. The elimination process is completely local to the five equations, and any row interchanging does not carry through to the main matrix.

The second step is to deal with the 3×3 block in a similar manner. Hence the equations are now

$$
\begin{bmatrix}
6 & 2 & 0 \\
3 & 7 & 2 \\
0 & 3 & 8
\end{bmatrix}
\hat{\mathbf{x}} =
\begin{bmatrix}
1 - 3x_4 - 2x_5 \\
1 \\
1 - 2x_9
\end{bmatrix}
$$

with again current x values being used on the right-hand side. The resulting x vector is now

$$\hat{x}_1 = -0.0759637, \qquad \hat{x}_2 = 0.156463$$
$$\hat{x}_3 = 0.063265$$

which replace the values of x_6, x_7 and x_8.

The final two steps are on the singleton blocks which of course reduce to pure Gauss–Seidel with the equations

$$9x_9^{(n+1)} = 1 - 3x_8^{(n+1)} - 3x_{10}^{(n)}$$
$$10x_{10}^{(n+1)} = 1 - 3x_9^{(n+1)}.$$

These iterations yield the two values

$$x_9 = 0.0890023, \qquad x_{10} = 0.0732993$$

The whole process is now repeated to convergence with the results shown in Table 4.3.

Table 4.3.

n	x_1	x_2	x_3	x_4	x_5
	x_6	x_7	x_8	x_9	x_{10}
1	0.263736	-0.0769231	-0.0549451	0.307692	0.109890
	-0.0759637	0.156463	0.0663265	0.0890023	0.0732993
2	0.495938	-0.117107	-0.175647	0.324154	0.0885382
	-0.0801148	0.165575	0.0406587	0.0731251	0.0780624
3	0.494135	-0.130027	-0.165116	0.322042	0.0956508
	-0.0811759	0.164814	0.0449133	0.0701192	0.0789642
4	0.494115	-0.126377	-0.164963	0.320729	0.0940476
	-0.0796631	0.163848	0.0460273	0.0694473	0.0791658
	$\vdots$				$\vdots$
30	0.494358	-0.126478	-0.166400	0.321640	0.0939059
	-0.0801235	0.164005	0.0461680	0.0693205	0.0792039
31	0.494358	-0.126478	-0.166400	0.321640	0.0939059
	-0.0801235	0.164005	0.0461680	0.0693205	0.0792039

Good convergence is observed and the lack of diagonal dominance has not proved a problem.

There are some algorithmic details which can be highlighted at this stage. The coefficients of the block matrices are always unchanged, it is only the x

values which are updated, and by implication the right-hand sides in each block. Hence there is no need to repeat the elimination steps every time, as long as the multipliers and interchanges are stored. These can then be applied to the relevant right-hand side and the stored upper triangular matrix used to get the updated x values. The storage requirement is roughly that of a further sparse matrix which doubles the requirement, but this is still far less than storing the full matrix for the application of straight Gaussian elimination with fill-in.

There are two major directions of progress in solving large sets of linear equations which will be covered only briefly here, with reference to more substantive texts for the reader wishing to pursue these topics further. There is a school of thought based round the extensive work of Duff, Reid and Erisman (1986) that the best approach to sparse systems is to abandon iterative techniques and return to elimination methods. A little taste of this approach has already appeared in the preliminary chapter with Thomas's algorithm, for the solution of a tridiagonal matrix by elimination.

For well-conditioned matrices, the elimination procedure can be carried out without pivoting, and it is the pivoting process which exchanges rows and causes the filling in. The Thomas algorithm can be extended to consider a matrix of bandwidth s. A matrix A has bandwidth s if $a_{i,j} = 0$ for every $i, j \in \{1, 2, \ldots, n\}$ for which $|i - j| > s$. By this definition $s = 1$ for tridiagonal and $s = 2$ for quindiagonal. For a full matrix, the operation count is $O(n^3)$, but a similar count for an s-banded matrix is $O(s^2 n)$, with substantial saving. This conclusion depends on s being substantially smaller than n, and this is not the case for partial differential equations. Clearly a loss in this type of algorithm is that any zeros inside the band structure could well be destroyed by the elimination process, and hence in this respect the iterative methods hold an advantage. However there is no need with direct methods to worry about SOR extrapolation factors or about low convergence rates and associated blocking possibilities, so like a great deal of numerical analysis there are losses and gains to be balanced in the choice of approach. Even if some restricted pivoting is required, the operation count maintains the same order of size quoted above.

The usual approach to the direct method is to use LU decomposition in which the matrix A is split into a lower triangle L and an upper triangle U (Wilkinson, 1965).

An advantage of the decomposition method lies in the simplifications which arise if A is symmetric. The decomposition then becomes

$$A = LL^T$$

and this is known as Cholesky decomposition. Such a decomposition always exists if A is positive definite. One of the fruitful developments in direct methods is in the recognition that the ordering of the equations can substantially affect the resulting bandwidth. Remember that the matrix elements will arise from the evaluation of computational molecules at each point of the finite difference grid. How these grid points should be numbered is clearly going to affect the spread of the bandwidth. With the cross-shaped computational molecule for

Laplace's operator, numbering along rows in the x direction will leave the three x derivative points lying consecutively, but the point above and the point below will differ by roughly the number of points in a row.

The ordering can be revealed by a link with graph theory. A graph is generated by drawing each node and linking these nodes to every point which appears as a non-zero element in the same equation. If the resulting graph is a rooted tree then it can be shown that the corresponding ordering allows of Cholesky factorisation. The pursuit of this topic will take far more space than can be utilised here and the reader is referred to Golumbic (1980).

Inevitably there have also been developments in the iterative work which speed up the solution process markedly. Again space will only permit a flavour of these advances and reference will be given to more detailed accounts.

The Gauss–Seidel method can be improved after some detailed analysis of its behaviour. The observation which allows this is that the residual $\mathbf{r}^{[k]}$ after the kth step defined by

$$\mathbf{r}^{[k]} = A\mathbf{x}^{[k]} - \mathbf{b} \tag{4.3.1}$$

appears to fall very rapidly before it levels out to the asymptotic level which is predicted by the spectral radius analysis of Chapter 1. For problems arising in finite differences (or finite elements), the errors in a typical five-point finite difference representation satisfy the five-term recurrence relation

$$e_{i-1,j}^{[k+1]} + e_{i,j-1}^{[k+1]} + e_{i+1,j}^{[k]} + e_{i,j+1}^{[k]} - 4e_{i,j}^{[k+1]} = 0 \tag{4.3.2}$$

obtained by simply subtracting the five-point finite difference scheme from the equivalent Gauss–Seidel scheme where

$$e_{i,j}^{[k]} = \phi_{i,j}^{[k]} - \phi_{i,j}. \tag{4.3.3}$$

By taking a bivariate Fourier transform of the error $e_{i,j}^{[k]}$ an analysis can be made of how various frequencies in the error are reduced or attenuated by the Gauss–Seidel process. The result is that the amplitude of the oscillatory forms of the error are halved at each iteration. This explains the earlier observation. The rapid reduction occurs during the early iterates when the oscillatory error terms are reduced. The slower asymptotic attenuation then takes over. The high frequency error oscillations depend on the grid size. Hence to take advantage of this phenomenon the Gauss–Seidel method needs to be applied to a range of grid sizes. The process is called a *multigrid* technique. Take the finest grid as the grid on which the problem is ultimately required to be solved. A grid hierarchy can be set up with a sequence of coarser grids such as a 64×64, 32×32 and so on. Hence in the coarse grids the solution is a subset of the original equations.

The algorithm involves first applying a few Gauss–Seidel steps to the finest grid to knock down the high frequency components for this grid. The process of *restriction* is then applied to reduce the problem onto the coarser grid, and the process repeated to the coarsest grid. The process of *prolongation* is then applied to move back up the grid sequence to the finest. This up-and-down process (termed the V-cycle) is repeated so that the fine grid is the ultimate finishing point.

The process can be applied to the residues once the initial set of residues is found. This follows by considering the full problem

$$A\mathbf{x} = \mathbf{b} \tag{4.3.4}$$

with the residue defined in (4.3.1). Then if

$$\mathbf{x} = \mathbf{x}^{[k]} + \delta\mathbf{x}^{[k]}, \tag{4.3.5}$$

$\delta\mathbf{x}^{[k]}$ satisfies

$$A(\mathbf{x}^{[k]} + \delta\mathbf{x}^{[k]}) = \mathbf{b} \tag{4.3.6}$$

and substituting $A\mathbf{x}^{[k]} = \mathbf{r}^{[k]} + \mathbf{b}$ from (4.3.1) and cancelling $\mathbf{b}$ gives the residual equation

$$A\delta\mathbf{x}^{[k]} = -\mathbf{r}^{[k]}. \tag{4.3.7}$$

Let the fine grid be represented by a subscript f and the coarse grid by subscript c for any two consecutive grids in the sequence. Hence for the solution of (4.3.1), define the residual for the coarse grid as related to that of the fine grid by

$$\mathbf{r}_c = R\mathbf{r}_f \tag{4.3.8}$$

where R is called the restriction matrix. The smoothing then continues on the coarser grid according to

$$A_c\mathbf{v}_c = -\mathbf{r}_c \tag{4.3.9}$$

where the elements of A_c are those of A_f restricted to the coarse grid elements and the vector $\mathbf{v}$ is the iterated residue.

This new problem has its higher frequencies removed. The reverse problem uses the prolongation matrix P where

$$\mathbf{s}_f = P\mathbf{v}_c \tag{4.3.10}$$

and then the required $\mathbf{x}$'s are updated according to

$$\mathbf{x}_f^{\text{new}} = \mathbf{x}_f^{\text{old}} + \mathbf{s}_f. \tag{4.3.11}$$

The choice of matrices R and P will be made shortly. The residues in the new $\mathbf{x}$'s are related to the old residues by

$$\mathbf{r}_f^{\text{new}} = (I - A_f P A_c^{-1} R)\mathbf{r}_f^{\text{old}} \tag{4.3.12}$$

hence the new residue has both coarse and fine grid frequencies removed.

The choice of restriction matrix might be to take

$$r_{i,j}^c = r_{2i,2j}^f \tag{4.3.13}$$

or, more involved, take a weighted sum of neighbouring residuals such as

$$r_{i,j}^c = \frac{1}{4}r_{2i,2j}^f + \frac{1}{8}\left(r_{2i-1,2j}^f + r_{2i+1,2j}^f + r_{2i,2j-1}^f + r_{2i,2j+1}^f\right)$$

$$+ \frac{1}{16}\left(r_{2i-1,2j-1}^f + r_{2i+1,2j-1}^f + r_{2i-1,2j+1}^f r_{2i+1,2j+1}^f\right) \quad (4.3.14)$$

and the obvious choice for the prolongation transformation is to use linear interpolation:

$$r_{2i,2j}^f = r_{i,j}^c \qquad\qquad i,j = 1,2,\ldots,n$$

$$r_{2i-1,2j}^f = \tfrac{1}{2}(r_{i,j}^c + r_{i,j+1}^c) \qquad \begin{aligned}&i = 1,2,\ldots,n-1,\\&j = 1,2,\ldots,n\end{aligned}$$

$$r_{2i,2j-1}^f = \tfrac{1}{2}(r_{i,j}^c + r_{i+1,j}^c) \qquad \begin{aligned}&i = 1,2,\ldots,n,\\&j = 1,2,\ldots,n-1\end{aligned} \qquad (4.3.15)$$

$$r_{2i-1,2j-1}^f = \tfrac{1}{4}(r_{i,j}^c + r_{i+1,j}^c + r_{i,j+1}^c + r_{i+1,j+1}^c) \qquad \begin{aligned}&i = 1,2,\ldots,n,\\&j = 1,2,\ldots,n-1.\end{aligned}$$

As a practical example, the solution of Laplace's equation with the boundary conditions:

$$\begin{aligned}
y &= 0, & \phi &= 0, & 0 &\leq x \leq 1\\
y &= 1, & \phi &= 1, & 0 &\leq x \leq 1\\
x &= 0, & \phi &= y, & 0 &\leq y \leq 1\\
x &= 1, & \phi &= y, & 0 &\leq y \leq 1
\end{aligned} \qquad (4.3.16)$$

on an 8×8 grid was considered. There are 49 unknowns in the fine grid with the matrix A having the elements corresponding to the computational molecule of (4.1.2) and Figure 4.1. Hence the diagonal elements have the value 4 and the other four non-zero values will be unity. If the equations are numbered along the rows so that element (1,1) of the grid is variable 1, (1,2) is variable 2, on to element (1,7) as variable 7 and then (2,1) is variable 8 and so on to (7,7) as 49, then the ith equation will have element (i,i) as 4, $(i, j-1)$ as 1, $(i, j+1)$ as 1, and then for the element in the grid row below, $(i, j-7)$ is 1 and the grid row above will force $(i, j+7)$ to be unity. Where one of these points is a boundary point then the corresponding boundary value is required and the resulting number appears on the right-hand side. Care is needed to always be clear that in the finite difference grid with 7×7 unknown elements at each interior point, each yields an equation of the type above to give a 49×49 matrix.

The direct use of Gauss–Seidel on this set of 49 equations yields only about 3-figure accuracy after 25 steps, but the rapid decrease of the residue occurs in the first few iterates alone. By taking say five iterates, the residue is reduced to $1.0(-2)$ on average. The restriction operator is then applied using formula (4.3.14) on the residue. The nine unknowns of the coarser grid are then iterated.

These equations are 9, 11 and 13 corresponding to grid elements (2,2), (2,4) and (2,6): then 23, 25 and 27 corresponding to grid elements (4,2), (4,4) and (4,6) and finally 37, 39 and 41 which correspond to (6,2), (6,4) and (6,6). In a program small routines can be written to convert from equation numbers to grid points and to convert to the relevant equation numbers of the coarser grids.

The resulting residues for these nine points are now reduced to around $1.0(-3)$ and a further restriction operator is applied to the single equation 25 corresponding to the central grid element (4,4) with a residue of $-3.6(-5)$.

Now the reverse process using the prolongation operators was applied, followed by a Gauss–Seidel application of just five steps. The resulting residues were then just of order $1.0(-6)$ on the nine points at this intermediate level. Prolongation again to get back to the finest level and the original 49 points saw an error ranging between $1.0(-7)$ and $1.0(-8)$ in the final values.

In terms of workload, this quite accurate outcome was the result of effectively just ten iterations, as the work load for the intermediate nine points and the coarsest grid with a single point can effectively be ignored. The original Gauss–Seidel method was at only 3 figures after 25 iterations, hence the power of the technique is very apparent

In summary, there are many advances being made in this area, and the reader is advised that for a detailed account the work of either Bramble (1993) or Hackbusch (1985) should be consulted. A simple approach is given in Briggs (1987).

EXERCISES

4.19 As a very simple example of a block iterative approach consider the problem

$$\begin{pmatrix} 5 & 2 & -1 \\ -1 & 3 & 1 \\ -1 & 2 & 4 \end{pmatrix} \begin{pmatrix} \phi_1 \\ \phi_2 \\ \phi_3 \end{pmatrix} = \begin{pmatrix} 2 \\ 5 \\ 2 \end{pmatrix}$$

with a single first element block and a 2×2 second block. Experiment with the convergence rate by reducing the element $(3,3)$ to 3, 2, 1 and 0 to remove the diagonal dominance in stages.

4.20 Apply a simple block iterative method using 2×2 blocks to the linear equations:

$$\begin{pmatrix} 4 & -1 & -1 & 0 \\ -1 & 4 & 0 & -1 \\ -1 & 0 & 4 & -1 \\ -1 & -1 & 0 & 4 \end{pmatrix} \begin{pmatrix} \phi_1 \\ \phi_2 \\ \phi_3 \\ \phi_4 \end{pmatrix} = \begin{pmatrix} 2/3 \\ 5/3 \\ 5/3 \\ 2 \end{pmatrix}$$

from Section 4.1 and compare the convergence rate with that of Gauss–Seidel. It has been shown (by Varga) that there is a gain of around $\sqrt{2}$ over using single blocks.

4.21 Apply block iteration to the (6×6) matrix

$$
\begin{pmatrix}
3 & -1 & -1 & 0 & 0 & 0 \\
-1 & 1 & 0 & -1 & 0 & 0 \\
-1 & 0 & 4 & -1 & 0 & 0 \\
0 & -1 & 0 & 4 & 1 & 0 \\
-1 & 0 & 3 & -1 & 5 & 1 \\
-1 & -1 & 0 & 4 & 1 & 6
\end{pmatrix}
\begin{pmatrix}
\phi_1 \\
\phi_2 \\
\phi_3 \\
\phi_4 \\
\phi_5 \\
\phi_6
\end{pmatrix}
=
\begin{pmatrix}
3 \\
3 \\
5 \\
2 \\
1 \\
6
\end{pmatrix}.
$$

Use three (2×2) blocks, and then try two (3×3) blocks.

4.22 This exercise certainly moves into the programming regime. Use a suitable arrangement of blocks for the (12×12) problem

$$
\begin{bmatrix}
2 & 3 & 4 & 5 & 0 & 0 & 0 & 0 & 0 & 0 & 0 & 0 \\
1 & 2 & 1 & 3 & 2 & 0 & 0 & 0 & 0 & 0 & 0 & 0 \\
3 & 2 & 3 & 2 & 1 & 2 & 0 & 0 & 0 & 0 & 0 & 0 \\
0 & 3 & 2 & 4 & 1 & 2 & 3 & 0 & 0 & 0 & 0 & 0 \\
0 & 0 & 3 & 2 & 5 & 1 & 2 & 3 & 0 & 0 & 0 & 0 \\
0 & 0 & 0 & 3 & 2 & 6 & 2 & 0 & 0 & 0 & 0 & 0 \\
0 & 0 & 0 & 0 & 0 & 3 & 7 & 2 & 0 & 0 & 0 & 0 \\
0 & 0 & 0 & 0 & 0 & 0 & 3 & 8 & 2 & 0 & 0 & 0 \\
0 & 0 & 0 & 0 & 0 & 0 & 1 & 3 & 9 & 3 & 1 & -1 \\
0 & 0 & 0 & 0 & 0 & 0 & 0 & 0 & 3 & 10 & 2 & 4 \\
0 & 0 & 0 & 0 & 0 & 0 & 1 & 0 & 1 & 3 & 3 & 1 \\
0 & 0 & 0 & 0 & 1 & 0 & 0 & 3 & 7 & 3 & 2 & 13
\end{bmatrix}
\mathbf{x} =
\begin{bmatrix}
1 \\
1 \\
1 \\
1 \\
1 \\
1 \\
1 \\
1 \\
1 \\
1 \\
1 \\
1
\end{bmatrix}.
$$

5

Finite Element Method for Ordinary Differential Equations

5.1 Introduction

Finite element methods depend on a much more enlarged background of mathematics than the finite difference methods of the previous chapters. Most of these concepts can be applied to the solution of ordinary differential equations, and it is expedient to introduce these ideas through this medium. By this means the reader is less likely to become disorientated in the discussion on partial differential equations in the next chapter, as the underlying concepts will be clear.

To take a practical approach, consider the typical problem of finding the approximate solution of

$$\frac{d^2u}{dx^2} + u = -x, \quad u(0) = 0,\ u(1) = 0. \tag{5.1.1}$$

The finite difference approach is to approximate the derivative using finite differences and solve the resulting difference equation either step-by-step or as a set of linear algebraic equations. In this way, the actual unknown u is computed at certain grid points. An alternative is to seek an approximate solution of the form

$$u_N = \sum_{i=1}^{N} \alpha_i \beta_i(x) = \alpha_1 \beta_1(x) + \alpha_2 \beta_2(x) + \cdots + \alpha_N \beta_N(x) \tag{5.1.2}$$

where the functions $\beta_i(x)$ are called the *basis* functions. Now the object is to compute the coefficients in some sense, from which the unknown u_N can be found by summation. This approach is a very old one and various ways of achieving the goal will be discussed: the objective is to choose the coefficients to minimise the resulting error in the computed solution.

If u_N is an approximate solution to the above example, then a measure of the error E is

$$E = \frac{d^2 u_N}{dx^2} + u_N + x. \tag{5.1.3}$$

Clearly if u_N is the exact solution then $E = 0$. In the following paragraphs three methods are employed to fix the coefficients α_i in (5.1.2).

There are many problems with the approach suggested here. For example, the choice of basis function will be crucial. Just how this choice can be made to yield a method with high accuracy for modest workload is not clear. The finite element method overcomes these difficulties by reducing the dependence of the computations on the basis functions. The region of interest is broken into elements, and the fitting process is localised to a specific element. Relatively simple approximations may now be used within the element, such as linear or quadratic forms, and the convergence of the process becomes dependent primarily on the 'gauge' of the elements.

To make the process work across a set of elements it is important to ensure continuity conditions across element boundaries, and this constitutes an important part of the setting-up process. Moreover, the shape of the elements may be chosen to be triangles or rectangles (in two dimensions) or tetrahedrons (in three dimensions) for example, and this allows curved boundaries to be handled fairly easily and naturally.

The final general issue is the choice of criterion for the determination of the coefficients. Approaches include making the error zero at a chosen set of points to give the collocation method, minimising the integral of the square of the error in the interval of interest to yield the least squares method, and using weighted errors to give the weighted residual method. If the weights are the basis functions themselves, the Galerkin method is generated. In effect the differential equation is being replaced with a variational formulation in which an integral is minimised by the solution function. This minimisation condition allows linear equations to be set up for the local coefficients in the basis set approximation, and hence a solution is found. For partial differential equations, the conventional variational principle is often used as a starting point, and the classical Euler–Lagrange equations yield the original partial differential equation. These issues are pursued in the next chapter.

5.2 The Collocation Method

In this method, the values of $\alpha_1, \ldots, \alpha_N$ are chosen so that $E = 0$ at a set of given points. To illustrate this approach consider solving (5.1.1) with a solution

of the form
$$u_N = x(1-x)(\alpha_1 + \alpha_2 x + \cdots + \alpha_N x^{N-1}) \qquad (5.2.1)$$
which is chosen so that the boundary conditions are automatically satisfied for any choice of α_i. The basis functions are
$$\beta_1 = x(1-x), \quad \beta_2 = x^2(1-x), \quad \beta_3 = x^3(1-x), \quad \ldots \qquad (5.2.2)$$
and for the purpose of this example attention will be restricted to just the two basis functions
$$u_2 = \alpha_1\beta_1 + \alpha_2\beta_2 = \alpha_1 x(1-x) + \alpha_2 x^2(1-x). \qquad (5.2.3)$$

The corresponding error E is
$$
\begin{aligned}
E &= \frac{d^2 u_2}{dx^2} + u_2 + x \\
&= \alpha_1[-2 + x - x^2] + \alpha_2[2 - 6x + x^2 - x^3] + x. \qquad (5.2.4)
\end{aligned}
$$

In order to determine α_1 and α_2, E is set to zero at say $x = 1/4$ and $x = 1/2$, so that (5.2.4) gives
$$
\begin{aligned}
\text{at} \quad x &= \frac{1}{4}, \quad \frac{29}{16}\alpha_1 - \frac{35}{64}\alpha_2 = \frac{1}{4}, \\
\text{at} \quad x &= \frac{1}{2}, \quad \frac{7}{4}\alpha_1 - \frac{7}{8}\alpha_2 = \frac{1}{2}.
\end{aligned}
$$

Solving for α_1 and α_2 gives
$$\alpha_1 = \frac{6}{31} \qquad \text{and} \qquad \alpha_2 = \frac{40}{217}$$

with the corresponding solution
$$u_2 = \frac{x(1-x)}{217}[42 + 40x]. \qquad (5.2.5)$$

This solution can be compared with the exact analytic solution in Table 5.1, and it is encouraging that so simple an approach yields quite reasonable values.

Table 5.1.

x	u_{exact}	u_2
0	0	0
0.25	0.044014	0.044931
0.5	0.069747	0.071429
0.75	0.060056	0.062212
1	0	0

An alternative to this method is now considered in which the integral of the squared error is minimised, but first there are a few exercises on the above method.

EXERCISES

5.1 Consider the initial value problem $u'' + u = x$ with conditions $u(0) = 0$ and $u'(0) = 2$. Find an approximate solution of the form

$$u_N = \alpha_0 + \alpha_1 x + \alpha_2 x^2 + \alpha_3 x^3 + \cdots$$

(i) Choose α_0 and α_1 to satisfy the given conditions.

(ii) Define $\epsilon = u_N'' + u_N - x$, and find ϵ in this case.

Restricting u_N to be a cubic, use $x = 1/2$ and $x = 1$ as collocation points to find an approximate solution.

5.2 Use the collocation method to find an approximate solution to $u'' + 2u + x^2 = 0$ with $u(0) = 0$ and $u(1) = 0$,

(a) using two collocation points $x = 1/3$ and $x = 2/3$;

(b) using three collocation points $x = 1/4$, $x = 1/2$ and $x = 3/4$.

5.3 Use the collocation method on the equation

$$y'' + 4y' + 3y = 10e^{-2x}$$

with boundary conditions $y = 0$ on $x = 0$ and $y = 1$ on $x = 1$. Given that the general solution is

$$y = Ae^{-x} + Be^{-3x} - 10e^{-2x}$$

find A and B and compare your approximate results with the true solution. Try a series of increasingly accurate collocations. In this example this approach will need some modification. The $x = 0$ condition can indeed be used to fix α_0, but the second condition will constitute an extra linear equation.

5.4 Repeat Exercise 5.4 with the equation

$$y'' + 4y' + 4y = \cos x$$

which has the general solution

$$y = (A + Bx)e^{-2x} + 1/25(3\cos x + 4\sin x)$$

and the same boundary conditions.

5.5 Extend the process of collocation to the third order equation

$$y''' - 6y'' + 12y' - 8y = 0$$

with boundary conditions $y = 1$ and $y' = 0$ at $x = 0$, and $y = e^2$ when $x = 1$. For comparison purposes the general solution is

$$y = (1 - 2x + 2x^2)e^{2x}.$$

Use the collocation method to obtain approximate solutions and make comparisons with the exact solution.

5.6 Consider the collocation method on the non-constant coefficient equation

$$x^2 y'' + 6xy' + 6y = 1/x^2$$

with boundary conditions $y = 1$ when $x = 1$ and $y = 2$ when $x = 2$. Again for comparison purposes the general solution is

$$y = \frac{A}{x^2} + \frac{B}{x^3} + \frac{\ln x}{x^2}.$$

Use the collocation method on this problem and also find the values of A and B to allow a comparison to be made with the exact solution.

5.7 As a final example consider

$$(1 - x^2)\frac{d^2 y}{dx^2} - x\frac{dy}{dx} = 0$$

with boundary conditions $y = 0$ when $x = 0$ and $y = \pi/2$ when $x = 1$ with general solution

$$y = A \arcsin x + B.$$

5.3 The Least Squares Method

In this second alternative, an approximate solution of the form

$$u_N = \sum_{i=1}^{N} \alpha_i \beta_i \tag{5.3.1}$$

is found by choosing the parameters so that

$$F = \int_a^b E^2 \, dx \tag{5.3.2}$$

is minimised where $[a, b]$ is the range of interest. Once again the solution of (5.1.1) is attempted and the approximation (5.2.3) is utilised. Now as before

$$E = \alpha_1[-2 + x - x^2] + \alpha_2[2 - 6x + x^2 - x^3] + x \tag{5.3.3}$$

and using partial derivatives of (5.3.2) to obtain conditions that F should be minimised by the choice of α_1 and α_2 gives

$$\frac{\partial F}{\partial \alpha_1} = \int_0^1 2E\frac{\partial E}{\partial \alpha_1} \, dx = 2\int_0^1 E(-2 + x - x^2) \, dx = 0 \tag{5.3.4}$$

and

$$\frac{\partial F}{\partial \alpha_2} = \int_0^1 2E \frac{\partial E}{\partial \alpha_2} \, dx = 2 \int_0^1 E(2 - 6x + x^2 - x^3) \, dx = 0. \qquad (5.3.5)$$

Expanding and evaluating the above integrals gives

$$202\alpha_1 + 101\alpha_2 \;=\; 55$$
$$707\alpha_1 + 1572\alpha_2 \;=\; 399$$

with solution

$$\alpha_1 = 0.1875, \qquad \alpha_2 = 0.1695$$

which yields

$$u_2 = x(1 - x)(0.1875 + 0.1695x).$$

For comparison purposes Table 5.2 gives numerical values from this approximation.

Table 5.2.

x	u_n	u_{exact}
0	0	0
0.25	0.043	0.044
0.5	0.068	0.070
0.75	0.059	0.060
1	0	0

As with the previous two methods, this quite crude approximation yields approximately two-figure accuracy, and the use of more basis functions in the approximation would be expected to give increasingly better accuracy. Some exercises are now provided on this section.

EXERCISES

5.8 Consider the initial value problem $u'' + u = x$ with conditions $u(0) = 0$ and $u'(0) = 2$ from Exercise 5.1. Again find an approximate solution of the form

$$u_N = \alpha_0 + \alpha_1 x + \alpha_2 x^2 + \alpha_3 x^3 + \cdots$$

but now use the method of least squares to find an approximate solution. Compare this solution with that of Exercise 5.1.

5.9 Use the least squares method to find an approximate solution to $u'' + 2u + x^2 = 0$ with $u(0) = 0$ and $u(1) = 0$. Use a quadratic and a cubic solution and compare the results with Exercise 5.2.

5.10 An approximate solution to the boundary value problem $u''+u'+u = x$ with $u(1) = 1$ and $u(2) = 3$ is sought. Take

$$u_N = \alpha_1 + \alpha_2 x + (1-x)(2-x)(\alpha_3 + \alpha_4 x)$$

and find α_1 and α_2 so that the boundary conditions are satisfied. Determine α_3 and α_4 by using the least squares method.

5.11 Consider Exercise 5.3 to 5.7 using least squares and compare the performances of the two methods. In this comparison consider the resulting accuracy and the workload employed to attain this goal.

5.4 The Galerkin Method

The third of the methods in this chapter is a *weighted residual method*. Taking the same example from (5.1.1) with E defined in (5.1.3), the defining equations are obtained by forcing the weighted error given by

$$\int_0^1 Ev\,dx \tag{5.4.1}$$

to be zero where v is called the weight function or the test function. The approximate solution given in (5.1.2) is again used and in the Galerkin method, the test functions are taken to be identical to the basis functions. The Galerkin approximation is found by calculating α_i such that

$$\int_0^1 E\beta_i\,dx = 0, \quad i = 1,\ldots,N \tag{5.4.2}$$

where at this stage it is convenient that each β_i should satisfy the boundary conditions. We are forcing E to be perpendicular to the basis functions β_i, $i = 1(1)N$ in function space terminology.

Hence proceeding with the example, the trial solution

$$u_N = \alpha_1 x(1-x) + \alpha_2 x^2(1-x) \tag{5.4.3}$$

with

$$\beta_1 = x(1-x) \quad \text{and} \quad \beta_2 = x^2(1-x) \tag{5.4.4}$$

which both satisfy the boundary conditions gives the equations

$$\int_0^1 E\beta_1 \, dx = 0 \quad \text{and} \quad \int_0^1 E\beta_2 \, dx = 0$$

which with

$$E = \alpha_1[-2 + x - x^2] + \alpha_2[2 - 6x + x^2 - x^3] + x$$

gives the simultaneous equations

$$\frac{3}{10}\alpha_1 + \frac{3}{20}\alpha_2 = \frac{1}{12}$$

and

$$\frac{3}{10}\alpha_1 + \frac{13}{105}\alpha_2 = \frac{1}{20}$$

with solution

$$\alpha_1 = \frac{71}{369} \quad \text{and} \quad \alpha_2 = \frac{7}{41}.$$

The approximate solution is therefore

$$u_N = x(1 - x)\left[\frac{71}{369} + \frac{7}{41}x\right]. \tag{5.4.5}$$

A comparison with the analytic solution appears in Table 5.3 and two-figure agreement is again observed.

Table 5.3.

x	u_n	u_{exact}
0	0	0
0.25	0.0440	0.0440
0.5	0.0689	0.0697
0.75	0.0600	0.0601
1	0	0

This process can be extended very easily to a partial differential equation such as Poisson's equation

$$\frac{\partial^2 u}{\partial x^2} + \frac{\partial^2 u}{\partial y^2} = f(x, y), \quad (x, y) \in (0, a) \times (0, b). \tag{5.4.6}$$

Suppose now the object is to find an approximate solution using only one basis function when $f(x, y) = c$ (constant) and with boundary conditions $u = 0$

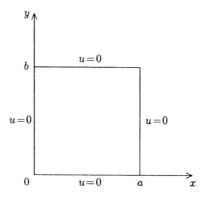

Fig. 5.1.

when $x = 0$, $x = a$, $y = 0$ and $y = b$, as illustrated in Figure 5.1. Then the trial solution now has the form

$$u_n = \alpha\beta(x,y). \tag{5.4.7}$$

Also $\beta(x,y)$ must satisfy all the boundary conditions and the integral must be taken over the rectangular domain R, so a double integral is required.

Hence try

$$u_1 = \alpha x(x-a)y(y-b) \tag{5.4.8}$$

as an appropriate solution, with E defined by

$$E = \frac{\partial^2 u_1}{\partial x^2} + \frac{\partial^2 u_1}{\partial y^2} - c = 2\alpha[y^2 - by + x^2 - ax] - c. \tag{5.4.9}$$

The Galerkin method requires

$$\int_R E\beta\, dxdy = 0 \tag{5.4.10}$$

or

$$\int_0^b \int_0^a \{2\alpha[x^2 - ax + y^2 - by] - c\}x(x-a)y(y-b)dxdy = 0 \tag{5.4.11}$$

which yields the value

$$\alpha = -\frac{5c}{2(a^2 + b^2)}. \tag{5.4.12}$$

The resulting approximate solution is

$$u_1 = \frac{-5c}{2(a^2 + b^2)}x(x-a)y(y-a).$$

More will be said about the solution of partial differential equations in the next chapter, but this simple example illustrates the ease with which the ideas in this section can be extended to such equations. Some exercises are now provided on Galerkin's method.

EXERCISES

5.12 An approximate solution to $-u'' + u - x = 0$ with boundary conditions $u(0) = u(1) = 0$ is to be found using the Galerkin method. Use

$$u_3 = \sum_{i=1}^{3} \alpha_i \beta_i(x)$$

with $\phi_i(x) = \sin i\pi x$ to find the appropriate constants α_i.

5.13 Apply the Galerkin method to Exercise 5.3 namely

$$y'' + 4y' + 3y = 10e^{-2x}$$

using as basis functions terms such as $\phi_i = x^i$, each of which satisfies the boundary condition $y = 0$ at $x = 0$, and $y = 1$ at $x = 1$ can be satisfied as a separate equation as in the earlier methods.

5.14 Apply the Galerkin method to the third order example in Exercise 5.5 namely

$$y''' - 6y'' + 12y' - 8y = 0$$

with boundary conditions $y = 1$ and $y' = 0$ at $x = 0$, and $y = e^2$ when $x = 1$. Suggested basis functions are $\phi_i(x) = x^i$ for $i \geq 2$ to give $y = 1 + \alpha_2\phi_2 + \cdots$ which will fix the two conditions at $x = 0$ for all i, and leave the requirement that $f(1) = e^2$.

5.15 Attempt Exercises 5.6 and 5.7 in the same spirit, using $\phi_i = x^i$ as basis functions.

5.5 Symmetric Variational Formulation

For this method, which is a variation of the previous technique, the test function is generalised to

$$v = \sum_{i=1}^{N} \gamma_i \beta_i \tag{5.5.1}$$

so that v is now a linear combination of the basis functions, β_i, which are chosen to be linearly independent, and satisfy the boundary conditions. This choice of test function ensures that u_n and v have similar forms. The Galerkin statement is now

$$\int_a^b Ev\,dx = 0, \quad \forall v \in \text{span}\{\beta_1, \ldots, \beta_N\} \tag{5.5.2}$$

and since the β_i are independent then this actually gives N conditions as each coefficient of β must be zero. These conditions are used to solve for the N unknowns α_i in the solution

$$u_N = \sum_{i=1}^N \alpha_i \beta_i. \tag{5.5.3}$$

Consider again the ordinary differential equation

$$u'' + u + x = 0, \quad u(0) = 0,\ u(1) = 0 \tag{5.5.4}$$

then as before the basis functions β_i are chosen to satisfy

$$\beta_i(0) = 0, \quad \beta_i(1) = 0.$$

The objective is to find values of α_i for which

$$\int_0^1 Ev\,dx = \int_0^1 (u_N'' + u_N + x)v\,dx = 0 \quad \forall v \in \text{span}\{\beta_1, \ldots, \beta_N\} \tag{5.5.5}$$

where

$$u_N = \sum_{i=1}^N \alpha_i \beta_i(x).$$

This is known as the variational formulation of the equation. The similarity of u_N and v are now employed to advantage as

$$\int_0^1 u_N'' v\,dx = [u_N' v]_0^1 - \int_0^1 u_N' v'\,dx$$

$$= -\int_0^1 u_N' v'\,dx \quad \forall v \in \text{span}\{\beta_1, \ldots, \beta_N\} \tag{5.5.6}$$

using integration by parts and the boundary conditions $v(0) = v(1) = 0$ (built up of the β_i). Hence

$$\int_0^1 (u_N'' + u_N + x)v\,dx = \int_0^1 (-u_N' v' + u_N v + xv)\,dx \quad \forall v \in \text{span}\{\beta_1, \ldots, \beta_N\}$$

and

$$u'' + u + x = 0, \qquad u(0) = u(1) = 0$$

can be replaced by

$$\int_0^1 (-u_N' v' + u_N v + xv)\, dx = 0, \quad \forall v \in \text{span}\{\beta_1, \ldots, \beta_N\}. \qquad (5.5.7)$$

This is the symmetric variational formulation. The same order of derivative of trial function (u_N) and test function (v) occur in the integrand and the following four points are to be noted:

(1) The order of the derivative in the variational formulation is less than in the original problem.

(2) The derivation places restrictions on v' and so we are not allowed to place any additional restrictions on v'. For example we cannot require $v'(0)$ or $v'(1)$ to have a specific value.

(3) If the boundary conditions are of the form $u = u_0$ at $x = 0$ and $u = u_1$ at $x = 1$ the test function v still satisfies the homogeneous boundary conditions ($v = 0$ at $x = 0, 1$). These are called essential boundary conditions.

(4) If the boundary conditions involve derivatives of u (called natural boundary conditions) then we do not specify the appropriate v, that is if $u_0'(0)$ is involved then $v(0)$ is not specified.

A formal method for finding the α_i is now illustrated using example (5.5.4). which is replaced by the variational formulation to find u_N such that

$$\int_0^1 (-u_N' v' + u_N v + xv)\, dx = 0 \qquad (5.5.8)$$

with

$$u_N(0) = u_N(1) = 0 \qquad \text{and} \qquad v(0) = v(1) = 0.$$

Let

$$v = \sum_{i=1}^N \gamma_i \beta_i, \qquad u_N = \sum_{j=1}^N \alpha_j \beta_j \qquad \text{and} \qquad \beta_i(0) = 0,\ \beta_i(1) = 0,\ i = 1, N$$

then (5.5.8) yields

$$\sum_i \gamma_i \left(\sum_j \left\{ \int_0^1 (-\beta_i' \beta_j' + \beta_i \beta_j)\, dx \right\} \alpha_j + \int_0^1 x\beta_i\, dx \right) = 0.$$

Define

$$K_{ij} = \int_0^1 -\beta_i'\beta_j' + \beta_i\beta_j \, dx \qquad (5.5.9)$$

and

$$F_i = -\int_0^1 x\beta_i \, da.x. \qquad (5.5.10)$$

Then

$$\sum_{i=1}^N \gamma_i \left(\sum_{j=1}^N K_{ij}\alpha_j - F_i \right) = 0$$

and since the γ_i are arbitrary then the following set of N equations arises to determine the required constants α_i.

$$\sum_{j=1}^N K_{ij}\alpha_j = F_i \quad \text{for } i = 1, 2, \dots, N \qquad (5.5.11)$$

or

$$K\alpha = \mathbf{F} \qquad (5.5.12)$$

where

$$K = \begin{pmatrix} K_{11} & K_{12} & \cdots & K_{1N} \\ K_{21} & K_{22} & \cdots & K_{2N} \\ \vdots & \cdots & \cdots & \vdots \\ K_{N1} & K_{N2} & \cdots & K_{NN} \end{pmatrix}$$

$$\alpha = \begin{pmatrix} \alpha_1 \\ \alpha_2 \\ \vdots \\ \alpha_N \end{pmatrix} \quad \text{and} \quad \mathbf{F} = \begin{pmatrix} F_1 \\ F_2 \\ \vdots \\ F_N \end{pmatrix}.$$

K is called the *stiffness* matrix, and $\mathbf{F}$ is called the *load* vector. In this case, K is symmetric, but this will not always be the case.

Stiffness matrices and load vectors of this type will arise in the finite element method as applied to partial differential equations which is described in the next chapter, and it is familiarity with such concepts which is the point of this diversion into ordinary differential equations.

As a second example, consider the differential equation

$$\frac{d^2u}{dx^2} + u = x^2 \quad \text{for } 0 < x < 1 \qquad (5.5.13)$$

subject to

(i) $u(0) = 0$, $u(1) = 0$

(ii) $u(0) = 0$, $u'(1) = 1$.

The solution in both cases starts the same way, and E is defined by

$$E = u_N'' + u_N - x^2 \tag{5.5.14}$$

and so, for any sufficiently smooth function v,

$$
\begin{aligned}
\int_0^1 Ev\,dx &= \int_0^1 (u_N'' + u_N - x^2)v\,dx \\
&= u_N'v\Big|_0^1 - \int_0^1 u_N'v'\,dx + \int_0^1 u_N v\,dx - \int_0^1 x^2 v\,dx \\
&= \int_0^1 (-u_N'v' + u_N v)\,dx + u_N'v\Big|_0^1 - \int_0^1 x^2 v\,dx.
\end{aligned}
$$

The boundary conditions now come into play, hence considering (i), which are essential conditions, and so require $v(0) = 0$, $v(1) = 0$, and allow $u_N(1) = 0$ and $u_N(1) = 0$ to be chosen, gives

$$\int_0^1 Ev\,dx = \int_0^1 (-u_N'v' + u_N v)\,dx - \int_0^1 vx^2\,dx = 0. \tag{5.5.15}$$

Now set

$$u_N = \sum_{j=1}^N \alpha_j \beta_j(x)$$

where

$$\beta_j(0) = \beta_j(1) = 0$$

and to satisfy this choose

$$\beta_1 = x(1-x), \qquad \beta_2 = x^2(1-x), \qquad \ldots, \qquad \beta_N = x^N(1-x). \tag{5.5.16}$$

Further take

$$v = \sum_{i=1}^N \gamma_i \beta_i(x)$$

and substitute into (5.5.15) to give

$$\int_0^1 \left(-\left(\sum_j \alpha_j \beta_j'\right)\left(\sum_j \gamma_i \beta_i'\right) + \left(\sum_j \alpha_j \beta_j\right)\left(\sum_j \gamma_i \beta_i\right) \right) dx$$
$$- \int_0^1 x^2 \left(\sum_i \gamma_i \beta_i\right) dx = 0$$

which reduces to

$$\sum_i \gamma_i \left(\sum_j K_{ij}\alpha_j - F_i \right) = 0 \qquad (5.5.17)$$

where

$$K_{ij} = \int_0^1 \left(-\beta_i'\beta_j' + \beta_i\beta_j \right) dx \qquad (5.5.18)$$

and

$$F_i = \int_0^1 x^2 \beta_i \, dx$$

and as before the equations can be written succinctly as

$$\sum_j K_{ij}\alpha_j = F_i \qquad (5.5.19)$$

as γ_i is arbitrary. Now using the β's defined in (5.5.16) gives

$$
\begin{aligned}
K_{ij} &= \int_0^1 \Big(-(ix^{i-1} - (i+1)x^i)(jx^{j-1} - (j+1)x^j) \\
&\quad + (x^i - x^{i+1})(x^j - x^{j+1}) \Big) dx \\
&= -\frac{2ij}{(i+j)((i+j)^2 - 1)} + \frac{2}{(i+j+1)(i+j+2)(i+j+3)}.
\end{aligned}
$$

and

$$F_i = \int_0^1 x^2 x^i (1 - x) \, dx = \frac{1}{(i+3)(i+4)}$$

Taking $N = 1$ then $u_1 = \alpha_1 \beta_1 = \alpha_1 x(1 - x)$ and $K_{11} = -\frac{3}{10}$ with $F_1 = \frac{1}{20}$. Hence as $K_{11}\alpha_1 = F_1$ then $\alpha_1 = -\frac{1}{6}$ yielding the approximate solution

$$u_1 = -\frac{1}{6}x(1 - x).$$

Taking $N = 2$ gives

$$u_2 = \alpha_1 \beta_1 + \alpha_2 \beta_2 = \alpha_1 x(1 - x) + \alpha_2 x^2(1 - x)$$

with $K_{11} = -\frac{3}{10}$, $K_{12} = K_{21} = -\frac{3}{20}$, $K_{22} = -\frac{13}{105}$, $F_1 = \frac{1}{20}$ and $F_2 = \frac{1}{30}$ which give the equations

$$
\begin{aligned}
-\frac{3}{10}\alpha_1 - \frac{3}{20}\alpha_2 &= \frac{1}{20} \\
-\frac{3}{20}\alpha_1 - \frac{13}{105}\alpha_2 &= \frac{1}{30}
\end{aligned}
$$

and the solution
$$\alpha_1 = -0.0813, \qquad \alpha_2 = -0.1707.$$

Hence the required function is
$$u_2 = (-0.0813)x(1-x) + (-0.1707)x^2(1-x).$$

Taking $N = 3$ and repeating the analysis gives
$$u_3 = -0.0952x(1-x) - 0.1005x^2(1-x) - 0.0702x^3(1-x).$$

For comparison, the exact solution is
$$u = \frac{\sin x + 2\sin(1-x)}{\sin(1)} + x^2 - 2$$

and Table 5.4 illustrates the above results in compact form. These results are also shown graphically in Figure 5.2.

Table 5.4.

x	$N = 1$	$N = 2$	$N = 3$	Exact
0.0	0.0	0.0	0.0	0.0
0.1	0.1500	0.0885	0.0954	0.0955
0.2	0.2667	0.1847	0.1890	0.1890
0.3	0.3500	0.2783	0.2766	0.2763
0.4	0.4000	0.3590	0.3520	0.3518
0.5	0.4167	0.4167	0.4076	0.4076
0.6	0.4000	0.4410	0.4340	0.4342
0.7	0.3500	0.4217	0.4200	0.4203
0.8	0.2667	0.3486	0.3529	0.3530
0.9	0.1500	0.2115	0.2183	0.2182
1.0	0.0	0.0	0.0	0.0

Taking the second set of boundary conditions $u(0) = 0$, $u'(1) = 1$, then $u(0) = 0$ is an essential condition and homogeneous, which therefore requires $v(0) = 0$. Further let u_N be such that $u_N(0) = 0$, $u'_N(1) = 1$ so that the boundary conditions are satisfied by u_N. This gives

$$\int_0^1 (-u'_N v' + u_N v)dx + v(1) - \int_0^1 x^2 v\, dx = 0. \qquad (5.5.20)$$

In this case, the β_i should be selected to satisfy only the essential boundary condition that is $\beta_i(0) = 0$. The function $\beta_i = x^i$ satisfies this condition. Let

$$v = \sum_{i=1}^{N} \gamma_i \beta_i(x)$$

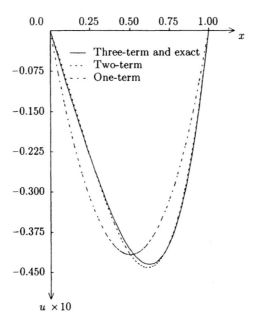

Fig. 5.2.

and

$$u_N = \sum_{j=1}^{N} \alpha_j \beta_j(x).$$

Substituting into (5.5.20) gives

$$\int_0^1 \left(-\left(\sum_j \alpha_j \beta_j' \right) \left(\sum_i \gamma_i \beta_i' \right) + \left(\sum_j \alpha_j \beta_j \right) \left(\sum_i \gamma_i \beta_i \right) \right) dx$$

$$+ \sum_{i=1}^{N} \gamma_i - \int_0^1 x^2 \left(\sum_i \gamma_i \beta_i \right) dx = 0$$

or

$$\sum_i \gamma_i \left(\sum_j K_{ij} \alpha_j - F_i \right) = 0 \qquad (5.5.21)$$

where

$$K_{ij} = \int_0^1 (-\beta_i' \beta_j' + \beta_i \beta_j)\, dx \qquad \text{and} \qquad F_i = \int_0^1 x^2 \beta_i\, dx - 1$$

to give as before

$$\sum_j K_{ij}\alpha_j = F_i \tag{5.5.22}$$

as the γ_i are arbitrary. In the particular case if $\beta_i = x^i$,

$$K_{ij} = \int_0^1 \left(-(ix^{i-1}jx^{j-1}) + x^i x^j\right) dx = -\frac{ij}{i+j-1} + \frac{1}{i+j+1}$$

and

$$F_i = \int_0^1 x^2 x^i\, dx - 1 = \frac{1}{i+3} - 1.$$

Taking $N = 1$, then $u_1 = \alpha_1\beta_1 = \alpha_1 x$ and $K_{11} = -\frac{2}{3}$ with $F_1 = -\frac{3}{4}$ to yield $\alpha_1 = \frac{9}{8}$ and hence $u_1 = \frac{9}{8}x$. Taking $N = 2$ requires $u_2 = \alpha_1\beta_1 + \alpha_2\beta_2 = \alpha_1 x + \alpha_2 x^2$ to yield in this case $K_{11} = -\frac{2}{3}$, $K_{12} = K_{21} = -\frac{3}{4}$, $K_{22} = -\frac{17}{15}$, $F_1 = -\frac{3}{4}$ and $F_2 = -\frac{4}{5}$. with the equations

$$-\frac{2}{3}\alpha_1 - \frac{3}{4}\alpha_2 = -\frac{3}{4}, \qquad -\frac{3}{4}\alpha_1 - \frac{17}{15}\alpha_2 = -\frac{4}{5}$$

to give $\alpha_1 = 1.295$ and $\alpha_2 = -0.1511$ and an approximate solution

$$u_2 = 1.295x - 0.1511x^2.$$

Taking $N = 3$ and repeating the analysis gives

$$u_3 = 1.283x - 0.1142x^2 - 0.02462x^3$$

whereas the exact solution is

$$u(x) = \frac{2\cos(1-x) - \sin x}{\cos 1} + x^2 - 2.$$

These results are again compared in Table 5.5.

As a final example of how to set up a variational solution consider

$$u'' + u' - 2u = x^2 + x \tag{5.5.23}$$

with either the boundary conditions

(a) $u(1) = 2$, $u(2) = 5$, or

(b) $u'(1) + u(1) = 3$, $2u'(2) - 3u(2) = 4$.

A variational statement of the problem will be derived for each case.

The variational statement is

$$\int_1^2 Ev\, dx = 0 \tag{5.5.24}$$

Table 5.5.

x	$N = 1$	$N = 2$	$N = 3$	Exact
0.0	0.0	0.0	0.0	0.0
0.1	0.1125	0.1280	0.1271	0.1262
0.2	0.2250	0.2529	0.2518	0.2513
0.3	0.3375	0.3749	0.3740	0.3742
0.4	0.4500	0.4938	0.4934	0.4943
0.5	0.5625	0.6067	0.6099	0.6112
0.6	0.6750	0.7226	0.7234	0.7244
0.7	0.7875	0.8324	0.8337	0.8340
0.8	0.9000	0.9393	0.9407	0.9402
0.9	1.0120	1.043	1.044	1.043
1.0	1.125	1.144	1.144	1.144

which can be written as

$$\int_1^2 [u_N'' + u_N' - 2u_N - (x^2 + x)]v \, dx = 0 \qquad (5.5.25)$$

or

$$\int_1^2 [-u_N'v' + u_N'v - 2u_N v - (x^2 + x)v]dx + [u_N'v]_1^2 = 0. \qquad (5.5.26)$$

If the boundary conditions are as in (a), then these are essential boundary conditions and in this case it is necessary to set $v(1) = v(2) = 0$ and the variational statement becomes

$$\int_1^2 [-u_n'v' + u_N'v - 2u_N v - (x^2 + x)v] \, dx = 0 \qquad (5.5.27)$$

as the highest derivatives of both u_N and v are the same. If the boundary conditions are as in (b), then both are natural boundary conditions and we cannot specify $v(1)$ or $v(2)$, but instead let u_N be such that

$$u_N'(1) + u_N(1) = 3$$

and

$$2u_N'(2) - 3u_N(2) = 4$$

to satisfy the boundary conditions.

Thus the variational statement is

$$\int_1^2 [-u_N'v' + u_N'v - 2u_N v - (x^2 + x)v]dx + u_N'(2)v(2) - u_N'(1)v(1) = 0 \quad (5.5.28)$$

which leads to

$$\int_1^2 [-u_N' v' + u_N' v - 2u_N v - (x^2 + x)v]dx$$

$$+(2 + \frac{3}{2}u_N(2))v(2) - (3 - u_N(1))v(1) = 0. \qquad (5.5.29)$$

As the boundary conditions in (a) are non-homogeneous essential boundary conditions, β_0 must be chosen so that these conditions are satisfied, that is $\beta_0(1) = 2$ and $\beta_0(2) = 5$, and $\beta_i(1) = \beta_i(2) = 0$, $i = 1(1)N$. Hence take

$$u_N = \sum_{j=1}^N \alpha_j \beta_j + \beta_0 \qquad \text{and} \qquad v = \sum_{i=1}^N \gamma_i \beta_i \qquad (5.5.30)$$

and substitute into the variational statement to give

$$\int_1^2 \left[-\left(\sum_j \alpha_j \beta_j' + \beta_0'\right)\left(\sum_i \gamma_i \beta_i'\right) + \left(\sum_j \alpha_j \beta_j' + \beta_0'\right)\left(\sum_i \gamma_i \beta_i\right) \right.$$

$$\left. -2\left(\sum_j \alpha_j \beta_j + \beta_0\right)\left(\sum_i \gamma_i \beta_i\right) - (x^2 + x)\left(\sum_i \gamma_i \beta_i\right) \right] dx = 0,$$

which can be rewritten as

$$\sum_i \gamma_i \left[\sum_j \left[\int_1^2 (-\beta_j' \beta_i' + \beta_j' \beta_i - 2\beta_j \beta_i) \, dx \right] \alpha_j \right.$$

$$\left. + \int_1^2 (-\beta_0' \beta_i' + \beta_0' \beta_i - 2\beta_j \beta_i - (x^2 + x)\beta_i)dx \right] = 0.$$

The usual matrix elements K_{ij} then become

$$K_{ij} = \int_1^2 (-\beta_j' \beta_i' + \beta_j' \beta_i - 2\beta_j \beta_i)dx$$

and

$$F_i = \int_1^2 (\beta_0' \beta_i' - \beta_0' \beta_i + 2\beta_0 \beta_i + (x^2 + x)\beta_i) \, dx$$

with

$$\sum_i \gamma_i \left[\sum_j K_{ij} \alpha_j - F_i \right] = 0$$

yielding the usual linear equations since the γ_i are arbitrary

$$\sum_j K_{ij}\alpha_j = F_i. \tag{5.5.31}$$

On the other hand, as both the boundary conditions in (b) are natural boundary conditions they do not restrict the choice of β_i , and hence

$$u_N = \sum_{j=1}^N \alpha_j \beta_j, \qquad v = \sum_{i=1}^N \gamma_i \beta_i \tag{5.5.32}$$

to give in this case

$$K_{ij} = \int_1^2 [-\beta'_j\beta'_i + \beta'_j\beta_i - 2\beta_j\beta_i]dx$$
$$+ \frac{3}{2}\beta_j(2)\beta_i(2) + \beta_j(1)\beta_i(1)$$

and

$$F_i = \int_1^2 (x^2 + x)\beta_i\,dx + 2\beta_i(2) - 3\beta_i(1)$$

to give the required matrices.

This section is concluded with some further exercises before attention is turned to overcoming the shortcomings of the methods by introducing the finite element approach in the form it will be used in the next chapter for partial differential equations.

EXERCISES

5.16 An approximate solution to $-u'' + u - x = 0$ with boundary conditions $u(0) = u(1) = 0$ is found using the symmetric variational formulation:

$$\int_0^1 u'_N v' + u_N v - xv\,dx = 0 \qquad \text{and} \qquad u_N(0) = u_N(1) = 0$$

where

$$u_N = \sum_{i=1}^N \alpha_i \phi_i.$$

Justify this formulation.

(i) Let $N = 3$ and choose $\phi_i(x) = \sin i\pi x$. Calculate the stiffness matrix and load vector. Solve for the coefficients α_i and construct an approximate solution. Plot the exact and approximate solutions and comment on the accuracy of the approximation.

(ii) Repeat (i) but take $\phi_i(x)$ as polynomials of degree $(i + 1)$ and satisfy the given boundary conditions.

5.17 It is required to find the approximate solution of the equation $u'' + 2u' - u = x$ where the boundary conditions may be either

(a) $u(0) = 1$ and $u(3) = 4$, or

(b) $u'(0) = 2$ and $u'(1) + 3u(1) = 5$.

Derive a symmetric variational formulation of the problem for each case. Also derive integral expressions for the elements of the stiffness matrix and the load vector. For the first boundary condition use the basis set $\phi_i = x^i(3 - x)$ with a suitable ϕ_0 and in the second case use $\phi_i = x^{i-1}$ for $i = 1, 2, \ldots$.

5.6 Finite Element Method

The quality of the Galerkin approximation is completely determined by the choice of the basis functions β_i. Once these have been chosen the determination of the coefficients α_i is purely computational. One weakness of the Galerkin method is that there is no systematic way of constructing reasonable basis functions. Apart from satisfying the homogeneous form of the essential boundary conditions these functions are arbitrary. In addition, the resulting system of linear algebraic equations has a full matrix.

These problems are overcome by using the *finite element* method. The finite element method provides a general and systematic technique for constructing basis functions for Galerkin approximations of boundary value problems. The main idea is that the basis functions β_i can be defined piecewise over subregions of the domain. Over any subregion, the β_i can be chosen to be a very simple functions such as low order polynomials. The subregions are called *finite elements*. In effect the work in this section is the one-dimensional version of the finite element method as it will be applied to partial differential equations, and hence it acts as a simple introduction to the full approach in the next chapter.

Suppose that the domain is the interval $[0, 1]$. The first step in the finite element method is to divide the domain into finite elements.

For example in Figure 5.3 there are four elements. The 'crosses' are called *nodes* or *nodal points*. The main features to note are

(i) The elements do not have to have the same length.

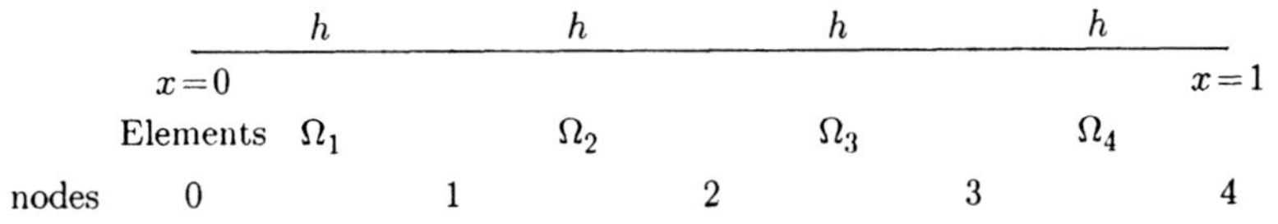

$$
\begin{array}{c|cccc}
 & h & h & h & h \\
\hline
x=0 & & & & x=1 \\
\text{Elements} \quad \Omega_1 & \Omega_2 & \Omega_3 & \Omega_4 & \\
\text{nodes} \quad 0 & 1 & 2 & 3 & 4
\end{array}
$$

Fig. 5.3.

(ii) An element may contain more than two nodes.

The basis functions need to be considered next. The following criteria are used for their construction:

(a) The basis functions are constructed from simple functions defined piecewise – element by element. The simple functions defined on the elements are called *shape functions*.

(b) The basis functions are smooth enough so that any integral appearing in the variational formulation exists.

(c) Each basis function has the property that its value is 1 at one node and 0 at all the other nodes. If x_j is the x-coordinate of node j then

$$
\beta_i(x_j) = \begin{cases} 1, & i = j, \\ 0, & i \neq j. \end{cases} \tag{5.6.1}
$$

It follows immediately from (iii) that $\alpha_i = u_N(x_i)$; this means that the coefficient α_i has precisely the value of u_N at the nodal point x_i.

An example of basis functions satisfying these points is

$$
\beta_i(x) = \begin{cases} \frac{x-x_{i-1}}{h_i} & x_{i-1} \leq x \leq x_i \\ \frac{x_{i+1}-x}{h_{i+1}} & x_i \leq x \leq x_{i+1} \quad i = 1,2,3,\ldots \\ 0 & \text{otherwise} \end{cases} \tag{5.6.2}
$$

where the x_i are the coordinates of the nodes, and $h_i = x_i - x_{i-1}$, and $h_{i+1} = x_{i+1} - x_i$. This basis function is illustrated in Figure 5.4.

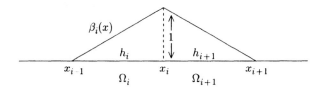

Fig. 5.4.

Note that the lengths of the elements need not be equal. The shape functions from which the functions β_i are constructed are the lines

$$
\Psi_i = \frac{x - x_{i-1}}{h_i} \tag{5.6.3}
$$

on element Ω_i and

$$\Psi_{i+1} = \frac{x_{i+1} - x}{h_{i+1}} \tag{5.6.4}$$

on element Ω_{i+1}.

There will in general be many shape functions defined on each element. The shape functions are defined only across a single element. It is critical that upon patching together the shape functions defined on elements to form the basis functions, the shape functions match perfectly at common nodes. The β_i will then be continuous. To illustrate the basis functions consider the domain $[0, 1]$ divided into four elements, then the basis functions β_1, β_2, β_3 are shown in Figure 5.5.

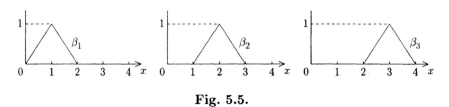

Fig. 5.5.

An important advantage of the finite element method is that the stiffness matrix K and the load vector $\mathbf{F}$ can be computed for each element. The stiffness matrix and load vector for the full domain is the sum of the stiffness matrices and load vectors for each element. This fact can be easily verified in the simple cases being considered.

Suppose the domain is $[0, 1]$ with four equal elements then using

$$K_{ij} = \int_0^1 (\beta_i' \beta_j' + \beta_i \beta_j)\, dx \tag{5.6.5}$$

gives

$$K_{ij} = \int_0^h (\beta_i' \beta_j' + \beta_i \beta_j)\, dx + \int_h^{2h} (\beta_i' \beta_j' + \beta_i \beta_j)\, dx$$

$$+ \int_{2h}^{3h} (\beta_i' \beta_j' + \beta_i \beta_j)\, dx + \int_{3h}^{1} (\beta_i' \beta_j' + \beta_i \beta_j)\, dx$$

$$= \sum_{e=1}^{4} \int_{\Omega_e} (\beta_i' \beta_j' + \beta_i \beta_j)\, dx$$

where $\int_{\Omega_e}$ means integrate across element Ω_e. Writing

$$K_{ij}^e = \int_{\Omega_e} (\beta_i' \beta_j' + \beta_i \beta_j)\, dx \tag{5.6.6}$$

to denote the element stiffness matrix for element Ω_e, then

$$K_{ij} = \sum_{e=1}^{4} K_{ij}^e \qquad (5.6.7)$$

and

$$F_i = \sum_{e=1}^{4} F_i^e \qquad (5.6.8)$$

where F_i^e is the element load vector for element Ω_e.

Computationally this simplifies the problem as we need only to calculate K_{ij}^e for a typical element and then sum.

To illustrate the above ideas, consider the solution of the following problem:

$$-u'' + u = x, \quad u(0) = u(1) = 0 \qquad (5.6.9)$$

with

$$u_N(x) = \sum_{j=1}^{N} \alpha_j \beta_j \quad \text{and} \quad v = \sum_{i=1}^{N} \gamma_i \beta_i \qquad (5.6.10)$$

where $\beta_i(0) = \beta_i(1) = 0$ as the boundary conditions are essential. The resulting linear equations are:

$$K\alpha = F \qquad (5.6.11)$$

where

$$K_{ij} = \int_0^1 (\beta_i' \beta_j' + \beta_i \beta_j)\, dx \quad \text{and} \quad F_i = \int_0^1 x\beta_i\, dx. \qquad (5.6.12)$$

The finite element method with four equal elements is employed and N is set as $N = 3$. The β_i are illustrated in Figure 5.5.

Consider element 1 in the first instance, then

$$K_{ij}^1 = \int_0^h (\beta_i' \beta_j' + \beta_i \beta_j)\, dx \qquad (5.6.13)$$

and on element 1, $\beta_2 = \beta_3 = 0$ so only K_{11}^1 is non-zero and has the value

$$K_{11}^1 = \int_0^h (\beta_1')^2 + (\beta_1)^2\, dx. \qquad (5.6.14)$$

The 'part' of β_1 on element 1 is just the shape function $\Psi_1 = x/h$ and hence

$$K_{11}^1 = \int_0^h \left(\frac{1}{h}\right)^2 + \left(\frac{x}{h}\right)^2\, dx$$

$$= \left[\left(\frac{1}{h^2} \right) x + \left(\frac{1}{h^2} \right) \frac{x^3}{3} \right]_0^h \tag{5.6.15}$$

$$= \frac{1}{h} + \frac{h}{3}.$$

But $h = \frac{1}{4}$ to give $K_{11}^1 = \frac{49}{12}$ and the element stiffness matrix for element 1 is

$$K^1 = \begin{pmatrix} \frac{49}{12} & 0 & 0 \\ 0 & 0 & 0 \\ 0 & 0 & 0 \end{pmatrix}. \tag{5.6.16}$$

The procedure is now repeated for element 2 with

$$K_{ij}^2 = \int_h^{2h} (\beta_i' \beta_j' + \beta_i \beta_j) \, dx. \tag{5.6.17}$$

Hence for element 2 part of β_1 and part of β_2 are pertinent as $\beta_3 = 0$ and $K_{11}, K_{12}, K_{21}, K_{22}$ will be non-zero for this element with $K_{13}, K_{31}, K_{32}, K_{23}, K_{33}$ zero. It is useful (and easier) to write the shape functions in a local coordinate, say ρ as defined in Figure 5.6.

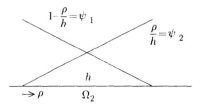

Fig. 5.6.

Hence with $\rho = x - h$ it is easy to show

$$K_{ij}^2 = \int_0^h \Psi_i' \Psi_j' + \Psi_i \Psi_j \, d\rho \quad i = 1, 2, \, j = 1, 2, \tag{5.6.18}$$

$$K_{11}^2 = \int_0^h \left(-\frac{1}{h} \right) \left(-\frac{1}{h} \right) + \left(1 - \frac{\rho}{h} \right) \left(1 - \frac{\rho}{h} \right) d\rho = \frac{1}{h} + \frac{h}{3}, \tag{5.6.19}$$

$$K_{12}^2 = \int_0^h \Psi_1' \Psi_2' + \Psi_1 \Psi_2 \, d\rho$$

$$= \int_0^h -\frac{1}{h^2} + \left(1 - \frac{\rho}{h} \right) \left(\frac{\rho}{h} \right) d\rho = -\frac{1}{h} + \frac{h}{6}. \tag{5.6.20}$$

From symmetry $K_{12} = K_{21}$

$$K_{22}^2 = \int_0^h \Psi_2'\Psi_2 + \Psi_2\Psi_2 \, d\rho = \frac{1}{h} + \frac{h}{3} \qquad (5.6.21)$$

and the element stiffness matrix for element 2 is

$$K^2 = \frac{1}{24}\begin{pmatrix} 98 & -95 & 0 \\ -95 & 98 & 0 \\ 0 & 0 & 0 \end{pmatrix}. \qquad (5.6.22)$$

The process is now repeated for element 3 so that here part of β_2 and part of β_3 are used. Also $\beta_1 = 0$ on Ω_3. In this case $K_{22}, K_{23}, K_{32}, K_{33}$ are non-zero on element 3 and $K_{11}, K_{12}, K_{13}, K_{21}, K_{31}$ are zero on element 3 The relevant local coordinates are $\rho = x - 2h$ and the same integrals as for element 2 have to be evaluated. At this stage it is clear that elements 2 and 3 are typical elements, whereas elements 1 and 4 at the ends are not typical. Hence the element stiffness matrix for element 3 is

$$K^3 = \frac{1}{24}\begin{pmatrix} 0 & 0 & 0 \\ 0 & 98 & -95 \\ 0 & -95 & 98 \end{pmatrix}. \qquad (5.6.23)$$

By similar argument it is left as an exercise to show that element 4 has the stiffness matrix

$$K^4 = \begin{pmatrix} 0 & 0 & 0 \\ 0 & 0 & 0 \\ 0 & 0 & \frac{49}{12} \end{pmatrix}. \qquad (5.6.24)$$

The stiffness matrices are now assembled to give

$$K = K^1 + K^2 + K^3 + K^4 = \frac{1}{24}\begin{pmatrix} 196 & -95 & 0 \\ -95 & 196 & -95 \\ 0 & -95 & 196 \end{pmatrix}. \qquad (5.6.25)$$

For the load vectors, the following derivations apply: firstly for element 1,

$$F_1^1 = \int_0^h x\beta_1 \, dx = \int_0^h x\frac{x}{h} \, dx = \frac{h^2}{3} = \frac{1}{48} \qquad (5.6.26)$$

and both F_2^1 and F_3^1 are zero, and then for element 2

$$F_1^2 = \int_h^{2h} x\beta_1 \, dx = \int_h^{2h}\left[2x - \frac{x^2}{h}\right] dx = \frac{2}{48}$$

$$F_2^2 = \int_h^{2h} x\beta_2 \, dx = \int_h^{2h} x\left[\frac{x}{h} - 1\right] dx = \frac{5}{96}, \qquad (5.6.27)$$

and

$$F_3^2 = \int_h^{2h} x\beta_3 \, dx = 0$$

and for element 3

$$F_1^3 = \int_{2h}^{3h} x\beta_1 \, dx = 0$$

$$F_2^3 = \int_{2h}^{3h} x\beta_2 \, dx = \int_{2h}^{3h} x\left[3 - \frac{x}{h}\right] dx = \frac{7}{96} \tag{5.6.28}$$

$$F_3^3 = \int_{2h}^{3h} x\beta_3 \, dx = \int_{2h}^{3h} x\left[-2 + \frac{x}{h}\right] dx = \frac{1}{12},$$

and finally for element 4

$$F_1^4 = \int_{3h}^{4h} x\beta_1 \, dx = 0, \qquad F_2^4 = \int_{3h}^{4h} x\beta_2 \, dx = 0 \tag{5.6.29}$$

$$F_3^4 = \int_{3h}^{4h} x\beta_3 \, dx = \int_{3h}^{4h} x\left[x - \frac{x}{h}\right] dx = \frac{5}{48}$$

which give the vectors:

$$\mathbf{F}^1 = \begin{pmatrix} \frac{1}{48} \\ 0 \\ 0 \end{pmatrix}, \qquad \mathbf{F}^2 = \begin{pmatrix} \frac{2}{48} \\ \frac{5}{96} \\ 0 \end{pmatrix},$$

$$\mathbf{F}^3 = \begin{pmatrix} 0 \\ \frac{7}{96} \\ \frac{1}{12} \end{pmatrix}, \qquad \mathbf{F}^4 = \begin{pmatrix} 0 \\ 0 \\ \frac{5}{48} \end{pmatrix}. \tag{5.6.30}$$

Assembly then yields:

$$\mathbf{F} = \mathbf{F}^1 + \mathbf{F}^2 + \mathbf{F}^3 + \mathbf{F}^4 = \frac{1}{16}\begin{pmatrix} 1 \\ 2 \\ 3 \end{pmatrix} \tag{5.6.31}$$

and

$$K\boldsymbol{\alpha} = \mathbf{F}. \tag{5.6.32}$$

Since $\alpha_i = u_N(x_i)$, it is common to replace α_i by u_i and write

$$K\mathbf{u} = \mathbf{F} \quad \text{where} \quad \mathbf{u} = \begin{pmatrix} u_1 \\ u_2 \\ u_3 \end{pmatrix}. \tag{5.6.33}$$

Finally solving the linear equations:

$$\frac{1}{24}\begin{pmatrix} 196 & -95 & 0 \\ -95 & 196 & -95 \\ 0 & -95 & 196 \end{pmatrix}\begin{pmatrix} u_1 \\ u_2 \\ u_3 \end{pmatrix} = \frac{1}{16}\begin{pmatrix} 1 \\ 2 \\ 3 \end{pmatrix} \quad (5.6.34)$$

gives

$$\alpha_1 = u_1 = 0.0353, \qquad \alpha_2 = u_2 = 0.0569, \qquad \alpha_3 = u_3 = 0.0505$$

and the solution:

$$u_N = 0.0353\beta_1 + 0.0569\beta_2 + 0.0505\beta_3.$$

A comparison of the exact and computed values is shown in Table 5.6.

Table 5.6.

x	u_n	u_{exact}
0	0	0
0.25	0.0353	0.0350
0.5	0.0569	0.0566
0.75	0.0505	0.0503
1	1	1

In the above example, only homogeneous boundary conditions have been considered, namely, $u(0) = u(1) = 0$. Consider now the more general form:

$$A(x)u''(x) + B(x)u'(x) + c(x)u(x) = g(x) \quad (5.6.35)$$

which can be rewritten as

$$(-a(x)u')' + b(x)u' + c(x)u = g(x) \quad (0 < x < l) \quad (5.6.36)$$

with boundary conditions:

$$r_0 \frac{du}{dx}(0) + s_0 u(0) = t_0$$

and

$$r_l \frac{du}{dx}(l) + s_l u(l) = t_l$$

where r_0, r_l, s_0, s_l, t_0 and t_l are specified constants.

The differential equation is first reset into a variational form. It would be convenient to use

$$E = -(a(x)u'_N)' + b(x)u'_N + c(x)u_N - g(x) \quad (5.6.37)$$

where E is the usual error term, but u'_N is discontinuous and u''_N does not exist. Hence the weak formulation is required equivalent to

$$\int_0^l Ev \, dx = 0 \qquad (5.6.38)$$

where v is a suitably defined test function. Formally integrating by parts gives

$$\int_0^l (a(x)u'_N)'v \, dx = [a(x)u'_N v]_0^l - \int_0^l au'_N v' \, dx$$

and leads to the weak formulation

$$\int_0^l [au'_N v' + bu'_N v + cu_N v] \, dx = \int_0^l g(x)v dx + [au'_N v]_0^l \quad \forall v \in \text{span}\{\beta_1, \ldots, \beta_N\}$$

$$(5.6.39)$$

and u_N satisfies the boundary conditions.

$$r_0 u'_N(0) + s_0 u_N(0) = t_0, \qquad r_l u'_N(l) + s_l u_N(l) = t_l. \qquad (5.6.40)$$

The boundary conditions are made part of the variational statement by setting

$$\int_0^l [au'_N v' + bu'_N v + cu_N v] \, dx$$

$$= \int_0^l gv \, dx + a(l)\left[\frac{t_l - s_l u_N(l)}{r_l}\right]v(l) - a(0)\left[\frac{t_0}{} - \frac{s_0 u_N(0)}{r_0}\right]v(0).$$

If the boundary conditions do not involve derivative terms, such as

$$s_0 u(0) = t_0, \qquad s_l u(l) = t_l \qquad (5.6.41)$$

then $v(0) = v(l) = 0$ (essential boundary conditions) should be specified. In this case the variational problem is

$$\int_0^l [au'_N v' + bu'_N v + cu_N v] \, dx = \int_0^l gv \, dx \quad \forall v \in \text{span}\{\beta_1, \ldots, \beta_N\} \quad (5.6.42)$$

where

$$u_N(0) = \frac{t_0}{s_0}, \qquad u_N(l) = \frac{t_l}{s_l}, \qquad v(0) = v(l) = 0.$$

Boundary conditions such as these (not involving derivatives) are called essential and do not appear in the variational statement. Suppose the boundary conditions also contain derivative terms (natural boundary conditions), since we have already placed restrictions on v' we cannot insist on any more conditions involving v'. The variational statement is now to find u_N such that

$$\int_0^l [au_N'v' + bu_N'v + cu_Nv]\,dx$$

$$= \int_0^l gv\,dx + a(l)\left[\frac{t_l - s_lu_N(l)}{r_l}\right]v(l) - a(0)\left[\frac{t_0 - s_0u_N(0)}{r_0}\right]v(0)$$

$$\forall v \in \text{span}\{\beta_1, \ldots, \beta_N\}. \tag{5.6.43}$$

In the analysis which follows only linear shape functions will be utilised, although greater accuracy can be gained using higher order polynomials but this increases the difficulty of the computations. Further advantage is taken of the finite element scheme allowing attention to be confined to a single typical element. Hence consider the variational statement on an element Ω_e defined as the interval (s_1, s_2), for consecutive s_1 and s_2. Then

$$\int_{s_1}^{s_2} (au_N'v' + bu_N'v + cu_Nv)\,dx$$

$$= \int_{s_1}^{s_2} gv\,dx + a(s_2)u_N'(s_2)v(s_2) - a(s_1)u_N'(s_1)v(s_1). \tag{5.6.44}$$

This statement does not depend on the boundary conditions existing at $x = 0$ and $x = l$. We seek an approximate solution, u_N^e, on this element and work will proceed with two nodes per element, which means that there are two different shape functions $\Psi_1(x)$, $\Psi_2(x)$ in the same way as in Figure 5.4. Then

$$u_N^e(x) = \sum_{j=1}^{2} \alpha_j^e \Psi_j^e(x) = \sum_{j=1}^{2} u_j^e \Psi_j^e(x) \tag{5.6.45}$$

where the e reminds us that we are refering to an element. Also

$$v(x) = \sum_{i=1}^{2} \gamma_i^e \Psi_i^e(x). \tag{5.6.46}$$

Substituting the expressions for u_N and v into the variational statement gives

$$\sum_{j=1}^{2} k_{ij}^e \alpha_j^e = g_i^e + a(s_2)u_N'(s_2)\Psi_i^e(s_2) - a(s_1)u_N'(s_1)\Psi_i^e(s_1) \tag{5.6.47}$$

where $i = 1, 2$ and

$$k_{ij}^e = \int_{s_1}^{s_2} \left[a\Psi_i'\Psi_j' + b\Psi_i\Psi_j' + c\Psi_i\Psi_j \right] dx \qquad (5.6.48)$$

with

$$g_i^e = \int_{s_1}^{s_2} g\Psi_i \, dx. \qquad (5.6.49)$$

Hence the element stiffness matrix, K^e and the element load vector, $\mathbf{G}^e$, can be calculated.

To find the (global) stiffness matrix, the element stiffness matrices are summed or assembled. There are two equations per element.

$$\text{for} \quad i \;=\; 1, \quad k_{11}^e\alpha_1 + k_{12}^e\alpha_2 = g_1^e - a(s_1)u_N'(s_1) \qquad (5.6.50)$$
$$\text{and for } i \;=\; 2, \quad k_{21}^e\alpha_1 + k_{22}^e\alpha_2 = g_2^e + a(s_2)u_N'(s_2). \qquad (5.6.51)$$

The subscripts 1 and 2 refer to node 1 and node 2 of a typical element. When assembling the matrices, the subscripts will have to be relabelled to take account of the actual element. Hence for example, if the 6th element is considered, then the equations are

$$k_{11}^6\alpha_6 + k_{12}^6\alpha_7 \;=\; g_1^6 - a(s_1)u_N'(s_1), \qquad (5.6.52)$$
$$k_{21}^6\alpha_6 + k_{22}^6\alpha_7 \;=\; g_2^6 + a(s_2)u_N'(s_2). \qquad (5.6.53)$$

The stiffness matrix and load vector may now be assembled. Suppose there are N nodes ($N - 1$ elements). The stiffness matrix is of dimension $N \times N$ and the load vector $N \times 1$, and the complete set of nodes is shown in Figure 5.7.

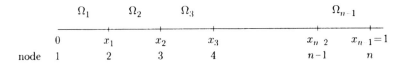

Fig. 5.7.

For Ω_1 :

$$\text{node 1} : \quad k_{11}^1\alpha_1 + k_{12}^1\alpha_2 = g_1^1 - a(0)u_N'(0)$$
$$\text{node 2} : \quad k_{21}^1\alpha_1 + k_{22}^1\alpha_2 = g_2^1 + a(x_1)u_N'(x_1)$$

and for Ω_2 :

$$\text{node 2} : \quad k_{11}^2\alpha_2 + k_{12}^2\alpha_3 = g_1^2 - a(x_1)u_N'(x_1)$$
$$\text{node 3} : \quad k_{21}^2\alpha_2 + k_{22}^2\alpha_3 = g_2^2 + a(x_2)u_N'(x_2).$$

Assembling these two sets of equations gives

$$
\begin{aligned}
k_{11}^1\alpha_1 \quad &+k_{12}^1\alpha_2 && &&= g_1^1 - a(0)u_N'(0) \\
k_{21}^1\alpha_1 \quad &+(k_{22}^1 + k_{11}^2)\alpha_2 &&+k_{12}^2\alpha_3 &&= g_2^1 + g_1^2 \\
&\quad k_{21}^2\alpha_2 &&+k_{22}^2\alpha_3 &&= g_2^2 + a(x_2)u_N'(x_2).
\end{aligned} \tag{5.6.54}
$$

Now set up the equations for Ω_3 :

$$
\begin{aligned}
\text{node } 3 \,:\, k_{11}^3\alpha_3 + k_{12}^3\alpha_4 &= g_1^3 - a(x_2)u_N'(x_2) \\
\text{node } 4 \,:\, k_{21}^3\alpha_3 + k_{22}^3\alpha_4 &= g_2^3 + a(x_3)u_N'(x_3)
\end{aligned}
$$

and adding these gives

$$
\begin{aligned}
k_{11}^1\alpha_1 \quad &+k_{12}^1\alpha_2 && && &&= g_1^1 - a(0)u_N'(0) \\
k_{21}^1\alpha_1 \quad &+(k_{22}^1 + k_{11}^2)\alpha_2 &&+k_{12}^2\alpha_3 && && = g_2^1 + g_1^2 \\
&\quad k_{21}^2\alpha_2 &&+(k_{22}^2 + k_{11}^3)\alpha_3 &&+k_{12}^3\alpha_4 && = g_2^2 + g_1^3 \\
&&&\quad k_{21}^3\alpha_3 &&+k_{22}^3\alpha_4 && = g_2^3 + a(x_3)u_N'(x_3).
\end{aligned} \tag{5.6.55}
$$

Finally consider the nodes $N-2$, $N-1$ and N. For Ω_{N-2} :

$$
\begin{aligned}
\text{node } N-2 \,:\, \quad k_{11}^{N-2}\alpha_{N-2} + k_{12}^{N-2}\alpha_{N-1} &= g_1^{N-2} - a(x_{N-3})u_N'(x_{N-3}) \\
\text{node } N-1 \,:\, \quad k_{21}^{N-2}\alpha_{N-2} + k_{22}^{N-2}\alpha_{N-1} &= g_2^{N-2} + a(x_{N-2})u_N'(x_{N-2})
\end{aligned}
$$

and for Ω_{N-1} :

$$
\begin{aligned}
\text{node } N-1 \,:\, \quad k_{11}^{N-1}\alpha_{N-1} + k_{12}^{N-1}\alpha_N &= g_1^{N-1} - a(x_{N-2})u_N'(x_{N-2}) \\
\text{node } N \,:\, \quad k_{21}^{N-1}\alpha_{N-1} + k_{22}^{N-1}\alpha_N &= g_2^{N-1} + a(l)u_N'(l).
\end{aligned}
$$

Finally assembling the stiffness matrix yields:

$$
K = \begin{pmatrix}
k_{11}^1 & k_{12}^1 & & & & & \\
k_{21}^1 & k_{22}^1 + k_{11}^2 & k_{12}^2 & & & & \\
& k_{21}^2 & k_{22}^2 + k_{11}^3 & k_{12}^3 & & & \\
& & \cdots & \cdots & \cdots & & \\
& & & k_{21}^{N-2} & k_{22}^{N-2} + k_{11}^{N-1} & k_{12}^{N-1} \\
& & & & k_{21}^{N-1} & k_{22}^{N-1}
\end{pmatrix} \tag{5.6.56}
$$

and the corresponding load vector is

$$
\mathbf{G} = \begin{pmatrix}
g_1^1 - a(0)u_N'(0) \\
g_2^1 + g_1^2 \\
g_2^2 + g_1^3 \\
g_2^3 + g_1^4 \\
\vdots \\
g_2^{N-2} + g_1^{N-1} \\
g_2^{N-1} + a(l)u_N'(l)
\end{pmatrix}. \tag{5.6.57}
$$

Note that $u'_N(0)$ and $u'_N(l)$ are found from the boundary conditions.

The derivation of K and $\mathbf{G}$ has been independent of the boundary conditions prescribed at $x = 0$ and $x = l$. Consider next how K and $\mathbf{G}$ will be modified by these boundary conditions. The entries $(1, 1)$ and (N, N) of the stiffness matrix and the first and last entries of the load vector will be altered by the boundary conditions. Two cases will be considered: firstly the natural boundary conditions given by

$$\begin{aligned} r_0 u'(0) + s_0 u(0) &= t_0 \\ r_l u'(l) + s_l u(l) &= t_l. \end{aligned} \tag{5.6.58}$$

The approximate solution must satisfy

$$r_0 u'_N(0) + s_0 u_N(0) = t_0 \tag{5.6.59}$$

and

$$r_l u'_N(l) + s_l u_N(l) = t_l. \tag{5.6.60}$$

Hence the first equation from $K\boldsymbol{\alpha} = \mathbf{G}$ now becomes

$$\begin{aligned} k_{11}^1 \alpha_1 + k_{12}^1 \alpha_2 &= g_1^1 - a(0) u_N^1(0) \\ &= g_1' - a(0) \left[\frac{t_0 - s_0 u_N(0)}{r_0} \right]. \end{aligned} \tag{5.6.61}$$

Further $\alpha_j = u_N(x)$ evaluated at the jth node to give $\alpha_1 = u_N(0)$ and replacing $u_N(0)$ by α_1, gives

$$\left(k_{11}^1 - \frac{a(0)s_0}{r_0} \right) \alpha_1 + k_{12}^1 \alpha_2 = g_1^1 - \frac{a(0)t_0}{r_0}. \tag{5.6.62}$$

Similarly $\alpha_N = u_N(l)$ and the last equation becomes

$$k_{21}^{N-1} \alpha_{N-1} + \left(k_{22}^{N-1} + \frac{a(l)s_l}{r_l} \right) \alpha_N = g_2^{N-1} + \frac{a(l)t_l}{r_l}. \tag{5.6.63}$$

Hence to summarise, with natural boundary conditions, K and $\mathbf{G}$ are modified thus

k_{11}^1	becomes	$k_{11}^1 - \dfrac{a(0)s_0}{r_0}$
k_{22}^{N-1}	becomes	$k_{22}^{N-1} + \dfrac{a(l)s_l}{r_l}$
$g_1^1 - a(0)u'_N(0)$	becomes	$g_1^1 - \dfrac{a(0)t_0}{r_0}$
$g_2^{N-1} + a(l)u'_N(l)$	becomes	$g_2^{N-1} + \dfrac{a(l)t_l}{r_l}.$

$$\tag{5.6.64}$$

Secondly with the essential boundary conditions

$$u(0) = \frac{t_0}{s_0}, \qquad u(l) = \frac{t_l}{s_l} \tag{5.6.65}$$

the approximate solution, u_N, needs to satisfy

$$u_N(0) = \frac{t_0}{s_0}, \qquad u_N(l) = \frac{t_l}{s_l}. \tag{5.6.66}$$

Since $\alpha_1 = u_N(0)$, $\alpha_N = u_N(l)$ there are only $N - 2$ values to find namely $\alpha_2, \alpha_3, \ldots, \alpha_{N-1}$, hence removing the first and last equations from the system. The second equation is modified thus:

$$k_{21}^1 \alpha_1 + (k_{22}^1 + k_{11}^2)\alpha_2 + k_{12}^2 \alpha_3 = g_2^1 + g_1^2 \tag{5.6.67}$$

but

$$\alpha_1 = u_N(0) = \frac{t_0}{s_0}$$

to give

$$(k_{22}^1 + k_{11}^2)\alpha_2 + k_{12}^2 \alpha_3 = g_2^1 + g_1^2 - k_{21}^1 \frac{t_0}{s_0}. \tag{5.6.68}$$

In a similar way, the $(N-1)$-th equation becomes

$$k_{21}^{N-2} \alpha_{N-2} + (k_{22}^{N-2} + k_{11}^{N-1})\alpha_{N-1} + k_{12}^{N-1} \alpha_N = g_2^{N-1} + g_1^N \tag{5.6.69}$$

but

$$\alpha_N = u_N(l) = \frac{t_l}{s_l}$$

to give

$$k_{21}^{N-2} \alpha_{N-2} + (k_{22}^{N-2} + k_{11}^{N-1})\alpha_{N-1} = g_2^{N-1} + g_1^N - k_{21}^{N-1} \frac{t_l}{s_l}. \tag{5.6.70}$$

Hence the full set of equations is now

$$
\begin{pmatrix}
k_{22}^1 + k_{11}^2 & k_{12}^2 & & & & & \\
k_{21}^2 & k_{22}^2 + k_{11}^3 & k_{12}^3 & & & & \\
 & k_{21}^3 & k_{22}^3 + k_{11}^4 & k_{12}^4 & & & \\
 & & \ddots & \ddots & \ddots & & \\
 & & & k_{21}^{N-3} & k_{22}^{N-3} + k_{11}^{N-2} & k_{12}^{N-1} & \\
 & & & & k_{21}^{N-2} & k_{22}^{N-2} + k_{11}^{N-1}
\end{pmatrix}
$$

$$
\begin{pmatrix}
\alpha_2 \\
\alpha_3 \\
\vdots \\
\alpha_{N-1}
\end{pmatrix}
=
\begin{pmatrix}
g_2^1 + g_1^2 - \frac{k_{21}^1 t_0}{s_0} \\
g_2^2 + g_1^3 \\
g_3^1 + g_1^4 \\
\vdots \\
g_2^{N-1} + g_1^N - \frac{k_{12}^{N-1} t_l}{s_l}
\end{pmatrix}
\tag{5.6.71}
$$

from which $\alpha_2, \ldots, \alpha_{N-1}$ can be calculated.

Note also that the first and last equations of the full system $K\alpha = G$ are

$$k_{11}^1 \alpha_1 + k_{12}^1 \alpha_2 = g_1^1 - a(0)u_N'(0) \qquad (5.6.72)$$
$$k_{21}^{N-1} \alpha_{N-1} + k_{22}^{N-1} \alpha_N = g_2^{N-1} + a(l)u_N'(l). \qquad (5.6.73)$$

Since α_1, α_2 are given when α_2 and α_{N-1} have been found the approximate derivatives at the endpoints $u_N'(0)$ and $u_N'(l)$ can be calculated.

The exercises for this section will be presented after the next section of worked examples.

5.7 Some Worked Examples

To conclude this chapter some worked examples will be treated in detail using the finite element method in its one-dimensional form. For the first example consider the solution of

$$-u'' + u - (1 + x) = 0 \qquad (5.7.1)$$

subject to $u(0) = 1$, $u(1) = 1$ (essential boundary conditions).

The equation is in the form

$$(-a(x)u')' + b(x)u' + c(x)u = g(x) \qquad (5.7.2)$$

with

$$a(x) = 1, \qquad b(x) = 0, \qquad c(x) = 1, \qquad g(x) = 1 + x.$$

The relevant interval is $x = 0$ to $x = 1$, which will be divided into four equal finite intervals. Each element has two nodes ($h = 0.25$), and linear shape functions are used. A typical element is shown in Figure 5.8.

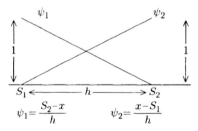

Fig. 5.8.

The approximate solutions, u_N, will satisfy the actual boundary conditions $u_N(0) = 1$ and $u_N(1) = 1$ and the basis functions, formed from the shape functions are shown in Figure 5.9.

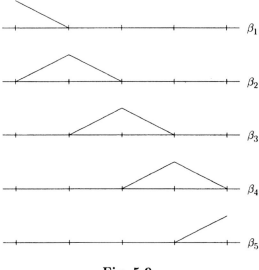

Fig. 5.9.

Note that the basis functions have a value of 1 at one particular node, and are zero at all other nodes. Hence taking a typical element Ω_e, say from s_1 to the consecutive element s_2, the variational statement is

$$\int_{s_1}^{s_2} u'_N v' + u_N v \, dx = \int_{s_1}^{s_2} (1+x)v \, dx + u'_N(s_2)v(s_2) - u'_N(s_1)v(s_1) \qquad (5.7.3)$$

as $a = 1$, $b = 0$, $c = 1$, $g = 1 + x$. Letting

$$u^e_N(x) = \sum_{j=1}^{2} \alpha^e_j \Psi^e_j(x) \qquad (5.7.4)$$

and

$$v(x) = \sum_{i=1}^{2} \gamma^e_i \Psi^e_i(x) \qquad (5.7.5)$$

gives

$$\sum_{j=1}^{2} k^e_{ij} \alpha^e_j = g^e_i + u'_N(s_2)\Psi^e_i(s_2) - u'_N(s_1)\Psi^e_i(s_1) \qquad (5.7.6)$$

where

$$k^e_{ij} = \int_{s_1}^{s_2} \Psi'_i \Psi'_j + \Psi_i \Psi_j \, dx$$

$$g_i^e = \int_{s_1}^{s_2} (1 + x)\Psi_i \, dx$$

$$k_{11}^e = \int_{s_1}^{s_2} \left(\frac{1}{h}\right)^2 + \left(\frac{s_2 - x}{h}\right)^2 dx = \frac{1}{h} + \frac{h}{3}$$

$$k_{12}^e = k_{21}^e = \int_{s_1}^{s_2} \left[-\frac{1}{h^2} + \frac{(s_2 - x)(x - s_1)}{h^2}\right] dx = \frac{h}{6} - \frac{1}{h}$$

and

$$k_{22}^e = \int_{s_1}^{s_2} \frac{1}{h^2} + \frac{(x - s_1)^2}{h^2} \, dx = \frac{1}{h} + \frac{h}{3}.$$

The stiffness matrix for a typical element is therefore

$$\begin{pmatrix} \frac{1}{h} + \frac{h}{3} & \frac{h}{6} - \frac{1}{h} \\ \frac{h}{6} - \frac{1}{h} & \frac{1}{h} + \frac{h}{3} \end{pmatrix}, \tag{5.7.7}$$

and in particular for element 1: $s_1 = 0$, $s_2 = 0.25$, $h = 0.25$ and

$$K^1 = \begin{pmatrix} 4.083333 & -3.958333 \\ -3.958333 & 4.083333 \end{pmatrix}.$$

For element 2: $s_1 = 0.25$, $s_2 = 0.5$, $h = 0.25$ and

$$K^2 = \begin{pmatrix} 4.083333 & -3.958333 \\ -3.958333 & 4.083333 \end{pmatrix}.$$

For element 3: $s_1 = 0.5$, $s_2 = 0.75$, $h = 0.25$ and

$$K^3 = \begin{pmatrix} 4.083333 & -3.958333 \\ -3.958333 & 4.083333 \end{pmatrix};$$

and for element 4: $s_1 = 0.75$, $s_2 = 1$, $h = 0.25$ and

$$K^4 = \begin{pmatrix} 4.083333 & -3.958333 \\ -3.958333 & 4.083333 \end{pmatrix}.$$

In each case here just the relevant 2×2 part of the full matrix is quoted. These element matrices are now assembled to form a global stiffness matrix.

$$K = \begin{pmatrix} 4.08333 & -3.95833 & 0 & 0 & 0 \\ -3.95833 & 8.16666 & -3.95833 & 0 & 0 \\ 0 & -3.95833 & 8.16666 & -3.95833 & 0 \\ 0 & 0 & -3.95833 & 8.16666 & -3.95833 \\ 0 & 0 & 0 & -3.95833 & 4.08333 \end{pmatrix}. \tag{5.7.8}$$

The element load vector is now calculated, the individual components being:

$$g_1^e = \int_{s_1}^{s_2} (1+x)\frac{(s_2 - x)}{h}\,dx = \frac{1}{h}\left[\frac{s_1^3 - s_2^3}{3} - \frac{(s_2^2 - s_1^2)(1 - s_2)}{2} + s_2 h\right]$$

and

$$g_2^e = \int_{s_1}^{s_2} (1+x)\frac{(x - s_1)}{h}\,dx = \frac{1}{h}\left[\frac{s_2^3 - s_1^3}{3} + \frac{(s_2^2 - s_1^2)(1 - s_1)}{2} - s_1 h\right]$$

Hence

$$\mathbf{G} = \begin{pmatrix} \frac{1}{h}\left[\frac{s_1^3 - s_2^3}{3} - \frac{(s_2^2 - s_1^2)(1-s_2)}{2} + s_2 h\right] \\ \frac{1}{h}\left[\frac{s_2^3 - s_1^3}{3} + \frac{(s_2^2 - s_1^2)(1-s_1)}{2} - s_1 h\right] \end{pmatrix}. \tag{5.7.9}$$

Hence the individual elements are, for element 1:

$$s_1 = 0.0, \qquad s_2 = 0.25, \qquad h = 0.25, \qquad \mathbf{G}^1 = \begin{pmatrix} 0.1354168 \\ 0.1458333 \end{pmatrix}.$$

For element 2:

$$s_1 = 0.25, \qquad s_2 = 0.5, \qquad h = 0.25, \qquad \mathbf{G}^2 = \begin{pmatrix} 0.16666666 \\ 0.1770832 \end{pmatrix}$$

For element 3:

$$s_1 = 0.5, \qquad s_2 = 0.75, \qquad h = 0.25, \qquad \mathbf{G}^3 = \begin{pmatrix} 0.1979168 \\ 0.2083333 \end{pmatrix}$$

and for element 4:

$$s_1 = 0.75, \qquad s_2 = 1.0, \qquad h = 0.25, \qquad \mathbf{G}^4 = \begin{pmatrix} 0.2291666 \\ 0.2395832 \end{pmatrix}.$$

The element load vectors are now assembled to form the global load vector

$$\mathbf{G} = \begin{pmatrix} 0.1354168 - a(0)u_N'(0) \\ 0.1458333 + 0.1666666 \\ 0.1770832 + 0.1979168 \\ 0.2083333 + 0.2291666 \\ 0.2395832 + a(1)u_N'(1) \end{pmatrix} \tag{5.7.10}$$

remembering the contributions from the boundary conditions. Now $a(0) = a(1) = 1$ to give

$$\mathbf{G} = \begin{pmatrix} 0.1354168 - u_N'(0) \\ 0.3125 \\ 0.375 \\ 0.4375 \\ 0.2395832 + u_N'(1) \end{pmatrix}. \tag{5.7.11}$$

So far the given boundary conditions have not been used and since the boundary conditions are essential we require $u_N(0) = 1$ and $u_N(1) = 1$. Thus $\alpha_1 = 1$ and $\alpha_5 = 1$. Since α_1 and α_5 are now specified, the first and the last rows of $K\alpha = G$ are removed and the effects of these values on the other rows are taken account of. In particular the second equation becomes

$$8.166666\alpha_2 - 3.958333\alpha_3 = 4.270833$$

and the fourth equation becomes

$$-3.958333\alpha_3 + 8.166666\alpha_4 = 4.395833.$$

The final set of linear equations is therefore

$$\begin{pmatrix} 8.166666 & -3.958333 & 0 \\ -3.958333 & 8.166666 & -3.9583333 \\ 0 & -3.958333 & 8.166666 \end{pmatrix} \begin{pmatrix} \alpha_2 \\ \alpha_3 \\ \alpha_4 \end{pmatrix} = \begin{pmatrix} 4.2708333 \\ 0.375 \\ 4.3958333 \end{pmatrix} \quad (5.7.12)$$

which solves to give

$$\begin{aligned} \alpha_2 &= 1.03521 = u_2 \\ \alpha_3 &= 1.05686 = u_3 \\ \alpha_4 &= 1.050521 = u_4 \end{aligned}$$

and

$$u_N(x) = \beta_1(x) + 1.03521\beta_2(x) + 1.05686\beta_3(x) + 1.050521\beta_4(x) + \beta_5(x). \quad (5.7.13)$$

A tabular comparison of the exact and calculated values appears in Table 5.7.

Table 5.7.

x	u_n	u_{exact}
0	1	1
0.25	1.03521	1.03505
0.5	1.05686	1.05659
0.75	1.050521	1.050276
1	1	1

For the second worked example, consider the problem

$$-u'' + u - (1 + x) = 0 \quad (5.7.14)$$

subject to boundary conditions

$$u'(0) + 2u(0) = 2.149, \qquad 2u'(1) - u(1) = -1.626$$

which are natural boundary conditions.

Using four elements and five nodes the stiffness matrix and load vector are calculated as above in equations (5.7.8) and (5.7.10). The natural boundary conditions now need to be treated, namely

$$u'_N(0) + 2u_N(0) = 2.149, \qquad 2u'_N(1) - u_N(1) = -1.626. \qquad (5.7.15)$$

The first equation is

$$
\begin{aligned}
4.083333\alpha_1 - 3.958333\alpha_2 &= 0.1354168 - u'_N(0) \\
&= -2.01358 + 2\alpha_1
\end{aligned}
$$

which becomes

$$2.083333\alpha_1 - 3.958333\alpha_2 = -2.01358. \qquad (5.7.16)$$

The last equation is similarly

$$-3.958333\alpha_4 + 3.583333\alpha_5 = -0.5734178. \qquad (5.7.17)$$

With these modifications, the system becomes

$$
\begin{pmatrix}
2.083333 & -3.958333 & 0 & 0 & 0 \\
-3.958333 & 8.166666 & -3.958333 & 0 & 0 \\
0 & -3.958333 & 8.166666 & -3.958333 & 0 \\
0 & 0 & -3.958333 & 8.166666 & -3.958333 \\
0 & 0 & 0 & -3.958333 & 3.583333
\end{pmatrix}
$$

$$
\begin{pmatrix}
\alpha_1 \\ \alpha_2 \\ \alpha_3 \\ \alpha_4 \\ \alpha_5
\end{pmatrix}
=
\begin{pmatrix}
-2.01358 \\ 0.3125 \\ 0.375 \\ 0.4375 \\ -0.5734168
\end{pmatrix}
\qquad (5.7.18)
$$

which solves to give

$$u_5(x) = 0.9985\beta_1 + 1.0342\beta_2 + 1.0563\beta_3 + 1.0504\beta_4 + 1.0003\beta_5. \qquad (5.7.19)$$

Table 5.8 shows a comparison of the computed and exact values for a range of x.

In conclusion, the work of this chapter has introduced all the major concepts of a finite element method. A suitable variational form is first set up, and the region of interest is subdivided into elements. The approximate solution is expressed locally in each element and the resulting stiffness matrix and load vector are found. These matrices and vectors for each element are then summed or assembled to give the final set of linear algebraic equations which are solved for the coefficients of the numerical solution. In the next chapter this work is extended to partial differential equations.

Exercises on the previous two sections are now presented.

Table 5.8.

x	u_5	u_{exact}
0	0.9985	1
0.25	1.0342	1.0351
0.5	1.0563	1.0566
0.75	1.0504	1.0503
1	1.0003	1

EXERCISES

5.18 Using a linear finite element method with four elements, calculate the stiffness matrix K and the load vector $\mathbf{F}$ for the boundary value problem $-u'' = 1 - x^2$ for $0 < x < 1$ and $u(0) = u(1) = 0$. Solve for u_N and compare u_N and u.

5.19 Using a linear finite element method with four elements, calculate the stiffness matrix K and load vector $\mathbf{F}$ for the boundary value problem $-(1 + x)u'' + u = x^2 - 2x$ subject to

(a) $u'(0) = 1$ and $u(1) + u'(1) = 2$

(b) $u(0) = 1$ and $u(1) = 2$

(c) $u'(0) = 1$ and $u(1) = 3$.

Solve for u_N.

6
Finite Elements for Partial Differential Equations

6.1 Introduction

Finite element methods are essentially methods for finding approximate solutions to a problem, commonly a partial differential equation, in a finite region or domain. Generally the unknown in the problem varies continuously over the domain but the solution found by the finite element method will not possess the same degree of continuity. The basis of the method is to divide the region or domain of the problem into a number of *finite elements*, and then find a solution which minimises some measure of the error, and is continuous inside the elements where it is expressed in terms of simple functions. The solution may not be continuous where the elements fit together. The ultimate accuracy of the solution is dependent upon the number and size of the elements, and the types of approximate function used within the elements.

In the previous chapter the approach has been used to solve ordinary differential equations and a sequence of methods was shown which led to a full-scale finite element approach. The elements are just segments of a straight line in this degenerate case, and some generalisation becomes necessary when more than one dependent variable is involved. The reader will observe a close parallel with the work in Chapter 5 as the approach for partial differential equations unfolds. The stages are first converting the differential equation into a variational form, followed by choosing a set of finite elements and an appropriate local form of solution (often linear as in the previous chapter), and finally leading by assembly to solving a sparse set of linear algebraic equations.

Finite element methods have their origins in the *displacement method* used in the 1950s to solve structural problems in aeronautical and civil engineering. In these early applications the elements could be identified with physical components of the structure and the problem was formulated as a system of equations, usually linear, in the displacements of the nodes or interconnection points of the elements. Frequently the formulation of the problem involved expressing the strain energy of the structure in terms of the nodal displacements and then applying the minimum energy principle to find the equilibrium position. For a large structure the number of equations could be very large and the solution formed an early use of computer technology.

As the power of computers increased, the range and complexity of problems soluble by finite element methods also grew. Not surprisingly the early applications of the method were all in structural mechanics and many of the commercial finite element packages were originally designed to make the finite element solution of structural problems more readily available to engineers. It was soon realised that finite element methods could be combined with variational calculus and applied to almost any problem which could be formulated in terms of minimising an integral functional over the domain of the problem. In this way, finite element methods could be used to solve heat transfer problems or electrostatic potential problems. In these cases, the integral or variational principle has a role which corresponds precisely to that of strain energy in elastic structural problems. Other more general methods for finite element approximation have been developed which are related to the Rayleigh–Ritz and Galerkin methods which were introduced in the previous chapter.

During the past twenty years a large number of papers and books on finite element methods have appeared in the literature, and in a single chapter only the basic principles can be sensibly covered. Pointers to further reading will allow the reader to pursue this topic more fully in specialised texts.

By analogy with Chapter 5, a typical finite element problem is solved in four phases. In the first phase, the region of interest is divided into elements, and element types, and associated basis functions are selected. Next the contribution of each element to the problem formulation is computed and generally this involves evaluating element *stiffness matrices* as in Chapter 5, together with *generalised force vectors*. In the third phase, the contributions from the elements are *assembled* to produce a large system of equations for the solution. Finally these equations are solved to find the primary unknowns of the problem, using the sparse solution methods from Chapter 1. These unknowns are generally the values of some physical quantity at the nodes of the elements.

Because of the nature of the finite element method, all sections of the solution need to be completed before it is possible to try examples and exercises which are meaningful. Hence the convention of considering exercises after each substantial section will be abandoned here and the relevant exercises will appear at the end of the chapter.

6.2 Variational Methods

In Chapter 5, the development of a variational form for an ordinary differential equation is covered in detail, and a similar approach can be used here for partial differential equations. Consider as a working example Poisson's equation

$$\frac{\partial^2 \phi}{\partial x^2} + \frac{\partial^2 \phi}{\partial y^2} = f(x, y) \qquad (6.2.1)$$

which includes the special case of Laplace's equation when $f(x, y) = 0$. The normal requirement is to solve (6.2.1) over some two-dimensional region R with boundary ∂R, subject to certain boundary conditions given on ∂R. The two main types of boundary condition are:

(i) $\phi(x, y) = g(x, y)$, and

(ii) $\partial \phi / \partial n = h(x, y)$

where $g(x, y)$ and $h(x, y)$ are given functions on ∂R. Condition (i) is called a Dirichlet condition and condition (ii) a Neumann condition. The partial derivative in (ii) denotes the derivative of ϕ in the direction of the outward normal to ∂R. More generally, the boundary might have a type (i) condition on one part, say ∂R_ϕ and a type (ii) condition on the remainder, say ∂R_n.

Though the emphasis here is on the practical aspects of the subject, the rigorous aspects cannot be ignored. The classical solution of say Poisson's equation belongs to $C_b^2(\mathcal{R})$, the set of bounded functions with continuous second derivatives on $\mathcal{R}$. Physical problems demand a 'bigger' space of functions to encompass discontinuities for example, and theoretical problems like proofs of existence are easier in such spaces. The problem cannot then be formulated in terms of second derivatives (which may not exist throughout the solution space), and an integral form is used in which the solution belongs to the Sobolev space $H^1(\mathcal{R})$ with the underlying space being $L^2(\mathcal{R})$. The space $L^2(\mathcal{R})$ is defined as

$$L^2(\mathcal{R}) = \left\{ v : \int_{\mathcal{R}} |v(x, y)|^2 \, dx dy < \infty \right\}$$

and the Sobolev space $H^1(\mathcal{R})$ as

$$H^1(\mathcal{R}) = \left\{ v \in L^2(\mathcal{R}) : \frac{\partial v}{\partial x} \in L^2(\mathcal{R}), \frac{\partial v}{\partial y} \in L^2(\mathcal{R}) \right\}.$$

This is the weak problem of the type used in (5.6.37) for the same reason. For the details of these basic ideas, the reader is referred to Renardy and Rogers (1992).

The variational principle for Poisson's equation can be formulated in the following theorem.

Theorem 6.1

Of all the continuous functions $\phi(x, y)$ which satisfy $\phi = g(x, y)$ on ∂R_ϕ, the one which is the solution of Poisson's equation with the above generalised boundary conditions is that which minimises

$$J(\phi) = \frac{1}{2} \int_R \left\{ \left(\frac{\partial \phi}{\partial x} \right)^2 + \left(\frac{\partial \phi}{\partial y} \right)^2 \right\} dx\, dy + \int_R f(x, y) \phi\, dx\, dy$$

$$- \int_{\partial R_n} h(x, y) \phi\, ds \tag{6.2.2}$$

where ds is a line segment along the boundary of R. The functions $\phi = g(x, y)$ on ∂R are called *admissible functions*, and $\phi \in H^1(\mathcal{R})$. This solution is called the weak solution. The solution of the Poisson partial differential equation is the strong solution. In fact ϕ is often just a number which represents the potential energy in many physical applications. Physically the theorem is a minimum energy principle which is often used to form partial differential equations in the first place! Also, as J maps functions ϕ onto numbers it is a *functional*.

Proof

Let ϕ be the admissible function which minimises the functional $J(\phi)$. Also, suppose that $\eta(x, y)$ satisfies $\eta = 0$ on ∂R_ϕ. Set

$$\psi(x, y) = \phi(x, y) + \epsilon \eta(x, y) \tag{6.2.3}$$

where ϵ is an arbitrary constant. Then clearly $\psi = g(x, y)$ is another admissible function. Substituting ψ for ϕ in (6.2.2) gives

$$J(\phi) = \frac{1}{2} \int_R \left\{ \left(\frac{\partial \phi}{\partial x} + \epsilon \frac{\partial \eta}{\partial x} \right)^2 + \left(\frac{\partial \phi}{\partial y} + \epsilon \frac{\partial \eta}{\partial y} \right)^2 \right\} dx\, dy$$

$$+ \int_R f(x, y)(\phi + \epsilon \eta)\, dx\, dy - \int_{\partial R_n} h(x, y)(\phi + \epsilon \eta)\, ds. \tag{6.2.4}$$

Subtract (6.2.2) to give

$$J(\psi) - J(\phi) = \frac{1}{2} \int_R \left\{ \epsilon^2 \left(\frac{\partial \eta}{\partial x} \right)^2 + \epsilon^2 \left(\frac{\partial \eta}{\partial y} \right)^2 + 2\epsilon \frac{\partial \phi}{\partial x} \frac{\partial \eta}{\partial x} + 2\epsilon \frac{\partial \phi}{\partial y} \frac{\partial \eta}{\partial y} \right\} dx\, dy$$

$$+ \epsilon \int_R f(x, y) \eta\, dx\, dy - \epsilon \int_{\partial R_n} h(x, y) \eta\, ds$$

$$= \frac{1}{2}\epsilon^2 \int_R \left\{ \left(\frac{\partial \eta}{\partial x}\right)^2 + \left(\frac{\partial \eta}{\partial y}\right)^2 \right\} dx\, dy$$

$$+ \epsilon \left[\int_R \left(\frac{\partial \phi}{\partial x}\frac{\partial \eta}{\partial x} + \frac{\partial \phi}{\partial y}\frac{\partial \eta}{\partial y}\right) dx\, dy + \int_R f(x,y)\eta\, dx\, dy \right.$$

$$\left. - \int_{\partial R_n} h(x,y)\eta\, ds \right]$$

$$= \frac{1}{2}\epsilon^2 F_1(\eta) + \epsilon F_2(\phi,\eta). \qquad (6.2.5)$$

Since ϕ is the function which minimises $J(\phi)$, then $J(\psi) \geq J(\phi)$, whence

$$\frac{1}{2}\epsilon^2 F_1(\eta) + \epsilon F_2(\phi,\eta) \geq 0. \qquad (6.2.6)$$

Assuming $\epsilon > 0$, (6.2.6) implies

$$\frac{1}{2}\epsilon F_1(\eta) + F_2(\phi,\eta) \geq 0$$

which in the limit as $\epsilon \to 0$ implies $F_2(\phi,\eta) \geq 0$ for all η. On the contrary, assuming $\epsilon < 0$, then (6.2.6) implies

$$\frac{1}{2}\epsilon F_1(\eta) + F_2(\phi,\eta) \leq 0$$

which in the limit as $\epsilon \to 0$ gives $F_2(\phi,\eta) \leq 0$ for all η. Hence $F_2(\phi,\eta) = 0$, and then

$$\int_R \left(\frac{\partial \phi}{\partial x}\frac{\partial \eta}{\partial x} + \frac{\partial \phi}{\partial y}\frac{\partial \eta}{\partial y}\right) dx\, dy + \int_R f(x,y)\eta\, dx\, dy$$

$$- \int_{\partial R_n} h(x,y)\eta\, ds = 0 \qquad (6.2.7)$$

$$\forall\, \eta \in H^1(\mathcal{R}), \eta|_{\partial R_\phi} = 0.$$

For ϕ sufficiently smooth, it is easy to establish that ϕ is a weak solution of Poisson's equation as the first term of (6.2.7) can be written as

$$\int_R \left\{ \frac{\partial}{\partial x}\left(\eta\frac{\partial \phi}{\partial x}\right) + \frac{\partial}{\partial y}\left(\eta\frac{\partial \phi}{\partial y}\right) \right\} dx\, dy - \int_R \eta\left\{ \frac{\partial^2 \phi}{\partial x^2} + \frac{\partial^2 \phi}{\partial y^2} \right\} dx\, dy \quad (6.2.8)$$

and the divergence theorem gives

$$\int_{\partial R_n} \eta\frac{\partial \phi}{\partial n}\, ds - \int_R \eta\left\{ \frac{\partial^2 \phi}{\partial x^2} + \frac{\partial^2 \phi}{\partial y^2} \right\} dx\, dy. \qquad (6.2.9)$$

Substituting into (6.2.7) gives

$$- \int_R \eta\left\{ \frac{\partial^2 \phi}{\partial x^2} + \frac{\partial^2 \phi}{\partial y^2} - f(x,y) \right\} dx\, dy + \int_{\partial R_n} \eta\left\{ \frac{\partial \phi}{\partial n} - h(x,y) \right\} ds$$

$$+ \int_{\partial R_\phi} \eta\frac{\partial \phi}{\partial \eta}\, ds = 0. \qquad (6.2.10)$$

But $\eta = 0$ on ∂R_ϕ, causing the last integral to vanish identically. Since η is otherwise arbitrary over R and ∂R_n, then the two remaining integrals must be identically zero. Hence

$$\frac{\partial^2 \phi}{\partial x^2} + \frac{\partial^2 \phi}{\partial y^2} = f(x, y) \quad \text{in} \quad R \qquad (6.2.11)$$

$$\phi = g(x, y) \quad \text{on} \quad \partial R_\phi, \qquad \frac{\partial \phi}{\partial n} = h(x, y) \quad \text{on} \quad \partial R_n \qquad (6.2.12)$$

which establishes that if ϕ minimises J then ϕ is a solution of the boundary value problem. To complete the theorem, the uniqueness of the solution of the boundary value problem has to be assumed. This is proved in detail in Renardy and Rogers (1992) and will add unnecessary detail to the development here.

Hence the variational formulation of the original Poisson problem is to find a function ϕ which minimises (6.2.4).

The finite element method now reduces to subdividing the region R into a finite number of subregions of simple geometric form. A local approximation is constructed to ϕ over each element separately in terms of ϕ values at a set of predetermined points or nodes. Linearity then allows the separate solutions to be summed over all the elements (assembly) and the resulting expression is minimised with respect to the nodal ϕ values. The ϕ values for which J is a minimum then yield the approximate solution.

6.3 Some Specific Elements

In the plane case, the domain R is subdivided into elements which are usually either triangles or rectangles. The nodes, or knots, are commonly taken to be the vertices of the element, though in more involved elements extra nodes may be used. The nodes are called boundary nodes if they lie on ∂R and interior nodes otherwise. An element is a boundary element if it possesses two or more boundary nodes, and otherwise is an interior element. Curved parts of ∂R are approximated by straight line segments joining boundary nodes.

The first element to be treated in detail is the triangular element shown in Figure 6.1.

The nodes have been numbered locally 1, 2 and 3, and it is conventional to number these nodes anticlockwise which standardises certain signs in formulae about to be derived. The individual elements also have a global node numbering system which will be used when all the elements are brought together in the assembly process. It is also convenient to give the element a number, say i in this case.

Consider therefore just a single element, then ϕ will be defined over this element in terms of the ϕ values at the three nodes. Hence a linear form for ϕ

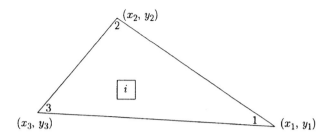

Fig. 6.1.

will have the correct number of degrees of freedom and hence

$$\phi(x,y) = \alpha_1 + \alpha_2 x + \alpha_3 y \tag{6.3.1}$$

in which the three unknowns α_1, α_2 and α_3 are fixed by forcing ϕ to take the values ϕ_1, ϕ_2 and ϕ_3 at the three nodes. The solution of the resulting set of linear equations can be written down in closed form, namely

$$\phi(x,y) = \phi_1 \beta_1(x,y) + \phi_2 \beta_2(x,y) + \phi_3 \beta_3(x,y) \tag{6.3.2}$$

where

$$\beta_1(x,y) = \frac{a_1 + b_1 x + c_1 y}{2\Delta} \tag{6.3.3}$$

and

$$a_1 = x_2 y_3 - x_3 y_2 \tag{6.3.4}$$
$$b_1 = y_2 - y_3 \tag{6.3.5}$$
$$c_1 = x_3 - x_2 \tag{6.3.6}$$

with

$$\Delta = \frac{1}{2}(x_2 y_3 - x_3 y_2 + x_3 y_1 - x_1 y_3 + x_1 y_2 - x_2 y_1)$$
$$= \frac{1}{2}\begin{vmatrix} 1 & 1 & 1 \\ x_1 & x_2 & x_3 \\ y_1 & y_2 & y_3 \end{vmatrix} \tag{6.3.7}$$

where Δ is the area of the triangle. The function β_1 has been constructed so as to be zero on the line joining (x_2, y_2) to (x_3, y_3), and to be unity at (x_1, y_1). Hence the function $\beta_1(x,y)$ defines the flat plane fitting the points $(x_1, y_1, 1)$, $(x_2, y_2, 0)$ and $(x_3, y_3, 0)$ in three dimensions. The functions $\beta_2(x,y)$ and $\beta_3(x,y)$ are defined by cyclically permuting the coefficients of (6.3.3) to (6.3.7). These will have the value unity at node 2 and node 3 respectively, and will be zero at the other two nodes.

The functions $\beta_1(x,y)$, $\beta_2(x,y)$ and $\beta_3(x,y)$ have various names in the finite element literature including *interpolation functions*, *shape functions* and *basis functions*, and have already been introduced in the context of ordinary differential equations in Section 5.6.

The second element to be treated in detail is the rectangular element shown in Figure 6.2.

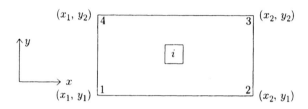

Fig. 6.2.

The assumption in the figure is that the sides of the rectangle are aligned with the x, y coordinate axes. Now there are four nodal values of ϕ and hence the approximating function equivalent to (6.3.1) requires four coefficients, and has the form

$$\phi(x, y) = \alpha_1 + \alpha_2 x + \alpha_3 y + \alpha_4 xy. \qquad (6.3.8)$$

This kind of function is called *bilinear* and is of course only one of many possible functions, but certainly one of the simplest.

As with the triangular element, the form in (6.3.8) is not used directly, as the requirement is to rewrite (6.3.8) with the terms ϕ_i rather than x and y being the controlling expressions. Hence a function such as

$$\beta_1(x, y) = \frac{(x - x_2)(y - y_2)}{(x_1 - x_2)(y_1 - y_2)} \qquad (6.3.9)$$

will have the property that it is zero on the sides of the rectangle $y = y_2$ and $x = x_2$ but unity at node 1. This is exactly like Lagrange's interpolation formula (Hildebrand, 1974). Hence the form for ϕ is

$$\phi(x, y) = \phi_1 \beta_1(x, y) + \phi_2 \beta_2(x, y) + \phi_3 \beta_3(x, y) + \phi_4 \beta_4(x, y) \qquad (6.3.10)$$

with β_2, β_3 and β_4 being defined by cyclic permuation.

For both triangular and rectangular elements as set up above, the variation of $\phi(x, y)$ along the boundaries is linear. Further, the particular linear function of x and y is dependent solely upon the ϕ values at the nodes at the two ends of the boundary concerned. It follows that if two adjacent elements have a common boundary as shown in Figure 6.3, then $\phi^{(i)}(x, y)$ and $\phi^{(j)}(x, y)$ are equal along this boundary. Hence $\phi(x, y)$ will automatically be continuous across all inter-element boundaries.

However with simple elements of this type, discontinuities of slope across inter-element boundaries cannot be avoided. Hence $\partial\phi/\partial x$ and $\partial\phi/\partial y$ will in general not be continuous.

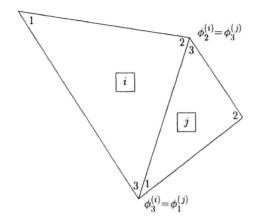

Fig. 6.3.

6.4 Assembly of the Elements

The ultimate aim is to minimise $J(\phi)$ defined in (6.2.2), and hence in terms of
the functionals for each element

$$J = \sum_{i=1}^{E} J^{(i)} \tag{6.4.1}$$

where

$$J^{(i)} = \frac{1}{2} \int_{R^{(i)}} \left\{ \left(\frac{\partial \phi^{(i)}}{\partial x} \right)^2 + \left(\frac{\partial \phi^{(i)}}{\partial y} \right)^2 \right\} dx\,dy + \int_{R^{(i)}} f(x,y)\phi^{(i)}\,dx\,dy$$

$$- \int_{\partial R_n^{(i)}} h(x,y)\phi^{(i)}\,ds \tag{6.4.2}$$

where $R^{(i)}$ is the region over which the ith element is defined, $\partial R_n^{(i)}$ is
that portion of its boundary over which $\partial \phi/\partial n$ is specified, and ϕ^i is the
approximating function defined over the ith element.

Equation (6.4.2) simplifies for all interior elements and all boundary elements having no boundaries on ∂R_n. In these two cases, $\partial R_n^{(i)} = 0$ and the
third integral vanishes.

It is not difficult to justify the replacement of (6.2.2) by (6.4.1) and (6.4.2).
The sum over all the elements of the first term of (6.4.2) is an integral over
the whole of R which is defined everywhere except on element boundaries,
where it has at worst finite jump discontinuities as ϕ is itself continuous across
these boundaries. Then this integral over R exists and is equal to the sum
of the components from which it is constructed by the linearity property of

the integral. Provided further that $f(x, y)$ and $h(x, y)$ are sufficiently well-behaved, the second integral may also be summed over all elements to obtain an integral over R. Similarly the third integral gives a sum over ∂R_n. The result of performing the summation is three integrals corresponding to the three terms in (6.2.2), in which ϕ should be regarded as a function which is admissible and which is composed of piecewise functions of the type defined in either (6.3.2) or (6.3.10).

Hence the procedure to set up the finite element equations is:

(i) The $J^{(i)}$ are evaluated using (6.3.2) or (6.3.10) according to whether triangular or rectangular elements are used.

(ii) The $J^{(i)}$ are summed to give J, the sum being expressed in terms of all the nodal values of ϕ.

(iii) Nodal values of ϕ on ∂R_ϕ are evaluated from the function $g(x, y)$.

(iv) Finally, J is minimised with respect to the unknown ϕ values at all internal nodes and boundary nodes on ∂R_n.

These steps are now dealt with in detail.

(i) This step involves the construction of the $J^{(i)}$. Both triangular and rectangular elements can be expressed in the form

$$\phi^{(i)}(x, y) = \sum_{k=1}^{K} \phi_k^{(i)} \beta_k(x, y) \tag{6.4.3}$$

where $K = 3$ for triangular elements and $K = 4$ for rectangular elements. In fact any two-dimensional element can be expressed as a sum of terms of this kind. Then

$$\frac{\partial \phi^{(i)}}{\partial x} = \sum_{k=1}^{K} \phi_k^{(i)} \frac{\partial \beta_k}{\partial x} \tag{6.4.4}$$

$$\frac{\partial \phi^{(i)}}{\partial y} = \sum_{k=1}^{K} \phi_k^{(i)} \frac{\partial \beta_k}{\partial y} \tag{6.4.5}$$

which may be expressed in convenient matrix terms as

$$\frac{\partial \phi^{(i)}}{\partial x} = \left(\phi_1^{(i)} \phi_2^{(i)} \ldots \phi_k^{(i)} \right) \begin{pmatrix} \frac{\partial \beta_1}{\partial x} \\ \frac{\partial \beta_2}{\partial x} \\ \vdots \\ \frac{\partial \beta_k}{\partial x} \end{pmatrix} = \left(\frac{\partial \beta_1}{\partial x} \frac{\partial \beta_2}{\partial x} \ldots \frac{\partial \beta_k}{\partial x} \right) \begin{pmatrix} \phi_1^{(i)} \\ \phi_2^{(i)} \\ \vdots \\ \phi_k^{(i)} \end{pmatrix} \tag{6.4.6}$$

with either representation being valid. Further

$$\left(\frac{\partial \phi^{(i)}}{\partial x}\right)^2 = (\phi_1^{(i)} \phi_2^{(i)} \ldots \phi_k^{(i)}) \begin{pmatrix} \frac{\partial \beta_1}{\partial x} \\ \frac{\partial \beta_2}{\partial x} \\ \vdots \\ \frac{\partial \beta_k}{\partial x} \end{pmatrix} (\frac{\partial \beta_1}{\partial x} \frac{\partial \beta_2}{\partial x} \ldots \frac{\partial \beta_k}{\partial x}) \begin{pmatrix} \phi_1^{(i)} \\ \phi_2^{(i)} \\ \vdots \\ \phi_k^{(i)} \end{pmatrix}$$

$$= \phi^{(i)^T} M^{(i)} \phi^{(i)}, \tag{6.4.7}$$

where $\phi^{(i)} = (\phi_1^{(i)}, \phi_2^{(i)}, \ldots, \phi_k^{(i)})^T$ and $M^{(i)}$ is the $K \times K$ matrix with elements

$$M_{kl}^{(i)} = \frac{\partial \beta_k}{\partial x} \frac{\partial \beta_l}{\partial x}. \tag{6.4.8}$$

Similarly

$$\left(\frac{\partial \phi^{(i)}}{\partial y}\right)^2 = \phi^{(i)^T} N^{(i)} \phi^{(i)} \tag{6.4.9}$$

where $N^{(i)}$ is the $K \times K$ matrix with elements

$$N_{kl}^{(i)} = \frac{\partial \beta_k}{\partial y} \frac{\partial \beta_l}{\partial y}. \tag{6.4.10}$$

Putting (6.4.3), (6.4.7) and (6.4.9) into (6.4.2) gives

$$\begin{aligned} J^{(i)} &= \frac{1}{2} \int_{R^{(i)}} \phi^{(i)^T} [M^{(i)} + N^{(i)}] \phi^{(i)} \, dx \, dy \\ &+ \int_{R^{(i)}} f(x,y) \left(\sum_{k=1}^{K} \phi_k^{(i)} \beta_k\right) dx \, dy \\ &- \int_{\partial R_n}^{(i)} h(x,y) \left(\sum_{k=1}^{K} \phi_k^{(i)} \beta_k\right) ds. \end{aligned} \tag{6.4.11}$$

The second and third terms can be expressed as

$$\sum_{k=1}^{K} \phi_k^{(i)} \left\{ \int_{R^{(i)}} f\beta_k \, dx \, dy - \int_{\partial R_n^{(i)}} h\beta_k \, ds \right\} = -\phi^{(i)^T} \mathbf{Q}^{(i)} \tag{6.4.12}$$

where $\mathbf{Q}^{(i)}$ is the column vector whose elements are

$$Q_k^{(i)} = -\int_{R^{(i)}} f(x,y)\beta_k(x,y) \, dx \, dy + \int_{\partial R_n^{(i)}} h(x,y)\beta_k(x,y) \, ds. \tag{6.4.13}$$

Equations (6.4.11) and (6.4.12) give together

$$J^{(i)} = \frac{1}{2} \phi^{(i)^T} K^{(i)} \phi^{(i)} - \phi^{(i)^T} \mathbf{Q}^{(i)} \tag{6.4.14}$$

in which

$$K^{(i)} = \int_{R^{(i)}} [M^{(i)} + N^{(i)}] \, dx \, dy \qquad (6.4.15)$$

or in other words, $K^{(i)}$ is the $K \times K$ matrix whose elements are given by

$$K_{kl}^{(i)} = \int_{R^{(i)}} \left(\frac{\partial \beta_k}{\partial x} \frac{\partial \beta_l}{\partial x} + \frac{\partial \beta_k}{\partial y} \frac{\partial \beta_l}{\partial y} \right) dx \, dy. \qquad (6.4.16)$$

Equation (6.4.15) defines $K^{(i)}$ which is called the *element stiffness matrix* analogous with the equivalent in Chapter 5. This name comes from solid mechanics. The vector $\phi^{(i)}$ is the *element generalised coordinate vector* and $\mathbf{Q}^{(i)}$ is the *element generalised force vector*. It will be shown later that each such term has its global counterpart which determines the behaviour of the system of elements as a whole. The element stiffness matrix $K^{(i)}$ has two important properties:

(a) The matrix is symmetric as $K_{kl}^{(i)} = K_{lk}^{(i)}$ from (6.4.16).

(b) The matrix is positive definite as $\phi^{(i)^T} K^{(i)} \phi^{(i)} > 0$ for any non-null vector $\phi^{(i)}$, because the first integral in (6.4.2) has an integrand which is a sum of squares.

These properties will have great importance when the computational problem is considered later.

The particular stiffness matrix for the triangular element has $K = 3$. The basis functions β in (6.4.16) are defined by (6.3.3), and hence $\partial \beta_k / \partial x = b_k / 2\Delta$ and $\partial \beta_k / \partial y = c_k / 2\Delta$ for $k = 1, 2, 3$. The stiffness matrix $K^{(i)}$ is then

$$\frac{1}{4\Delta} \begin{bmatrix} b_1^2 + c_1^2 & & \\ b_2 b_1 + c_2 c_1 & b_2^2 + c_2^2 & \\ b_3 b_1 + c_3 c_1 & b_3 b_2 + c_3 c_2 & b_3^2 + c_3^2 \end{bmatrix} \qquad (6.4.17)$$

where the upper right symmetric elements have been omitted. Effectively the k, l element is the element area Δ times the constant integrand (6.4.16).

For the rectangular element, $K = 4$. The partial derivatives in (6.4.16) are linear functions of x and y and the integration over the rectangular region $R^{(i)}$ is a simple matter. The resulting stiffness matrix is from equations (6.3.9) and (6.4.16)

$$\frac{1}{3ab} \begin{bmatrix} a^2 + b^2 & & & \\ -b^2 + \frac{1}{2}a^2 & a^2 + b^2 & & \\ -\frac{1}{2}b^2 - \frac{1}{2}a^2 & \frac{1}{2}b^2 - a^2 & a^2 + b^2 & \\ \frac{1}{2}b^2 - a^2 & -\frac{1}{2}b^2 - \frac{1}{2}a^2 & -b^2 + \frac{1}{2}a^2 & a^2 + b^2 \end{bmatrix} \qquad (6.4.18)$$

where again the symmetric elements have been omitted and the rectangle has sides of length $a = x_2 - x_1$ and $b = y_2 - y_1$ in the x and y directions respectively.

(ii) In this step the global stiffness matrix is assembled. In step (i), equation (6.4.2) is reformulated in matrix terms to give $J^{(i)}$. Now equation (6.4.1) will be used to get the overall functional J as

$$J = \sum_{i=1}^{E} J^{(i)} = \sum_{i=1}^{E} \left\{ \frac{1}{2} \phi^{(i)^T} K^{(i)} \phi^{(i)} - \phi^{(i)^T} \mathbf{Q}^{(i)} \right\}. \tag{6.4.19}$$

Some algebraic manipulation, which will follow shortly, allows (6.4.19) to be put into the form:

$$J = \frac{1}{2} \phi^T K \phi - \phi^T \mathbf{Q} \tag{6.4.20}$$

in which $\phi = [\phi_1, \phi_2, \ldots, \phi_N]^T$ and the $N \times N$ matrix K and the vector $\mathbf{Q}$ are respectively the *global stiffness matrix* and the *global generalised force vector*. Here N denotes the total number of nodes in the problem as a whole, and the components of ϕ are now labelled by their global node numbers.

(iii) The third major step in the process is the substitution of the Dirichlet boundary conditions. If these conditions hold on part of the boundary of R, then the ϕ values are known on this part of the boundary. Hence in principle, the components of ϕ can be ordered so that the unknown components occur first, and then (6.4.20) can be partitioned in the form

$$J = \frac{1}{2} [\phi_\alpha^T \quad \phi_\beta^T] \begin{bmatrix} K_{\alpha\alpha} & K_{\alpha\beta} \\ K_{\alpha\beta} & K_{\beta\beta} \end{bmatrix} \begin{bmatrix} \phi_\alpha \\ \phi_\beta \end{bmatrix} - [\phi_\alpha^T \quad \phi_\beta^T] \begin{bmatrix} \mathbf{Q}_\alpha \\ \mathbf{Q}_\beta \end{bmatrix}. \tag{6.4.21}$$

Minimisation of J with respect to the unknown components of ϕ, now the elements of ϕ_α, leads to the system of equations

$$K_{\alpha\alpha} \phi_\alpha = \mathbf{Q}_\alpha - \mathbf{K}_{\alpha\beta} \phi_\beta \tag{6.4.22}$$

in which the right-hand side is known. The unknown elements of ϕ_α may then be determined. In practice, the substitution of the known nodal values is performed slightly differently to avoid the need for reshuffling the order of the variables.

(iv) The final step involves finding the unknown nodal values of ϕ. This may be done either by solving a set of linear equations such as (6.4.22) or by a direct minimisation of J using (6.4.20). As the number of unknowns is large, sparse matrix methods from Chapter 1 are employed.

6.5 Worked Example

The four steps of the previous section are put into practice in the following worked example.

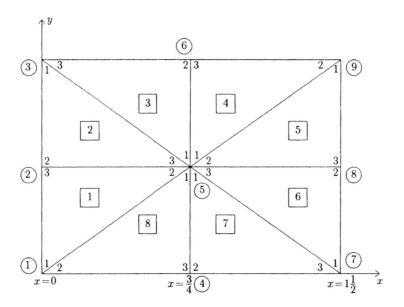

Fig. 6.4.

Solve

$$\frac{\partial^2 \phi}{\partial x^2} + \frac{\partial^2 \phi}{\partial y^2} = 2 \qquad (6.5.1)$$

with $0 \le x \le 3/2$ and $0 \le y \le 1$. The boundary conditions are $\phi = 3$ on $\phi = 0$, $\partial\phi/\partial y = 2x + 2$ on $y = 1$, $\partial\phi/\partial x = 2y$ on $x = 0$, and $\partial\phi/\partial x = 2y$ on $x = 3/2$. The triangular elements are shown in Figure 6.4. where the number enclosed in a square is the element number, that in a circle the global node number and the unenclosed numbers are the local node numbers, with anticlockwise labelling. Consider in detail the construction of the element stiffness matrix for element 2. Matrix (6.4.17) is required, and hence the coefficients b and c are required from (6.3.5) and (6.3.6). The a coefficient in (6.3.4) is also required and will be dealt with at the same time. Only a_1, b_1 and c_1 are quoted explicitly, as the others follow by cyclic permutation. Hence

$$
\begin{aligned}
a_1 &= x_2 y_3 - x_3 y_2 = -3/8, \\
b_1 &= y_2 - y_3 = 0, \qquad c_1 = x_3 - x_2 = 3/4, \\
a_2 &= x_3 y_1 - x_1 y_3 = 3/4, \\
b_2 &= y_3 - y_1 = -1/2, \qquad c_2 = x_1 - x_3 = -3/4, \qquad (6.5.2) \\
a_3 &= x_1 y_2 - x_2 y_1 = 0, \\
b_3 &= y_1 - y_2 = 1/2, \qquad c_3 = x_2 - x_1 = 0.
\end{aligned}
$$

Also

$$\Delta = \frac{1}{2}(x_2 y_3 - x_3 y_2 + x_3 y_1 - x_1 y_3 + x_1 y_2 - x_2 y_1) = 3/16.$$

Then from (6.4.17)

$$K^{(2)} = \frac{1}{4\Delta} \begin{bmatrix} b_1^2 + c_1^2 & b_1 b_2 + c_1 c_2 & b_1 b_3 + c_1 c_3 \\ b_2 b_1 + c_2 c_1 & b_2^2 + c_2^2 & b_2 b_3 + c_2 c_3 \\ b_3 b_1 + c_3 c_1 & b_3 b_2 + c_3 c_2 & b_3^2 + c_3^2 \end{bmatrix}$$

$$= \frac{16}{12} \begin{bmatrix} \frac{9}{16} & -\frac{9}{16} & 0 \\ -\frac{9}{16} & \frac{13}{16} & -\frac{1}{4} \\ 0 & -\frac{1}{4} & \frac{1}{4} \end{bmatrix} = \frac{1}{12} \begin{bmatrix} 9 & -9 & 0 \\ -9 & 13 & -4 \\ 0 & -4 & 4 \end{bmatrix}. \qquad (6.5.3)$$

Since elements 3, 6 and 7 are congruent to element 2, they will have the same stiffness matrix. On the other hand, elements 1, 4, 5 and 8 are congruent to a mirror reflection of element 2 and will have a stiffness matrix

$$\frac{1}{12} \begin{bmatrix} 9 & 0 & -9 \\ 0 & 4 & -4 \\ -9 & -4 & 13 \end{bmatrix}. \qquad (6.5.4)$$

Next the generalised force vector for element 2 is required and the components are given by (6.4.13),

$$Q_k^{(2)} = -\int_{R^{(2)}} f(x, y) \beta_k(x, y)\, dx\, dy + \int_{\partial R_n^{(2)}} h(x, y) \beta_k(x, y)\, ds. \qquad (6.5.5)$$

For this example, $f(x, y) = 2$ is the right-hand side of the Poisson equation, and $h(x, y)$ is a specified cross-boundary gradient. The part of ∂R_n which forms part of the boundary of this element is the interval $(1/2, 1)$ on the y-axis. Hence expressing the first integral as a double integral over the triangle gives

$$Q_k^{(2)} = -\int_{\frac{1}{2}}^{1} dy \int_0^{\frac{3}{2}(1-y)} 2\frac{a_k + b_k x + c_k y}{2\Delta}\, dx\, dy + \int_{\frac{1}{2}}^{1} (-2y)\frac{a_k + c_k y}{2\Delta}\, dy. \qquad (6.5.6)$$

In the second integral, $x = 0$ and since integration is along the left-hand boundary of the region, ds becomes just dy. Since $\Delta = 3/16$, then

$$Q_k^{(2)} = -\frac{16}{3} \int_{\frac{1}{2}}^{1} dy \left[a_k x + \frac{1}{2} b_k x^2 + c_k xy \right]_0^{\frac{3}{2}(1-y)} - \frac{16}{3} \int_{\frac{1}{2}}^{1} (a_k y + c_k y^2)\, dy$$

$$= -\frac{16}{3} \int_{\frac{1}{2}}^{1} \left(\frac{3}{2} a_k (1 - y) + \frac{9}{8} b_k (1 - y)^2 + \frac{3}{2} c_k y(1 - y) \right) dy$$

$$-\frac{16}{3} \left[\frac{1}{2} a_k y^2 + \frac{1}{3} c_k y^3 \right]_{\frac{1}{2}}^{1}.$$

Now put $1 - y = Y$ to give

$$Q_k^{(2)} = \frac{16}{3} \int_{\frac{1}{2}}^{0} \left(\frac{3}{2} a_k Y + \frac{9}{8} b_k Y^2 + \frac{3}{2} c_k (1 - Y)Y \right) dY - \frac{16}{3} \left(\frac{3}{8} a_k + \frac{7}{24} c_k \right)$$

$$
\begin{aligned}
&= -\frac{16}{3}\left[\frac{3}{4}a_k Y^2 + \frac{3}{8}b_k Y^3 + \frac{3}{4}c_k Y^2 - \frac{1}{2}c_k Y^3\right]_0^{\frac{1}{2}} - 2a_k - \frac{14}{9}c_k \\
&= -3a_k - \frac{1}{4}b_k - \frac{20}{9}c_k .
\end{aligned}
\tag{6.5.7}
$$

Substituting from (6.5.2) for a_k, b_k and c_k gives

$$
\begin{aligned}
\mathbf{Q}^{(2)} &= \left[Q_1^{(2)}, Q_2^{(2)}, Q_3^{(2)}\right]^T \\
&= \left[-\frac{13}{24}, -\frac{11}{24}, -\frac{1}{8}\right] .
\end{aligned}
\tag{6.5.8}
$$

In a similar manner the full list of eight force vectors can be found to yield:

$$
\begin{aligned}
\mathbf{Q}^{(1)} &= \left(-\frac{5}{24}, -\frac{1}{8}, -\frac{7}{24}\right)^T \\[4pt]
\mathbf{Q}^{(2)} &= \left(-\frac{13}{24}, -\frac{11}{24}, -\frac{1}{8}\right)^T \\[4pt]
\mathbf{Q}^{(3)} &= \left(-\frac{1}{8}, 1, \frac{13}{16}\right)^T \\[4pt]
\mathbf{Q}^{(4)} &= \left(-\frac{1}{8}, \frac{25}{16}, \frac{11}{8}\right)^T \\[4pt]
\mathbf{Q}^{(5)} &= \left(\frac{7}{24}, -\frac{1}{8}, \frac{5}{24}\right)^T \\[4pt]
\mathbf{Q}^{(6)} &= \left(-\frac{1}{24}, \frac{1}{24}, -\frac{1}{8}\right)^T \\[4pt]
\mathbf{Q}^{(7)} &= \left(-\frac{1}{8}, -\frac{1}{8}, -\frac{1}{8}\right)^T \\[4pt]
\mathbf{Q}^{(8)} &= \left(-\frac{1}{8}, -\frac{1}{8}, -\frac{1}{8}\right)^T .
\end{aligned}
\tag{6.5.9}
$$

Before assembling these matrices into the global matrix, the application of rectangular rather than triangular matrices is considered.

The elements and their numbering are shown in Figure 6.5, where the global numbering has been kept identical with the triangular element case. In this way the solutions from the two cases and from the analytic solution may be compared in due course.

Since the lengths of the element sides in the x and y directions are 3/4 and 1/2 respectively, then equations (6.4.17) give for the element stiffness matrix

$$
K^{(i)} = \frac{1}{36}\begin{bmatrix}
26 & 1 & -13 & -14 \\
1 & 26 & -14 & -13 \\
-13 & -14 & 26 & 1 \\
-14 & -13 & 1 & 26
\end{bmatrix}
\tag{6.5.10}
$$

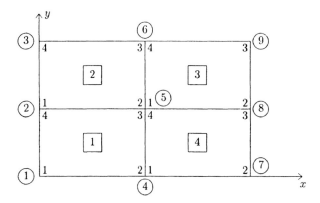

Fig. 6.5.

which will be identical for each of the four elements, since they have identical geometry.

The generalised force vector for element 1, again from (6.4.18), gives

$$Q_k^{(1)} = -\int_{R^{(1)}} f(x,y)\beta_k(x,y)\,dx\,dy + \int_{\partial R_n^{(1)}} h(x,y)\beta_k(x,y)\,ds \qquad (6.5.11)$$

and as before $f(x,y) = 2$ and $h(x,y) = -2y$ then using (6.3.9) for $\beta_1(x,y)$ gives

$$\begin{aligned}
Q_1 &= -2\int_0^{\frac{1}{2}} dy \int_0^{\frac{3}{4}} \frac{\left(x - \frac{3}{4}\right)\left(y - \frac{1}{2}\right)}{\left(\frac{3}{4}\right)\left(\frac{1}{2}\right)}\,dx + \int_0^{\frac{1}{2}} (-2y)\frac{-\frac{3}{4}\left(y - \frac{1}{2}\right)}{\left(\frac{3}{4}\right)\left(\frac{1}{2}\right)}\,dy \\
&= -\frac{16}{3}\int_0^{\frac{1}{2}}(y-1/2)\,dy \int_0^{\frac{3}{4}}(x-3/4)\,dx + 4\int_0^{\frac{1}{4}}(y^2 - y/2)\,dy \\
&= -\frac{3}{16} - \frac{1}{12} = -\frac{13}{48}. \qquad (6.5.12)
\end{aligned}$$

Similar computations give:

$$\begin{aligned}
\mathbf{Q}_1 &= \left(-\frac{13}{48}, -\frac{3}{16}, -\frac{3}{16}, \frac{17}{48}\right)^T \\[4pt]
\mathbf{Q}_2 &= \left(-\frac{25}{48}, -\frac{3}{16}, \frac{15}{16}, \frac{1}{3}\right)^T \\[4pt]
\mathbf{Q}_3 &= \left(-\frac{3}{16}, \frac{7}{48}, -\frac{23}{12}, -\frac{21}{16}\right)^T \qquad (6.5.13) \\[4pt]
\mathbf{Q}_4 &= \left(-\frac{3}{16}, -\frac{5}{48}, -\frac{1}{48}, -\frac{3}{16}\right)^T.
\end{aligned}$$

Again the assembly of these elements will be covered in the next section.

6.6 A General Variational Principle

A more general second order elliptic partial differential equation is

$$\frac{\partial}{\partial x}\left(a\frac{\partial \phi}{\partial x}\right) + \frac{\partial}{\partial x}\left(b\frac{\partial \phi}{\partial y}\right) + \frac{\partial}{\partial y}\left(b\frac{\partial \phi}{\partial x}\right) + \frac{\partial}{\partial y}\left(c\frac{\partial \phi}{\partial y}\right) + e\phi = f(x,y) \quad (6.6.1)$$

where a, b, c and e may be functions of x and y, and $b^2 - ac < 0$ throughout the region R of interest. The boundary conditions may be of the type considered earlier or may be of the more general type

$$\frac{\partial \phi}{\partial n} + m(x,y)\phi = h(x,y) \qquad (6.6.2)$$

on ∂R_n. The corresponding variational principle is that

$$
\begin{aligned}
J(\phi) \;=\; & \frac{1}{2}\int_R \left[a\left(\frac{\partial \phi}{\partial x}\right)^2 + 2b\frac{\partial \phi}{\partial x}\frac{\partial \phi}{\partial y} + c\left(\frac{\partial \phi}{\partial y}\right)^2 - e\phi^2 \right] dx\,dy \\
& + \frac{1}{2}\int_{\partial R_n} m(x,y)\phi^2\,ds + \int_R f(x,y)\phi\,dx\,dy \\
& - \int_{\partial R_n} h(x,y)\phi\,ds
\end{aligned}
\qquad (6.6.3)
$$

should be a minimum as ϕ varies over $H^1(\mathcal{R})$ with $\phi|_{\partial R_\phi} = g$. The procedure for setting up the element stiffness matrices and force vectors follows the same lines as given previously. The first two integrals contribute to the stiffness matrix, and the last two to the force vector. The integrations involved in evaluating the elements of the stiffness matrix and the force vector are more complicated than previously if the coefficients of the partial differential equation are variable, but these integrations are performed numerically in such cases. If individual elements are small, it is often sufficiently accurate to give the coefficients a constant value over each element, corresponding to its actual value at the centroid of the element for example. In the latter case the integrations can be performed analytically.

The procedure given earlier can also be generalised to more than two dimensions. For instance, in three dimensions, the simplest type of element is the tetrahedron whose vertices are four nodes at which the ϕ values are specified. The approximating function can then be written as

$$\phi(x,y,z) = \alpha_1 + \alpha_2 x + \alpha_3 y + \alpha_4 z \qquad (6.6.4)$$

or in terms of basis functions and the nodal ϕ values as

$$\phi(x,y,z) = \sum_{k=1}^{4} \phi_k \beta_k(x,y,z) \qquad (6.6.5)$$

A more complicated three-dimensional element is the hexahedron with eight vertices, for which $\phi(x, y, z)$ has the form

$$\phi(x, y, z) = \alpha_1 + \alpha_2 x + \alpha_3 y + \alpha_4 z + \alpha_5 xy + \alpha_6 yz + \alpha_7 zx + \alpha_8 xyz \quad (6.6.6)$$

which can be reformulated as

$$\phi(x, y, z) = \sum_{k=1}^{8} \phi_k \beta_k(x, y, z). \quad (6.6.7)$$

In setting up the element matrices, the foregoing procedures are generalised. Hence integrals over areas become integrals over volumes, and line integrals become surface integrals. The details are inevitably more complex than for the two-dimensional case, and very large systems of equations demand extremely efficient programs for their solution.

6.7 Assembly and Solution

After the stiffness matrices K_n, $n = 1, \ldots, N$, and the generalised force vectors $\mathbf{Q}_n$, $n = 1, \ldots, N$ have been calculated from the individual elements, the global stiffness matrix $\mathbf{K}$ and the global force vector $\mathbf{Q}$ must be computed.

The element stiffness matrix K_n defines the contribution of the nth element to the global matrix K and in a sense, K is nothing more than the sum of $K_1, K_2, \ldots, K_n$, but these matrices cannot be simply added together since the components of K_n are given as referred to the local node numbers of the nth element and before performing the addition, it is essential to give all components with reference to the global node numbers.

If, for example, the nth element has nodes with local numbers i and j and the corresponding global numbers are I and J then the element K_{ij} of K_n when included in K will be added into the element K_{IJ}.

In a similar way, element Q_i of $\mathbf{Q}_n$ will be assembled as part of element Q_I of $\mathbf{Q}$. The identification of corresponding pairs of local and global node numbers can be found from a diagram in simple cases or can be found using a computerised look-up table.

In practice, it is wasteful of computer time and storage to actually convert each stiffness matrix K_n into an $M \times M$ matrix and then add together all N of these large matrices to give K. The procedure usually adopted is to assemble each element stiffness matrix into K as it is computed, effectively generating a sequence

$$K^{(0)} = 0, \qquad K^{(n)} = K^{(n-1)} + K_n \quad (6.7.1)$$

finally producing $K^{(n)} = K$. The force vector $\mathbf{Q}$ is generated in a similar manner.

Once K and $\mathbf{Q}$ have been assembled, the remaining problem is to find the minimum of the functional

$$J = \frac{1}{2}\phi^T K \phi - \phi^T \mathbf{Q} \qquad (6.7.2)$$

where ϕ is an M-dimensional vector whose elements are the generalised coordinates. In some instances (6.7.2) may be minimised directly using a suitable optimisation technique, but it is more convenient to equate the partial derivatives of J with respect to each of the generalised coordinates to zero which is the analytic condition for a minimum. The result is the set of linear equations

$$K\phi = \mathbf{Q} \qquad (6.7.3)$$

which is solved using the sparse matrix methods of Chapter 1.

Equations (6.7.3) require the substitution of those unknowns predetermined by the boundary conditions before the numerical solution is attempted. The reduced number of equations is then written as

$$K_0\phi = \mathbf{Q}_0 \qquad (6.7.4)$$

where K_0 is usually positive definite and symmetric.

6.8 Solution of the Worked Example

Attention is now turned to assembling the final matrix problem from the element matrices for both the triangular and the rectangular cases. Beginning with the triangular case, the stiffness matrices fall into just two categories and are given by (6.5.3) and (6.5.4), with force vectors listed in (6.5.9). The numbering notation is given in Figure 6.4. The element matrices are added into their correct positions according to the rule in (6.7.1) to give the following sequence of matrices $K^{(i)}$, the numbering in Figure 6.4 being used to correctly position the local elements in their correct global position.

$$12K^{(1)} = \begin{bmatrix} 9 & -9 & 0 & 0 & 0 & 0 & 0 & 0 & 0 \\ -9 & 13 & 0 & 0 & -4 & 0 & 0 & 0 & 0 \\ 0 & 0 & 0 & 0 & 0 & 0 & 0 & 0 & 0 \\ 0 & 0 & 0 & 0 & 0 & 0 & 0 & 0 & 0 \\ 0 & -4 & 0 & 0 & 4 & 0 & 0 & 0 & 0 \\ 0 & 0 & 0 & 0 & 0 & 0 & 0 & 0 & 0 \\ 0 & 0 & 0 & 0 & 0 & 0 & 0 & 0 & 0 \\ 0 & 0 & 0 & 0 & 0 & 0 & 0 & 0 & 0 \\ 0 & 0 & 0 & 0 & 0 & 0 & 0 & 0 & 0 \end{bmatrix}$$

$$12K^{(2)} = \begin{bmatrix} 9 & -9 & 0 & 0 & 0 & 0 & 0 & 0 & 0 \\ -9 & 26 & -9 & 0 & -8 & 0 & 0 & 0 & 0 \\ 0 & -9 & 9 & 0 & 0 & 0 & 0 & 0 & 0 \\ 0 & 0 & 0 & 0 & 0 & 0 & 0 & 0 & 0 \\ 0 & -8 & 0 & 0 & 8 & 0 & 0 & 0 & 0 \\ 0 & 0 & 0 & 0 & 0 & 0 & 0 & 0 & 0 \\ 0 & 0 & 0 & 0 & 0 & 0 & 0 & 0 & 0 \\ 0 & 0 & 0 & 0 & 0 & 0 & 0 & 0 & 0 \\ 0 & 0 & 0 & 0 & 0 & 0 & 0 & 0 & 0 \end{bmatrix}$$

$$12K^{(3)} = \begin{bmatrix} 9 & -9 & 0 & 0 & 0 & 0 & 0 & 0 & 0 \\ -9 & 26 & -9 & 0 & -8 & 0 & 0 & 0 & 0 \\ 0 & -9 & 13 & 0 & 0 & -4 & 0 & 0 & 0 \\ 0 & 0 & 0 & 0 & 0 & 0 & 0 & 0 & 0 \\ 0 & -8 & 0 & 0 & 17 & -9 & 0 & 0 & 0 \\ 0 & 0 & -4 & 0 & -9 & 13 & 0 & 0 & 0 \\ 0 & 0 & 0 & 0 & 0 & 0 & 0 & 0 & 0 \\ 0 & 0 & 0 & 0 & 0 & 0 & 0 & 0 & 0 \\ 0 & 0 & 0 & 0 & 0 & 0 & 0 & 0 & 0 \end{bmatrix}$$

$$12K^{(4)} = \begin{bmatrix} 9 & -9 & 0 & 0 & 0 & 0 & 0 & 0 & 0 \\ -9 & 26 & -9 & 0 & -8 & 0 & 0 & 0 & 0 \\ 0 & -9 & 13 & 0 & 0 & -4 & 0 & 0 & 0 \\ 0 & 0 & 0 & 0 & 0 & 0 & 0 & 0 & 0 \\ 0 & -8 & 0 & 0 & 26 & -18 & 0 & 0 & 0 \\ 0 & 0 & -4 & 0 & -18 & 26 & 0 & 0 & -4 \\ 0 & 0 & 0 & 0 & 0 & 0 & 0 & 0 & 0 \\ 0 & 0 & 0 & 0 & 0 & 0 & 0 & 0 & 0 \\ 0 & 0 & 0 & 0 & 0 & -4 & 0 & 0 & 4 \end{bmatrix}$$

$$12K^{(5)} = \begin{bmatrix} 9 & -9 & 0 & 0 & 0 & 0 & 0 & 0 & 0 \\ -9 & 26 & -9 & 0 & -8 & 0 & 0 & 0 & 0 \\ 0 & -9 & 13 & 0 & 0 & -4 & 0 & 0 & 0 \\ 0 & 0 & 0 & 0 & 0 & 0 & 0 & 0 & 0 \\ 0 & -8 & 0 & 0 & 30 & -18 & 0 & -4 & 0 \\ 0 & 0 & -4 & 0 & -18 & 26 & 0 & 0 & -4 \\ 0 & 0 & 0 & 0 & 0 & 0 & 0 & 0 & 0 \\ 0 & 0 & 0 & 0 & -4 & 0 & 0 & 13 & -9 \\ 0 & 0 & 0 & 0 & 0 & -4 & 0 & -9 & 13 \end{bmatrix}$$

$$12K^{(6)} = \begin{bmatrix} 9 & -9 & 0 & 0 & 0 & 0 & 0 & 0 & 0 \\ -9 & 26 & -9 & 0 & -8 & 0 & 0 & 0 & 0 \\ 0 & -9 & 13 & 0 & 0 & -4 & 0 & 0 & 0 \\ 0 & 0 & 0 & 0 & 0 & 0 & 0 & 0 & 0 \\ 0 & -8 & 0 & 0 & 34 & -18 & 0 & -8 & 0 \\ 0 & 0 & -4 & 0 & -18 & 26 & 0 & 0 & -4 \\ 0 & 0 & 0 & 0 & 0 & 0 & 9 & -9 & 0 \\ 0 & 0 & 0 & 0 & -8 & 0 & -9 & 26 & -9 \\ 0 & 0 & 0 & 0 & 0 & -4 & 0 & -9 & 13 \end{bmatrix}$$

$$12K^{(7)} = \begin{bmatrix} 9 & -9 & 0 & 0 & 0 & 0 & 0 & 0 & 0 \\ -9 & 26 & -9 & 0 & -8 & 0 & 0 & 0 & 0 \\ 0 & -9 & 13 & 0 & 0 & -4 & 0 & 0 & 0 \\ 0 & 0 & 0 & 13 & -9 & 0 & -4 & 0 & 0 \\ 0 & -8 & 0 & -9 & 43 & -18 & 0 & -8 & 0 \\ 0 & 0 & -4 & 0 & -18 & 26 & 0 & 0 & -4 \\ 0 & 0 & 0 & -4 & 0 & 0 & 13 & -9 & 0 \\ 0 & 0 & 0 & 0 & -8 & 0 & -9 & 26 & -9 \\ 0 & 0 & 0 & 0 & 0 & -4 & 0 & -9 & 13 \end{bmatrix}$$

$$12K^{(8)} = \begin{bmatrix} 13 & -9 & 0 & -4 & 0 & 0 & 0 & 0 & 0 \\ -9 & 26 & -9 & 0 & -8 & 0 & 0 & 0 & 0 \\ 0 & -9 & 13 & 0 & 0 & -4 & 0 & 0 & 0 \\ -4 & 0 & 0 & 26 & -18 & 0 & -4 & 0 & 0 \\ 0 & -8 & 0 & -18 & 52 & -18 & 0 & -8 & 0 \\ 0 & 0 & -4 & 0 & -18 & 26 & 0 & 0 & -4 \\ 0 & 0 & 0 & -4 & 0 & 0 & 13 & -9 & 0 \\ 0 & 0 & 0 & 0 & -8 & 0 & -9 & 26 & -9 \\ 0 & 0 & 0 & 0 & 0 & -4 & 0 & -9 & 13 \end{bmatrix}$$

where $K^{(8)} = K$. Similarly the associated $\mathbf{Q}^{(i)}$ have the values:

$$12\mathbf{Q}^{(1)} = \begin{bmatrix} -5/2 \\ -7/2 \\ 0 \\ 0 \\ -3/2 \\ 0 \\ 0 \\ 0 \\ 0 \end{bmatrix}, \quad 12\mathbf{Q}^{(2)} = \begin{bmatrix} -5/2 \\ -9 \\ -13/2 \\ 0 \\ -3 \\ 0 \\ 0 \\ 0 \\ 0 \end{bmatrix}, \quad 12\mathbf{Q}^{(3)} = \begin{bmatrix} -5/2 \\ -9 \\ -13/4 \\ 0 \\ -9/2 \\ 12 \\ 0 \\ 0 \\ 0 \end{bmatrix}$$

$$12\mathbf{Q}^{(4)} = \begin{bmatrix} -5/2 \\ -9 \\ 13/4 \\ 0 \\ -6 \\ 57/2 \\ 0 \\ 0 \\ 75/4 \end{bmatrix}, \quad 12\mathbf{Q}^{(5)} = \begin{bmatrix} -5/2 \\ -9 \\ 13/4 \\ 0 \\ -15/2 \\ 57/2 \\ 0 \\ 5/2 \\ 89/6 \end{bmatrix}, \quad 12\mathbf{Q}^{(6)} = \begin{bmatrix} -5/2 \\ -9 \\ 13/4 \\ 0 \\ -9 \\ 57/2 \\ -1/2 \\ 3 \\ 89/4 \end{bmatrix}$$

$$12\mathbf{Q}^{(7)} = \begin{bmatrix} -5/2 \\ -9 \\ 13/4 \\ -3/2 \\ -21/2 \\ 57/2 \\ -2 \\ 3 \\ 89/4 \end{bmatrix}, \quad 12\mathbf{Q}^{(8)} = \begin{bmatrix} -4 \\ -9 \\ 13/4 \\ -3 \\ -12 \\ 57/2 \\ -2 \\ 3 \\ 89/4 \end{bmatrix}$$

and $\mathbf{Q}^{(8)} = \mathbf{Q}$. The predetermined boundary values are $\phi_1 = \phi_4 = \phi_7 = 3$ and hence the reduced system equivalent to (6.7.4) is

$$\begin{bmatrix} 26 & -9 & -8 & 0 & 0 & 0 \\ -9 & 13 & 0 & -4 & 0 & 0 \\ -8 & 0 & 52 & -18 & -8 & 0 \\ 0 & -4 & -18 & 26 & 0 & -4 \\ 0 & 0 & -8 & 0 & 26 & -9 \\ 0 & 0 & 0 & -4 & -9 & 13 \end{bmatrix} \begin{bmatrix} \phi_2 \\ \phi_3 \\ \phi_5 \\ \phi_6 \\ \phi_8 \\ \phi_9 \end{bmatrix} = \begin{bmatrix} 18 \\ 3.25 \\ 42 \\ 28.5 \\ 30 \\ 22.25 \end{bmatrix}$$

which has the form $A\phi^* = \mathbf{b}$ and is solved by any of the standard methods of Chapter 1 to give

$$\phi^{*T} = [3.4015, 4.3047, 3.9623, 5.5242, 4.6739, 6.6472].$$

The equivalent assembly using the rectangular mesh is now detailed. In this case the element matrices are given by (6.5.10), the force vectors are listed in (6.5.13) and the numbering notation is given in Figure 6.5. For this case the element matrices have been listed with their elements in their global positions.

$$36K_1 = \begin{bmatrix} 26 & -14 & 0 & 1 & -13 & 0 & 0 & 0 & 0 \\ -14 & 26 & 0 & -13 & 1 & 0 & 0 & 0 & 0 \\ 0 & 0 & 0 & 0 & 0 & 0 & 0 & 0 & 0 \\ 1 & -13 & 0 & 26 & -14 & 0 & 0 & 0 & 0 \\ -13 & 1 & 0 & -14 & 26 & 0 & 0 & 0 & 0 \\ 0 & 0 & 0 & 0 & 0 & 0 & 0 & 0 & 0 \\ 0 & 0 & 0 & 0 & 0 & 0 & 0 & 0 & 0 \\ 0 & 0 & 0 & 0 & 0 & 0 & 0 & 0 & 0 \\ 0 & 0 & 0 & 0 & 0 & 0 & 0 & 0 & 0 \end{bmatrix}$$

$$36K_2 = \begin{bmatrix} 0 & 0 & 0 & 0 & 0 & 0 & 0 & 0 & 0 \\ 0 & 26 & -14 & 0 & 1 & -13 & 0 & 0 & 0 \\ 0 & -14 & 26 & 0 & -13 & 1 & 0 & 0 & 0 \\ 0 & 0 & 0 & 0 & 0 & 0 & 0 & 0 & 0 \\ 0 & 1 & -13 & 0 & 26 & -14 & 0 & 0 & 0 \\ 0 & -13 & 1 & 0 & -14 & 26 & 0 & 0 & 0 \\ 0 & 0 & 0 & 0 & 0 & 0 & 0 & 0 & 0 \\ 0 & 0 & 0 & 0 & 0 & 0 & 0 & 0 & 0 \\ 0 & 0 & 0 & 0 & 0 & 0 & 0 & 0 & 0 \end{bmatrix}$$

$$36K_3 = \begin{bmatrix} 0 & 0 & 0 & 0 & 0 & 0 & 0 & 0 & 0 \\ 0 & 0 & 0 & 0 & 0 & 0 & 0 & 0 & 0 \\ 0 & 0 & 0 & 0 & 0 & 0 & 0 & 0 & 0 \\ 0 & 0 & 0 & 0 & 0 & 0 & 0 & 0 & 0 \\ 0 & 0 & 0 & 0 & 26 & -14 & 0 & 1 & -13 \\ 0 & 0 & 0 & 0 & -14 & 26 & 0 & -13 & 1 \\ 0 & 0 & 0 & 0 & 0 & 0 & 0 & 0 & 0 \\ 0 & 0 & 0 & 0 & 1 & -13 & 0 & 26 & -16 \\ 0 & 0 & 0 & 0 & -13 & 1 & 0 & -14 & 26 \end{bmatrix}$$

$$36K_4 = \begin{bmatrix} 0 & 0 & 0 & 0 & 0 & 0 & 0 & 0 & 0 \\ 0 & 0 & 0 & 0 & 0 & 0 & 0 & 0 & 0 \\ 0 & 0 & 0 & 0 & 0 & 0 & 0 & 0 & 0 \\ 0 & 0 & 0 & 26 & -14 & 0 & 1 & -13 & 0 \\ 0 & 0 & 0 & -14 & 26 & 0 & -13 & 1 & 0 \\ 0 & 0 & 0 & 0 & 0 & 0 & 0 & 0 & 0 \\ 0 & 0 & 0 & 1 & -13 & 0 & 26 & -14 & 0 \\ 0 & 0 & 0 & -13 & 1 & 0 & -14 & 26 & 0 \\ 0 & 0 & 0 & 0 & 0 & -4 & 0 & 0 & 4 \end{bmatrix}$$

which by adding leaves the full assembled matrix as

$$36K = \begin{bmatrix} 26 & -14 & 0 & 1 & -13 & 0 & 0 & 0 & 0 \\ -14 & 52 & -14 & -13 & 2 & -13 & 0 & 0 & 0 \\ 0 & -14 & 26 & 0 & -13 & 1 & 0 & 0 & 0 \\ 1 & -13 & 0 & 52 & -28 & 0 & 1 & -13 & 0 \\ -13 & 2 & -13 & -28 & 104 & -28 & -13 & 2 & -13 \\ 0 & -13 & 1 & 0 & -28 & 52 & 0 & -13 & 1 \\ 0 & 0 & 0 & 1 & -13 & 0 & 26 & -14 & 0 \\ 0 & 0 & 0 & -13 & 2 & -13 & -14 & 52 & -14 \\ 0 & 0 & 0 & 0 & -13 & 1 & 0 & -14 & 26 \end{bmatrix}$$

and the corresponding **Q** vectors are:

$$
36\mathbf{Q}_1 = \begin{bmatrix} -39/4 \\ -51/4 \\ 0 \\ -27/4 \\ -27/4 \\ 0 \\ 0 \\ 0 \\ 0 \end{bmatrix}, \qquad
36\mathbf{Q}_2 = \begin{bmatrix} 0 \\ -75/4 \\ 12 \\ 0 \\ -27/4 \\ -135/4 \\ 0 \\ 0 \\ 0 \end{bmatrix}, \qquad
36\mathbf{Q}_3 = \begin{bmatrix} 0 \\ 0 \\ 0 \\ 0 \\ -27/4 \\ -189/4 \\ 0 \\ 21/4 \\ 69 \end{bmatrix}
$$

$$
36\mathbf{Q}_4 = \begin{bmatrix} 0 \\ 0 \\ 0 \\ -27/4 \\ -27/4 \\ 0 \\ -15/4 \\ -3/4 \\ 0 \end{bmatrix}
$$

which sum to give

$$
36\mathbf{Q} = \begin{bmatrix} -39/4 \\ -63/2 \\ 12 \\ -27/2 \\ -27 \\ 81/4 \\ -150/4 \\ 9/2 \\ 69 \end{bmatrix}.
$$

In this case the predetermined boundary values are $\phi_1 = 3$, $\phi_4 = 3$ and $\phi_7 = 3$ which yields the reduced system

$$
\begin{bmatrix}
52 & -14 & 2 & -13 & 0 & 0 \\
-14 & 26 & -13 & 1 & 0 & 0 \\
2 & 26 & -13 & 1 & 0 & 0 \\
-13 & 1 & -28 & 52 & -13 & 1 \\
0 & 0 & 2 & -13 & 52 & -14 \\
0 & 0 & -13 & 1 & -14 & 26
\end{bmatrix}
\begin{bmatrix} \phi_2 \\ \phi_3 \\ \phi_5 \\ \phi_6 \\ \phi_8 \\ \phi_9 \end{bmatrix}
=
\begin{bmatrix} 49.5 \\ 12 \\ 135 \\ 81 \\ 85.5 \\ 69 \end{bmatrix}
$$

with solution by the methods of Chapter 1 giving $\phi_2 = 3.250$, $\phi_3 = 4.00$, $\phi_5 = 4.00$, $\phi_6 = 5.55001$, $\phi_8 = 4.7501$ and $\phi_9 = 7.0001$.

Hence the complete solution has been obtained using both the triangular element and the rectangular element approaches. It is clear from this example that the method is primarily a computational method and writing out the

stages by hand is quite long-winded. In order to complete the exercises for this
chapter it is recommended that the results of the above worked example are
obtained by coding the various stages, and then this code can be modified to
complete the exercises. Further details of the required codes are given in the
Appendix. In the next section a discussion of more complicated interpolation
functions will be made.

6.9 Further Interpolation Functions

In each finite element, the field variable, say $u(x)$ in the one-dimensional case,
is assumed to take a simple (usually polynomial) form. If continuity of $u(x)$ and
possibly its derivative, is required, then the approximating polynomials must be
matched at the boundaries between elements. The triangular and rectangular
elements used in the previous sections given continuity of $u(x)$.

Hence the function $u(x)$ might be approximated by a polynomial such as
$u(x) \simeq a_0 + a_1 x + a_2 x^2 + a_3 x^3$. By choosing four distinct points x_1, x_2, x_3 and
x_4, the coefficients a_i may be determined in terms of the values u_i by solving
the linear equations

$$\begin{aligned}
a_0 + a_1 x_1 + a_2 x_1^2 + a_3 x_1^3 &= u_1 \\
a_0 + a_1 x_2 + a_2 x_2^2 + a_3 x_2^3 &= u_2 \\
a_0 + a_1 x_3 + a_2 x_3^2 + a_3 x_3^3 &= u_3 \\
a_0 + a_1 x_4 + a_2 x_4^2 + a_3 x_4^3 &= u_4.
\end{aligned} \tag{6.9.1}$$

However the coefficients a_i are not simply related to the values u_i, as continuity
between this element and its neighbour defined on x_4, x_5, x_6 and x_7 will require
a constraint relation of the form

$$a_0 + a_1 x_4 + a_2 x_4^2 + a_3 x_4^3 = a_0' + a_1' x_4 + a_2' x_4^2 + a_3' x_4^3 \quad (= u_4). \tag{6.9.2}$$

Derivative continuity will impose a further condition. Hence minimisation of
the functional is now constrained.

However, if an approximation of the form

$$\bar{u} = \sum_{i=1}^{n} u_i f_i(x) \tag{6.9.3}$$

can be found in which the functions $f_i(x)$ satisfy

$$f_i(x_j) = \begin{cases} 1, & i = j \\ 0, & i \neq j. \end{cases} \tag{6.9.4}$$

then it follows that $\bar{u}(x_i) = u_i$, so that $\bar{u}(x)$ interpolates the values u_1, u_2,
u_3 and u_4. The functions $f_i(x)$ are called *interpolation functions*. Continuity

between the elements (x_1, x_2, x_3, x_4) and (x_4, x_5, x_6, x_7) is now automatic by choosing the value u_4 for both elements. This is a well-known technique in numerical analysis known as *Lagrangian interpolation* (Hildebrand, 1974).

If in addition continuity of the first derivative is required then the approximation has the form

$$\bar{u} = \sum_{i=1}^{n} \{u_i f_i(x) + u_i' g_i(x)\} \tag{6.9.5}$$

in which

$$
\begin{aligned}
f_i(x_j) &= \begin{cases} 1, & i = j, \\ 0, & i \neq j \end{cases} \\
f_i'(x_j) &= 0 \quad \text{for all} \quad i, j \\
g_i(x_j) &= 0 \quad \text{for all} \quad i, j \\
g_i'(x_j) &= \begin{cases} 1, & i = j, \\ 0, & 1 \neq j \end{cases}
\end{aligned}
\tag{6.9.6}
$$

for $i, j = 1, 2, \ldots, n$ and where u_i' is the value of $u'(x)$ at $x = x_i$.

The result is that $\bar{u}$ interpolates both the value and derivative of $u(x)$ at each x_i, $i = 1, 2, \ldots, n$. Hence continuity of both $\bar{u}(x)$ and its derivative are ensured at boundaries between elements by simply choosing a common value of u_i and u_i' at the common boundary point. This type of interpolation is *Hermite interpolation*.

In order to further standardise the treatment of elements of different dimensions, each element will be described in terms of a local system of coordinates in the range $(-1, 1)$. General coordinates are then obtained using coordinate transformations.

Hence an element defined on two nodes at $\xi_1 = -1$ and $\xi_2 = 1$ has linear interpolation functions

$$f_1(\xi) = \frac{1}{2}(1 - \xi) \quad \text{and} \quad f_2(\xi) = \frac{1}{2}(1 + \xi) \tag{6.9.7}$$

which are shown in Figure 6.6.

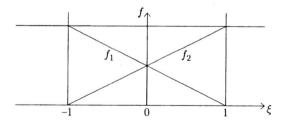

Fig. 6.6.

For an element with three interpolation points the local coordinates are $\xi_1 = -1$, $\xi_2 = 0$ and $\xi_3 = 1$ and the corresponding quadratic interpolation functions are

$$
\begin{aligned}
f_1(\xi) &= \frac{1}{2}\xi(\xi - 1) \\
f_2(\xi) &= (1 - \xi)(1 + \xi) \\
f_3(\xi) &= \frac{1}{2}\xi(\xi + 1)
\end{aligned}
\qquad (6.9.8)
$$

and these functions are shown in Figure 6.7.

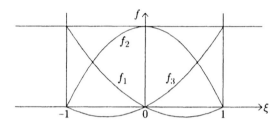

Fig. 6.7.

These functions are special cases of the general Lagrange interpolation formula

$$
f_i(x) = \frac{(x - x_1)(x - x_2)\ldots(x - x_{i-1})(x - x_{i+1})\ldots(x - x_{m+1})}{(x_i - x_1)(x_i - x_2)\ldots(x_i - x_{i-1})(x_i - x_{i+1})\ldots(x_i - x_{m+1})}. \qquad (6.9.9)
$$

For the Hermite interpolation functions, using equations and conditions (6.9.5) and (6.9.6), requires two conditions at each knot, and hence functions of degree 3 for two knots, and degree 5 for three. Hence the cubic on two knots is:

$$
\begin{aligned}
f_1(\xi) &= (2 + \xi)(1 - \xi)^2/4 \\
f_2(\xi) &= (2 - \xi)(1\xi)^2/4 \\
g_1(\xi) &= (1 + \xi)(1 - \xi)^2/4 \\
g_2(\xi) &= (\xi - 1)(1 + \xi)^2/4
\end{aligned}
\qquad (6.9.10)
$$

with the forms shown in Figure 6.8.

If continuity of the second derivative is also required, then the Hermite process can be generalised to the form

$$
\bar{u} = \sum_{i=1}^{n} \{u_i f_i(x) + u_i' g_i(x) + u_i'' h_i(x)\} \qquad (6.9.11)
$$

which is used to define a quintic on two knots x_1 and x_2 which has given values u_i, u_i' and u_i'' at the knots x_i for $i = 1, 2$, with $n = 2$. The conditions to be

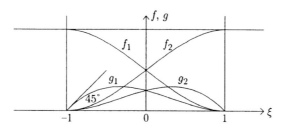

Fig. 6.8.

satisfied by the quintic interpolation polynomials are

$$f_i(x_j) = \delta_{ij}; \quad f_i'(x_j) = 0; \quad f_i''(x_j) = 0$$
$$g_i(x_j) = 0; \quad g_i'(x_j) = \delta_{ij}; \quad g_i''(x_j) = 0 \qquad (6.9.12)$$
$$h_i(x_j) = 0; \quad h_i'(x_j) = 0; \quad h_i''(x_j) = \delta_{ij}$$

where $\delta_{ij} = 1$ when $i = j$ and $\delta_{ij} = 0$ when $i \neq j$, and for a two-knot element the corresponding polynomials are

$$
\begin{aligned}
f_1(\xi) &= \{1 + 3(1+\xi)/2 + 3(1+\xi)^2/2\}(1-\xi)^3/8 \\
f_2(\xi) &= \{1 + 3(1-\xi)/2 + 3(1-\xi)^2/2\}(1+\xi)^3/8 \\
g_1(\xi) &= \{1 + 3(1+\xi)/2\}(1+\xi)(1-\xi)^3/8 \qquad (6.9.13) \\
g_2(\xi) &= \{1 + 3(1+\xi)/2\}(\xi - 1)(1+\xi)^3/8 \\
h_1(\xi) &= (1+\xi)^2(1-\xi)^3/16 \\
h_2(\xi) &= (1-\xi)^2(1+\xi)^3/16.
\end{aligned}
$$

It should be noted that the derivatives in (6.9.11) will be with respect to normalised coordinates ξ and the conditions of (6.9.4) will be satisfied.

The errors in the above approximations can be determined by expanding $u(x)$ as a Taylor series about a point x_0 in the form

$$u(x) = u(x_0) + (x - x_0)u'(x_0) + (x - x_0)^2 u''(x_0)/2! + \cdots \qquad (6.9.14)$$

from which we see that the linear approximation incurs an error of order h^2, and the quadratic approximation an error of h^3, where h is the largest value in absolute value of $x - x_0$.

Hence as an example, consider the linear interpolation with knots at $x = 2$ and $x = 5$, and $u(2) = 4$ and $u(5) = 7$. Then the linear interpolate in normalised variables has $\xi(2) = -1$ and $\xi(5) = 1$ and gives

$$\bar{u} = 4 \cdot \frac{1}{2}(1 - \xi) + 7 \cdot \frac{1}{2}(1 + \xi) = 11/2 + 3\xi/2 \qquad (6.9.15)$$

and in terms of x, let $\xi = ax + b$ to give

$$
\begin{aligned}
2a + b &= -1 \\
5a + b &= 1
\end{aligned}
$$

with solution $a = 2/3$ and $b = -7/3$ so that

$$\bar{u} = 11/2 + (2x - 7)/2 = x + 2. \tag{6.9.16}$$

The advantages of local coordinates lie in two- and three-dimensional problems where such coordinates are easy to work with. More will be said on coordinate transformations a little later.

Having considered linear elements in some detail, the same ideas can be used to look at two dimensions. Now there are two normalised coordinates ξ and η each lying between -1 and 1, in for example the use of bilinears on a rectangle. Consider as a first case that of a square with four knots. In order to provide a function $\bar{u}(\xi, \eta)$ which takes the prescribed values at $P_1(-1, -1)$, $P_2(1, -1)$, $P_3(1, 1)$ and $P_4(-1, 1)$, four interpolating functions which are zero at three corners of the square and take the value 1 at the fourth are needed. Hence for example $f_1(\xi, \eta)$ has to be zero at P_2, P_3 and P_4 and 1 at P_1. But all points on $P_2 P_3$ satisfy $\xi - 1 = 0$ and all points on $P_3 P_4$ satisfy $\eta - 1 = 0$. Hence P_2, P_3 and P_4 lie on the pair of straight lines $(\xi - 1)(\eta - 1) = 0$ and hence

$$f_1(\xi, \eta) = \frac{1}{4}(\xi - 1)(\eta - 1). \tag{6.9.17}$$

The contours for this function are plotted in Figure 6.9.

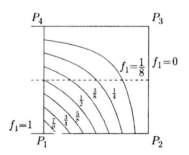

Fig. 6.9.

The complete set of functions for the square is

$$
\begin{aligned}
f_1(\xi, \eta) &= \frac{1}{4}(\xi - 1)(\eta - 1) \\
f_2(\xi, \eta) &= \frac{1}{4}(\xi + 1)(1 - \eta) \\
f_3(\xi, \eta) &= \frac{1}{4}(\xi + 1)(\eta + 1) \\
f_4(\xi, \eta) &= \frac{1}{4}(1 - \xi)(\eta + 1)
\end{aligned}
\tag{6.9.18}
$$

and if the four corner values are u_i for $i = 1, 4$ then

$$\bar{u} = u_1 f_1(\xi, \eta) + u_2 f_2(\xi, \eta) + u_3 f_3(\xi, \eta) + u_4 f_4(\xi, \eta) \tag{6.9.19}$$

which gives a bilinear interpolant at the four knots. On any side of the square the approximation is the linear Lagrangian given in (6.9.3) with $n = 2$, and is therefore continuous across the whole boundary between any two square elements.

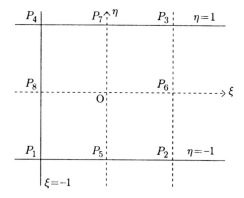

Fig. 6.10.

As an extension of the four-point square, suppose the interpolation is to include the midpoints, requiring quadratic interpolation along each side, then biquadratic functions are required. The arrangement is shown in Figure 6.10. and for example the point P_6 is distinguished by the fact that it is the only point not on either of $\eta = 1, \eta = -1$ or $\xi = -1$. Hence $f_6(\xi, \eta) = \frac{1}{2}(1-\eta^2)(1-\xi)$ and similar arguments yield the eight functions

$$
\begin{aligned}
f_1 &= -\frac{1}{4}(\eta - 1)(\xi - 1)(\xi + \eta + 1) \\
f_2 &= -\frac{1}{4}(\eta - 1)(\xi - 1)(\xi - \eta - 1) \\
f_3 &= -\frac{1}{4}(\eta + 1)(\xi - 1)(\xi + \eta - 1) \\
f_4 &= -\frac{1}{4}(\eta + 1)(\xi - 1)(\xi - \eta + 1) \\
f_5 &= \frac{1}{2}(\xi^2 - 1)(\eta - 1) \\
f_6 &= -\frac{1}{2}(\eta^2 - 1)(\xi + 1) \\
f_7 &= -\frac{1}{2}(\xi^2 - 1)(\eta + 1) \\
f_8 &= \frac{1}{2}(\eta^2 - 1)(\xi - 1)
\end{aligned}
\tag{6.9.20}
$$

and the interpolants will have the form

$$\bar{u} = a_1 + a_2\xi + a_3\eta + a_4\xi^2 + a_5\xi\eta + a_6\eta^2 + a_7\xi^2\eta + a_8\xi\eta^2. \tag{6.9.21}$$

As a second two-dimensional case consider a right-angled triangle. The specialisation here is intended as more general shapes can be obtained from these basic shapes by applying transformations. For the right-angled triangle, we use *triangular co-ordinates* (ξ_1, ξ_2, ξ_3) where $\xi_1 = \xi$, $\xi_2 = \eta$ and $\xi_3 = 1 - \xi_2 - \xi_3$. Hence the corners of the triangle with $(\xi, \eta) = (0,0), (1,0)$, and $(0,1)$ are $(0,0,1)$, $(1,0,0)$ and $(0,1,0)$ as shown in Figure 6.11.

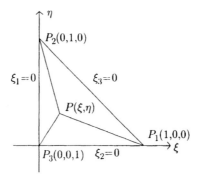

Fig. 6.11.

The point $P(\xi, \eta)$ is joined to the three vertices and the areas of the three triangles so formed are $\frac{1}{2}\xi_1$, $\frac{1}{2}\xi_2$ and $\frac{1}{2}\xi_3$. Hence these coordinates are often called *areal* coordinates.

For linear Lagrangian interpolation, areal coordinates provide a very simple form

$$\bar{u} = u_1\xi_1 + u_2\xi_2 + u_3\xi_3 \tag{6.9.22}$$

as $\xi_1 = 0$ at P_2 and P_3, $\xi_2 = 0$ at P_3 and P_1, and $\xi_3 = 0$ at P_1 and P_2.

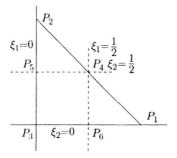

Fig. 6.12.

For quadratic interpolation, the points are shown in Figure 6.12 and now the midpoints of the sides of the triangle are included. Again by looking at the line pairs not including the point in question, the set of interpolation functions gives

$$
\begin{aligned}
f_1 &= \xi_1(2\xi_1 - 1) \\
f_2 &= \xi_2(2\xi_2 - 1) \\
f_3 &= \xi_3(2\xi_3 - 1) \\
f_4 &= 4\xi_1\xi_2 \\
f_5 &= 4\xi_2\xi_3 \\
f_6 &= 4\xi_3\xi_1.
\end{aligned}
\tag{6.9.23}
$$

Taking two intermediate points on each side of the triangle gives a ten-knot interpolation with cubic accuracy. The points are shown in Figure 6.13.

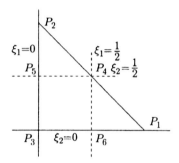

Fig. 6.13.

The set of functions is now:

$$
\begin{aligned}
f_1 &= \frac{1}{2}\xi_1(3\xi_1 - 1)(3\xi_1 - 2) \\
f_2 &= \frac{1}{2}\xi_2(3\xi_2 - 1)(3\xi_2 - 2) \\
f_3 &= \frac{1}{2}\xi_3(3\xi_3 - 1)(3\xi_3 - 2) \\
f_4 &= 9\xi_1\xi_2(3\xi_1 - 1)/2 \\
f_5 &= 9\xi_1\xi_2(3\xi_2 - 1)/2 \\
f_6 &= 9\xi_2\xi_3(3\xi_2 - 1)/2 \\
f_7 &= 9\xi_2\xi_3(3\xi_3 - 1)/2 \\
f_8 &= 9\xi_3\xi_1(3\xi_3 - 1)/2 \\
f_9 &= 9\xi_3\xi_1(3\xi_1 - 1)/2 \\
f_{10} &= \xi_1\xi_2\xi_3,
\end{aligned}
\tag{6.9.24}
$$

and these ten functions give a cubic approximation of the form

$$\begin{aligned}
\bar{u} \quad = \quad & a_1 + a_2\xi + a_3\eta + a_4\xi^2 + a_5\xi\eta \\
& +a_6\eta^2 + a_7\xi^3 + a_8\xi^2\eta + a_9\xi\eta^2 + a_{10}\eta^3.
\end{aligned} \qquad (6.9.25)$$

For a practical problem, more general shapes than the square or right-angled triangle will be used. The rectangle has already been obtained from the square with simple linear transformations on the two axes. More general transformations considered here allow the creation of general quadrilateral elements and curvilinear elements.

Let the transformation be

$$x = x(\xi, \eta), \qquad y = y(\xi, \eta) \qquad (6.9.26)$$

with corresponding inverse

$$\xi = \xi(x, y), \qquad \eta = \eta(x, y) \qquad (6.9.27)$$

as long as

$$\frac{\partial x}{\partial \xi}\frac{\partial y}{\partial \eta} \neq \frac{\partial x}{\partial \eta}\frac{\partial y}{\partial \eta}. \qquad (6.9.28)$$

The chain rule for differentiation can be written as

$$\begin{bmatrix} \frac{\partial u}{\partial \xi} \\ \frac{\partial u}{\partial \eta} \end{bmatrix} = \begin{bmatrix} \frac{\partial x}{\partial \xi} & \frac{\partial y}{\partial \xi} \\ \frac{\partial x}{\partial \eta} & \frac{\partial y}{\partial \eta} \end{bmatrix} \begin{bmatrix} \frac{\partial u}{\partial x} \\ \frac{\partial u}{\partial y} \end{bmatrix} = J \begin{bmatrix} \frac{\partial u}{\partial x} \\ \frac{\partial u}{\partial y} \end{bmatrix} \qquad (6.9.29)$$

where the matrix J is the Jacobian matrix of the transformation. The determinant $|J|$ is the Jacobian and the condition (6.9.28) is simply $|J| \neq 0$. Hence double integrals over a region R of the $(x - y)$ plane can be converted into an integration over the corresponding region P of the $(\xi - \eta)$ plane according to

$$\int\int_R u(x, y)\, dx\, dy = \int\int_P u[x(\xi, \eta), y(\xi, \eta)]\, |J|\, d\xi d\eta. \qquad (6.9.30)$$

Hence the transformation of a general rectangle at an angle to the axis as in Figure 6.14 requires

$$\begin{aligned}
x \quad = \quad & \frac{1}{4}x_1(\xi - 1)(\eta - 1) - x_2(\xi + 1)(\eta - 1) \\
& +x_3(\xi + 1)(\eta + 1) - x_4(\xi - 1)(\eta + 1)
\end{aligned} \qquad (6.9.31)$$

with a similar expression for y.

Differentiation gives

$$J = \begin{bmatrix} x_2 - x_1 & y_2 - y_1 \\ x_3 - x_2 & y_3 - y_2 \end{bmatrix} \qquad (6.9.32)$$

which reduces to just $J = ab/4$ where a and b are the lengths of the sides which can be used in (6.9.30).

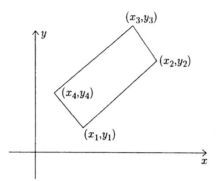

Fig. 6.14.

Transformation (6.9.31) also holds for a general quadrilateral, but now the Jacobian is no longer constant.

If $x(\xi,\eta)$ and $y(\xi,\eta)$ are defined by interpolation on the standard square using quadratic interpolating functions (6.9.20) then the side $\eta = 1$ becomes

$$x(\xi,-1) = \frac{1}{2}x_1\xi(\xi-1) - x_5(\xi^2-1) + \frac{1}{2}x_2\xi(\xi+1)$$

$$y(\xi,-1) = \frac{1}{2}y_1\xi(\xi-1) - y_5(\xi^2-1) + \frac{1}{2}y_2\xi(\xi+1) \qquad (6.9.33)$$

which are the parametric equations for the unique parabola passing through the points (x_1,y_1), (x_5,y_5) and (x_2,y_2). Hence the standard square is transformed into an element with sides which are parabolic arcs as shown in Figure 6.15.

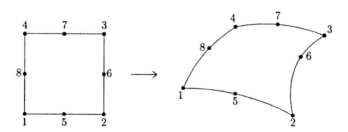

Fig. 6.15.

Cubic curves can be similarly generated with the twelve standard square as a starting point. For the right-angled triangle, areal coordinates may be applied to the general triangle with vertices (x_1,y_1), (x_2,y_2) and (x_3,y_3) using the transformation

$$x = x_1\xi_1 + x_2\xi_2 + x_3\xi_3$$

$$y = y_1\xi_1 + y_2\xi_2 + y_3\xi_3. \qquad (6.9.34)$$

An important consequence of the application of non-linear transformations to obtain curvilinear boundaries to the elements is that the degree of the expression in ξ and η is higher than that of the corresponding expression in x and y. Hence if (6.9.18) is used for interpolation, a linear function of x and y will be transformed into a quadratic function of ξ and η.

The problem now is that the linear functions of x and y have only three degrees of freedom, so that the corresponding quadratic functions of ξ and η cannot have more. However a *complete* quadratic function of ξ and η has six degrees of freedom, and hence the curvilinear element can only describe a restricted quadratic function of ξ and η. The interpolation is said to be not complete in ξ and η.

Hence the price paid for the use of curvilinear coordinate elements is that a higher degree polynomial must be used in ξ and η to ensure a given degree of completeness in x and y.

6.10 Quadrature Methods and Storage Considerations

The integrations required in the finite element process can only be made analytically in relatively simple cases such as that in the worked example earlier in the chapter. Numerical quadrature is a major topic in its own right and here only a simple introduction will be made. For further details see specialist books such as Evans (1993).

Integrals of the form

$$\int_{-1}^{1} f(\xi)\, d\xi \qquad (6.10.1)$$

are required in normalised coordinates. The simplest rule is obtained by replacing the curve $y = f(\xi)$ by a straight line passing through the two endpoints. The resulting integral is then just the area of a trapezium. Hence

$$\int_{-1}^{1} f(\xi)\, d\xi = f(-1) + f(1) - 2f''(\theta)/3 \qquad (6.10.2)$$

where the error term depends on θ with $-1 \le \theta \le 1$. The trapezoidal rule is

$$\int_{-1}^{1} f(\xi)\, d\xi \approx f(-1) + f(1).$$

If the curve $y = f(\xi)$ is approximated by a quadratic then the result is Simpson's rule given by

$$\int_{-1}^{1} f(\xi)\, d\xi \approx [f(-1) + 4f(0) + f(1)]/3 \qquad (6.10.3)$$

with error term $-u^{(4)}(\theta)/90$. Hence in x coordinates the trapezoidal error is order h^2 and the Simpson rule error is h^4. These rules may be used over any interval using linear transformations and may be used of a succession of consecutive subintervals to give increased accuracy. These rules are the first two in a family of rules called the Newton–Cotes rules. They have the principle property that the abscissae (the points at which the integrand is evaluated) are predetermined as being equally spaced across the interval of integration.

A large increase in accuracy for the same number of function evaluations arises if the abscissae are made free and these extra degrees of freedom are used to reduce the error still further. The result is the set of quadratures known as Gaussian formulae. There are many of these, but the relevant formulae in finite element work is the Gauss–Legendre rule

$$\int_{-1}^{1} f(\xi)\,d\xi = \sum_{i=1}^{m} w_i f(\xi_i) + E \qquad (6.10.4)$$

where the weights w_i and the abscissae $\xi_i \in (-1,1)$ are tabulated in standard tables such as Abramowitch and Stegun (1964) for orders from 2 to 256. The abscissae are the roots of the Legendre polynomials which are the relevant orthogonal polynomials for the interval $[-1,1]$ and the integral in (6.10.4). Hence the integration weights and abscissae require the solution of non-linear equations and are therefore normally stored.

The disadvantage of using Gaussian quadrature is that the evaluation points will be irregular and the working will require non-standard interpolation functions. Moreover the error in the finite element approximation implies that any quadratures need only be found to equivalent accuracy, hence high powered quadrature routines are not as useful as they might at first appear, though are often preferred in engineering practice.

In this concluding chapter, something should be said about the computational side of the process. The resulting matrices are banded and symmetric, though not in all cases, and hence only the upper triangle need be stored. Further only the non-zero elements need to be stored as long as a second array of integers is used to indicate the position along the row of each non-zero element. Hence if the non-zero elements for a given row are in columns 8, 10, 11, 12 and 17 say then the compressed matrix will have the full number of rows, say N, and the number of columns will be equal to the maximum number of non-zero elements in any row. The translation or integer array, say I, will then have elements:

$$I[1] = 8; \qquad I[2] = 10; \qquad I[3] = 11; \qquad I[4] = 12; \qquad I[5] = 17.$$

The elimination process without pivoting will preserve the band structure and only a limited amount of elimination is required to complete the process. The algorithm is easily modified to accept the compressed storage, as effectively every call to $a_{i,j}$ is replaced by a call to $a_{i,I[j]}$ and loop counters along rows will be correspondingly reduced. The actual band width is determined by how the nodes are numbered, and a significant reduction may be made by suitable renumbering.

This chapter will be concluded with a brief look at a related area, namely the boundary element method. This work is referred to in Evans, Blackledge and Yardley (1999, Chapter 5) on Green's functions.

6.11 Boundary Element Method

Many elliptic, steady state partial differential equations can be rewritten in an integral form, equating quantities defined only on the boundary or surface of a region. Examples include potential problems in electrostatics, fluid flow and in heat transfer. In order to obtain a solution to a boundary integral equation, it is usual to divide the boundary into elements and, in the same way as the finite element method, obtain the resulting integrals approximately on each element. The gain is that the number of elements and hence the number of equations is much reduced, but the loss is that the preliminary use of the fundamental solution is required, so limiting the process to certain problems only.

The boundary integral equation can be derived by considering the most commonly employed example of this technique, the solution of Laplace's equation.

$$\nabla^2 u \equiv \frac{\partial^2 u}{\partial x^2} + \frac{\partial^2 u}{\partial y^2} = 0 \qquad (6.11.1)$$

in the region shown in Figure 6.16.

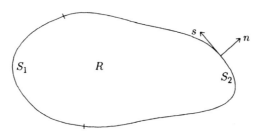

Fig. 6.16.

The boundary conditions are $u = \bar{u}$ on S_1 and $\partial u/\partial n = \bar{q}$ on S_2 where $S = S_1 + S_2$. A new function $v(x, y)$ is introduced which will be defined later, in terms of which Green's theorem gives

$$\int\int_R (u\nabla^2 v - v\nabla^2 u)\,dx\,dy = \int_S \left(u\frac{\partial v}{\partial n} - v\frac{\partial u}{\partial n}\right)\,dS. \qquad (6.11.2)$$

However $\nabla^2 u = 0$ and hence if $v(x, y)$ is also some special solution of Laplace's equation then the area integral can be completely eliminated. Hence choose v such that

$$\nabla^2 v = -\delta(x - x_i, y - y_i) = -\delta^i \qquad (6.11.3)$$

where δ is the usual Dirac delta function. Hence $\delta^i = \infty$ at the point (x_i, y_i) and $\delta^i = 0$ elsewhere. It is also essential that $(x_i, y_i) \in R$ and $(x_i, y_i) \notin S$. With this choice of v the left-hand side of (6.11.2) degenerates to

$$\int\int_R -u\delta^i \, dx \, dy = -u_i \qquad (6.11.4)$$

and (6.11.2) becomes

$$-u_i = \int_S \left(u \frac{\partial v}{\partial n} - v \frac{\partial u}{\partial n} \right) dS \qquad (6.11.5)$$

where u_i is the value of u at any point inside or importantly on the boundary of the region R. Hence if all the quantities on the right-hand side of (6.11.5) are known, then this equation may be used to find u at any interior point of R. Indeed this is the final step of the solution. Therefore a relation between the known and unknown boundary quantities is required and can be obtained by taking the point i to be a boundary point to give

$$c_i u_i = \int_S \left(v \frac{\partial u}{\partial n} - u \frac{\partial v}{\partial n} \right) dS \qquad (6.11.6)$$

where c_i is a constant depending on the boundary shape at the point i. If the boundary is smooth, $c_i = 1/2$. Equation (6.11.6) is often written succinctly as

$$c_i u_i + \int_S u q^* \, dS = \int_S u^* q \, dS \qquad (6.11.7)$$

where $q = \partial u/\partial n$, $u^* = v$ and $q^* = \partial v/\partial n$.

The fundamental solution $v(x, y)$ is the solution of (6.11.3) and for Laplace's equation in two dimensions has the form

$$v = \frac{1}{2\pi} \ln(1/r) \qquad (6.11.8)$$

where $r = \sqrt{(x - x_i)^2 + (y - y_i)^2}$ and (x, y) is any point in the region. Physically the fundamental solution gives the potential at any point (x, y) due to a unit potential at the point (x_i, y_i) in an infinite isotropic region. The determination of this fundamental solution in general is only possible in closed form for a limited class of equations.

The exterior problem in which the region of interest is outside the boundary presents no new difficulties. A circular boundary at infinity is introduced and the appropriate integrals are evaluated on this boundary. If these integrals tend to zero then the standard boundary element method can be used without modification.

It is clear that the boundary element method uses superposition of particular solutions to generate a full solution. Hence (6.11.7) becomes

$$c_a u_a + \int_{S_1} p^* u\, dS + \int_{S_2} p^* u\, dS = \int_{S_1} pu^*\, dS + \int_{S_2} pu^*\, dS \qquad (6.11.9)$$

for a given point a, and where the integrals are taken around the relevant sections of the boundary. The variables u and p both represent known and unknown boundary values. The boundary is now discretised and it is at this point that numerical errors are introduced, the equations being exact to this point. Suppose the boundary is divided into N elements and the quantities u and P are assumed to vary in a simple form over each element, then (6.11.9) is a summation of the integrals over each element. The discretisation is shown in Figure 6.17.

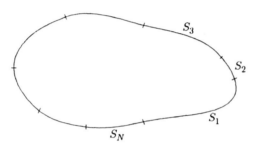

Fig. 6.17.

Hence (6.11.9) becomes

$$c_a u_a + \sum_{i=1}^{N} \int_{S_i} p^* u\, dS = \sum_{i=1}^{N} \int_{S_i} u^* p\, dS \qquad (6.11.10)$$

and a very simple form of u and p would be to assume they are constant over each element and then the function values are those at the midpoint of the straight line element. On each node, either u or p are known, and renumbering the elements so that the boundary point a becomes node 1, then

$$c_1 u_1 + \int_{S_1} p^* u_1\, dS + \int_{S_2} p^* u_2\, dS + \cdots + \int_{S_N} p^* u_N\, dS$$

$$= \int_{S_1} u^* p_1\, dS + \int_{S_2} u^* p_2\, dS + \cdots + \int_{S_N} u^* p_N\, dS \quad (6.11.11)$$

and a typical integral in this equation has the form

$$\int_S p^* u\, dS = \int_S \begin{bmatrix} p^*_{xx} & p^*_{xy} \\ p^*_{yx} & p^*_{yy} \end{bmatrix} \begin{bmatrix} u_x \\ u_y \end{bmatrix} dS \qquad (6.11.12)$$

which reduces to evaluating integrals of the form

$$I = \int p_{xx}^* u_x \, dS = u_x \int p^* \, dS \qquad (6.11.13)$$

where u_x is taken outside the integrals as it is constant.

These integrations are usually carried out exactly for constant elements and approximately using say Gaussian quadrature for more complex elements. It is of note that the fundamental solution often includes functions which are singular at some points of the range, and special integration routines are incorporated to deal with this problem (Evans, 1993). Defining the various integrals in (6.11.11) using

$$h_{11} = c_1 u_1 + \int_{S_1} p^* \, dS; \qquad h_{1i} = \int_{S_i} p^* \, dS; \qquad g_{1i} = \int_{S_i} u^* \, dS \quad (6.11.14)$$

allows equation (6.10.11) to be rewritten as

$$h_{11} u_1 + h_{12} u_2 + \cdots + h_{1N} u_N = g_{11} p_1 + g_{12} p_2 + \cdots + g_{1N} p_N \qquad (6.11.15)$$

in which it can be assumed that all the integrations have been carried out so that all the coefficients h and g are known. Hence for each of the N nodes an equation of the form (6.11.15) can be written to yield N equations in N unknowns. These equations are solved by standard means once the knowns and unknowns from u and p are put on opposite sides of the equation. Hence the boundary values are then all known and (6.10.5) is used to obtain the u values at any interior point.

This brief overview of the boundary element method can be extended by reference to Brebbia (1978), Brebbia and Dominguez (1992) and Brebbia and Mackerle (1988) for further reading.

Clearly in a single chapter there are limitations on what may be covered. Some leaders have been set in this chapter to more refined algorithms and further reading may be found in Davies (1980). Some exercises on this chapter are now given, and will require some programming activity to deal with the computational labour that the finite element method generates.

EXERCISES

6.1 The finite element procedure can be simplified somewhat when the boundary conditions are entirely of the Dirichlet type to force $h(x, y) = 0$ and when $f(x, y) = 0$ to give the solution of Laplace's equation rather than Poisson's. Hence it is instructive to try the worked example from Section 4.1 illustrated in Figure 4.2 using a finite element method with the same rectangular elements as used to form the grid in Chapter 4.

6.2 To test the simplified boundary conditions further, solve the same problem using triangular elements. Form these by dividing the rectangles of Exercise 6.1 from the bottom left-hand corners to the top right. In both cases, the results will compare directly with the earlier work. The numbering system used in the solution is given in the Appendix and may need to be referred to before coding these problems to allow the resulting matrices and force vectors to be easily compared.

6.3 Repeat the procedure with Exercise 4.2 which is a second Dirichlet problem.

6.4 Now the Neumann conditions can be tried with a non-zero $h(x, y)$ by attempting Exercise 4.3. Use the same rectangular grid as the new elements.

6.5 Triangulate as in Exercise 6.2 and repeat the solution of Exercise 4.3.

6.6 Curved boundaries do not require any special handling with the finite element method. Solve the non-transformed worked example from Section 4.2 given in Figure 4.7. Replace the curved grid lines by straight lines and triangulate the resulting quadrilaterals as in Exercise 6.2.

6.7 Solve Exercise 4.2 using finite elements, triangulating the grid of Figure 4.10.

6.8 To bring into the solution non-zero $f(x, y)$ use finite elements to solve Exercise 4.1 illustrated in Figure 4.9. Again triangulate as before.

A
Solutions to Exercises

In this appendix all the exercises will be solved. In cases where the solutions are close replicas of the worked examples in the text then only the new numerical values at the major steps, or the specific difference from the earlier work will be included. Standard methods will not be exhibited in detail, hence for example a solution which requires Gaussian elimination will simply quote the result, as the process itself will be considered a standard tool. The same will apply to iterative methods in chapters where these are again used as tools. In fact readers will find that access to simple computing facilities will make the problems in this volume very much easier, as software can be built up as the work proceeds.

On the other hand some of the exercises introduce new ideas for the first time to stretch the readers ingenuity. In these cases much fuller solutions are provided in this appendix.

Chapter 1

1.1
$$||A||_E^2 = \text{trace}(A^H A) = \sum_{i=1}^{n} \sigma_i^2$$

by the definitions of trace and singular value.

Hence $\sigma_1^2 \leq \sum_{i=1}^{n} \sigma_i^2 \leq n\sigma_1^2$ which is precisely the required inequality.

The matrix $A^H A$ is diagonal so no hard work here. Hence $||A||_2 = 2.33$, $||A||_E = 2.6$ and $n^{1/2}||A||_2 = 3.30$.

1.2 Only $||A||_2$ needs a bit of work. Direct use of the definitions gives

$||A||_1 = 5$, $||A||_\infty = 7$ and $||A||_E = 6$. For $||A||_2$

$$A^H A = \begin{bmatrix} 2 & 2 & 6 \\ 2 & 13 & 7 \\ 6 & 7 & 21 \end{bmatrix}.$$

The characteristic equation is

$$\lambda^3 - 36\lambda^2 + 252\lambda - 64 = 0$$

to yield the maximum eigenvalue of 26.626 by using Newton's method. A rough estimate of the largest root follows by arguing that for large λ the characteristic polynomial reduces to just the first two terms to yield a rough root of 36. Gerschgorin's theorems may also be used (see next exercise).

1.3 The characteristic polynomial for this matrix is

$$-\lambda^3 + 9\lambda^2 + 24\lambda - 332 = 0$$

with roots $\lambda_1 = -5.649627$ and $\lambda_{2,3} = 7.32481 \pm i2.26098$. Hence the complex pair has maximum modulus namely 7.6658. By comparison $||A||_1 = 14$ and $||A||_\infty = 12$, confirming the inequalities.

1.4 Because A is symmetric, κ is the ratio of the modulus of the largest to smallest eigenvalue of A itself. The characteristic equation is

$$-\lambda^3 + 127\lambda^2 - 230\lambda + 24 = 0$$

and its roots are 0.1111655, 1.72489 and 125.164 leaving $\kappa = 1125.92$. Hence one can expect to loose 3 digits in a numerical solution of this set of equations.

It is of note that κ can be estimated quite well without any recourse to solving polynomials. If one root is small, then λ^3 and λ^2 are negligible. Hence the small root will approximately satisfy $-230\lambda + 24 = 0$ to give 0.104 which is a good value. For a large root the reverse is true and $-\lambda^3 + 127\lambda^2 = 0$ to give 127.0. Hence an estimate of κ is 1221. As the order of this is the important feature, this estimate is quite adequate.

1.5 The relevant values for a range of c are given in Table A.1.

Table A.1.

c	λ_1	λ_n	κ
0.0	0.08333	1.3899	16.679
0.2	0.00268733	1.40832	524.06
0.4	0.283102	1.05909	3.741

The condition factor κ rises very sharply around the suggested problem area of $c = 1/5$.

1.6 The first and last rows give circles centre 2 and radius 1, and the middle rows have centre 2 and radius 2. The latter circle gives the range $[0, 4]$ as the eigenvalues will be real.

1.7 An estimate of the largest eigenvalue can be made from the third row with centre 120 and radius 30. The small eigenvalue is too vaguely determined to be useful, namely a circle centre 1 radius 8. It is the low value of this eigenvalue which is crucial to computing κ.

1.8 The situation remains as difficult for large order. Direct computation of the condition factors yields 1.8(6) for $n = 5$, and 3.4(13) for $n = 8$. Solving such a system by numerical means would result in a 13-figure loss of accuracy (Evans, 1995).

1.9 Progress can be made in this example because of the small circle associated with the first row. The first and last rows give the relevant circles with upper and lower bounds:

$$96 * 2/3 < \kappa < 104 * 2.$$

The actual eigenvalues are 0.975959, 7.56762, 10.2790 and 100.177.

1.10 The result of applying Jacobi's method is $x_1 = 0.115702479339$, $x_2 = 0.578512396694$ and $x_3 = 0.380165289256$ in 43 iterations. Gauss–Seidel with the same stopping criterion yields the same result in just 17 iterations. Here a factor of roughly 2 in the convergence rates is seen.

1.11 The $\mathbf{G}$ matrices are $\mathbf{D}^{-1}(\mathbf{L}+\mathbf{U})$ for Jacobi's method, and $(\mathbf{D}+\mathbf{L})^{-1}\mathbf{U}$ for Gauss–Seidel's. These matrices are:

$$\mathbf{G}_J = \begin{bmatrix} 0 & 1/2 & 1/4 \\ -1/3 & 0 & 1/3 \\ 1/8 & -1/4 & 0 \end{bmatrix}, \quad \mathbf{G}_{GS} = \begin{bmatrix} 0 & 1/2 & 1/4 \\ 0 & 1/6 & 5/12 \\ 0 & -1/48 & 7/96 \end{bmatrix}.$$

For $\mathbf{G}_J$ the characteristic equation is

$$-\lambda^3 - \frac{7}{32}\lambda + \frac{1}{24} = 0$$

with roots $\lambda_1 = 0.1685763$ and $\lambda_{2,3} = -0.084288 \pm i0.4899627$, hence making the modulus of the dominant eigenvalue 0.49715. Hence the error falls off like the powers of this factor. Hence for 12-figure accuracy the number of steps m satisfies:

$$\lambda^m = 10^{-12}$$

which gives $m = 39$. In practice, 44 steps were used, with the last iteration being used to confirm convergence. The convergence criterion was made fairly strict to make the theoretical iteration count very reasonable.

The equivalent characteristic equation for the Gauss–Seidel method is

$$-\lambda^3 + \frac{23}{96}\lambda^2 - \frac{1}{48}\lambda = 0$$

with roots $\lambda_1 = 0$ and $\lambda_{2,3} = 0.119792 \pm i0.08051888$. Hence the maximum modulus is 0.144 and then $m = 14$ for 12 figures, and 17 steps were used in practice.

1.12 The results are given in Table A.2.

Table A.2.

ω	Number of iterations
1.0	17
1.05	21
1.1	24
1.15	28
1.2	33

Clearly SOR is not improving the convergence rate and inspection of the Gauss–Seidel iterations shows that the error is not one-signed.

1.13 The **G** matrices are:

$$\mathbf{G_J} = \begin{bmatrix} 0 & 0 & 1 \\ -1 & 0 & 0 \\ -1/3 & -2/3 & 0 \end{bmatrix}, \qquad \mathbf{G_{GS}} = \begin{bmatrix} 0 & 0 & 1 \\ 0 & 0 & 1 \\ 0 & 0 & 1 \end{bmatrix}.$$

For $\mathbf{G_J}$ the characteristic equation has roots $\lambda_1 = 0.748$ and $\lambda_{2,3} = -0.374 \pm i0.868$, hence making the modulus of the dominant eigenvalue 0.945. Hence very slow convergence.

The equivalent (trivial) characteristic equation for the Gauss–Seidel method has roots $\lambda_1 = 0$ and $\lambda_{2,3} = 0$. Hence in this pathological case, Jacobi's method converges and Gauss–Seidel's does not.

1.14 To compute the optimum ω, $\mathbf{C_J}$ is

$$\begin{bmatrix} 0 & -1 & 0 & 0 \\ -1/3 & 0 & -1/3 & 0 \\ 0 & -1/6 & 0 & -1/6 \\ 0 & 0 & -1/11 & 0 \end{bmatrix}.$$

The characteristic polynomial of this matrix is

$$-\lambda^4 + \frac{40}{99}\lambda^2 - \frac{1}{198} = 0$$

with roots $x_{1,2} = \pm 0.113634$ and $x_{3,4} = \pm 0.625402$. Hence $\rho(C_J) = 0.25402$ and $\omega = 1.24$.

On the practical side the number of iterations for varying ω gives 31 at $\omega = 1.0$, 27 at 1.1 and 23 at 1.2. This rises to 28 at 1.3. Here we see agreement with the theoretical expectation.

1.15 With the small change in element (1,4) the new numbers of iterates are 31 at 1.0, 27 at 1.1, 23 at 1.2 and 28 at 1.3. These results show that the simple change makes little difference to the ω value.

1.16 For $n = 2$ the characteristic equation is

$$(a - \lambda)^2 - bc = 0$$

to give the roots $\lambda = a \pm \sqrt{bc}$. In the formula, $\cos s\pi/(n+1)$ gives the required factor of $\pm 1/2$ for agreement.

1.17 The characteristic equation factors rapidly to $(\lambda - 2)(4 - \lambda)(\lambda - 6)$. In the formula $a = 4$ and $bc = 2$. The cos has arguments $\pi/4$, $\pi/2$ and $3\pi/4$ to give the same values.

1.18 $x_1 = 0.1875$, $x_2 = 0.25$ and $x_3 = 0.625$.

The next few questions all depend on variations of the Gauss elimination algorithm. The pertinent parts of this are:

```
for(r = 1 to n − 1)            These are the n − 1 major steps
  for(i = r + 1 to n)          Find multiplying factors from row r + 1 to n
  w = a[i,r]/a[r,r]
  for(j = r to n)              Eliminate along the rows from r to n
    a[i,j] = a[i,j] − w * a[r,j]
```

The pivotal calculations involve no multiply or divide operations and are omitted. The back-substitution is of lower order and can be added in separately. The inner j loop could run from $r + 1$ to n as the first element must be zero in the new matrix by the construction.

1.19 Hence the count is:

$$\sum_{r=1}^{n-1} \left\{ \sum_{i=r+1}^{n} \left[1 + \sum_{j=r}^{n} 1 \right] \right\}$$

$$= \sum_{r=1}^{n-1} \left\{ \sum_{i=r+1}^{n} [n + 2 - r] \right\}$$

$$= \sum_{r=1}^{n-1} \left\{ (n + 2 - r)(n - r) \right\}$$

$$= n(n + 2)(n + 1) - \frac{(2n + 1)(n - 1)n}{2} + \frac{(n - 1)n(2n - 1)}{6}$$

$$= \frac{n^3}{3} + \frac{n^2}{2} - \frac{5n}{6}$$

where the given formulae for summing the natural numbers and their squares are used to evaluate the last two terms of the final summation.

For Thomas's algorithm the calculation degenerates as the i loop is just $i = r + 1$ and the j loop will just go from r to $r + 1$ as zero will be added to all succeeding elements. Hence the count is now:

$$\sum_{r=1}^{n-1} (1 + 2) = 3(n - 1).$$

1.20 One sweep of Gauss–Seidel gives a count of

$$1 + 2(n - 2) + 1 + 1 = 2n + 1$$

for a tridiagonal matrix. There is just one element to subtract off the RHS in the first and last rows giving two of the i's, and for the middle $(n - 2)$ rows there are 2 elements. There is a final division by the diagonal element to give the third 1. Hence from the point of view of workload, Thomas's algorithm will always be competitive.

1.21 To extend Thomas's algorithm to an upper Hessenberg matrix just means running the j loop from r to n as there will no longer be zero elements in the pivotal row.

1.22 For quindiagonal matrices the i loop needs to go from $r + 1$ to $r + 2$ to account for the extra non-zero element in each column and the j loop from r to $r + 2$.

It is the need for pivoting which will cause problems in general.

1.23 (a) elliptic

(b) hyperbolic

(c) parabolic

(d) hyperbolic for $x < 0$, elliptic for $x > 0$ and parabolic when $x = 0$.

1.24 This is a curve-sketching exercise. The equation is parabolic on $9x^4 y^4 = 4(x + y)$. The regions of hyperbolicity and ellipticity will be bounded by this curve. The salient features are as follows:

(i) $x + y$ can never be negative as $x^4 y^4$ is always positive. Hence no curve beneath and to the left of $x = -y$.

(ii) Convert to polars to get

$$r^7 = \frac{4(\cos\theta + \sin\theta)}{9\cos^4\theta\sin^4\theta}.$$

Hence asymptotes at $\theta = 0$ and $\theta = \pi/2$ on both sides of each line.

(iii) At $\theta = 3\pi/4$, $r = 0$ where the curve gradient is -1.

Hence the curve is as shown in Figure A.1.

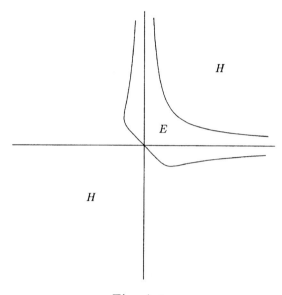

Fig. A.1.

1.25 The characteristics satisfy:

$$y'^2 - 2\left(x + \frac{1}{y}\right)y' + \frac{4x}{y} = 0$$

or

$$(y' - 2x)\left(y' - \frac{2}{y}\right) = 0$$

to leave the characteristics $y = x^2 + c$ and $y^2 = 4x + c$.

1.26 The characterics satisfy $y'^2 + 6y' + 9 = 0$ to give after factorisation characteristic transformation $\xi = y + 3x$ and $\eta = x$. Then $\xi_x = 3$, $\xi_y = 1$, $\eta_x = 1$ and $\eta_y = 0$. The required partial derivatives are then

$$\frac{\partial z}{\partial x} = 3\frac{\partial z}{\partial \xi} + \frac{\partial z}{\partial \eta}$$

$$\frac{\partial z}{\partial y} = \frac{\partial z}{\partial \xi}$$

$$\frac{\partial^2 z}{\partial x^2} = 9\frac{\partial^2 z}{\partial \xi^2} + 6\frac{\partial^2 z}{\partial \xi \partial \eta} + \frac{\partial^2 z}{\partial \eta^2}$$

$$\frac{\partial^2 z}{\partial y^2} = \frac{\partial^2 z}{\partial \xi^2}$$

$$\frac{\partial^2 z}{\partial x \partial y} = 3\frac{\partial^2 z}{\partial \xi^2} + \frac{\partial^2 z}{\partial \xi \partial \eta}.$$

Substitute into the original partial differential equation to get

$$\frac{\partial^2 z}{\partial \eta^2} = \frac{\partial z}{\partial \xi}$$

which is a conventional heat equation.

1.27   Using the quadratic solution formula the characteristic transformation is $\xi = y + kx$ and $\eta = y + lx$ with $k = \left(-3 - \sqrt{5}\right)/2$ and $l = \left(-3 + \sqrt{5}\right)/2$. The change of variable then gives:

$$\frac{\partial z}{\partial x} = k\frac{\partial z}{\partial \xi} + l\frac{\partial z}{\partial \eta}$$

$$\frac{\partial z}{\partial y} = \frac{\partial z}{\partial \xi} + \frac{\partial z}{\partial \eta}$$

$$\frac{\partial^2 z}{\partial x^2} = k^2\frac{\partial^2 z}{\partial \xi^2} + 2kl\frac{\partial^2 z}{\partial \xi \partial \eta} + l^2\frac{\partial^2 z}{\partial \eta^2}$$

$$\frac{\partial^2 z}{\partial y^2} = \frac{\partial^2 z}{\partial \xi^2} + 2\frac{\partial^2 z}{\partial \xi \partial \eta} + \frac{\partial^2 z}{\partial \eta^2}$$

$$\frac{\partial^2 z}{\partial x \partial y} = k\frac{\partial^2 z}{\partial \xi^2} + (k + l)\frac{\partial^2 z}{\partial \xi \partial \eta} + l\frac{\partial^2 z}{\partial \eta^2}.$$

Substitute and remember that k and l satisfy $m^2 - 3m + 1 = 0$, and $kl = 1$ (product of roots) and $k + l = 3$ (minus sum of roots), to give briskly

$$13\frac{\partial^2 z}{\partial \xi \partial \eta} = 0$$

which integrates twice to give

$$z = f(y + kx) + g(y + lx)$$

with f and g being general functions. (See Evans, Blackledge and Yardley, 1999.)

1.28   This follows the same lines as Exercise 1.27 with $k = -1 - i\sqrt{2}$ and $l = -1 + i\sqrt{2}$, to give the same cross derivative.

$$13\frac{\partial^2 z}{\partial \xi \partial \eta} = z.$$

To get a real form put $\xi = \alpha + i\beta$ and $\eta = \alpha - i\beta$. Hence $\alpha = \xi/2 + \eta/2$ and $\beta = \xi/2i - \eta/2i$. Repeat the transformation procedure to leave the real form:

$$\frac{1}{4}\left(\frac{\partial^2 z}{\partial \alpha^2} + \frac{\partial^2 \eta}{\partial \beta^2}\right) = z.$$

Chapter 2

2.1
$$\frac{\partial \phi}{\partial x} - \frac{\phi_{i+1,j} - \phi_{i,j}}{h}$$

$$= \frac{\partial \phi}{\partial x} + \phi_{i,j}/h - \frac{1}{h}\left(\phi_{i,j} + h\frac{\partial \phi}{\partial x} + \frac{h^2}{2!}\frac{\partial^2 \phi}{\partial x^2} + \cdots\right)$$

$$= \frac{h}{2}\frac{\partial^2 \phi}{\partial x^2}$$

where all expressions are evaluated at (i,j).

2.2
$$\frac{\partial^2 \phi}{\partial x^2} - \frac{\phi_{i+1,j} - 2\phi_{i,j} + \phi_{i-1,j}}{h}$$

$$= \frac{\partial^2 \phi}{\partial x^2} + 2\phi_{i,j}/h$$

$$- \frac{1}{h^2}\left(\phi_{i,j} + h\frac{\partial \phi}{\partial x} + \frac{h^2}{2!}\frac{\partial^2 \phi}{\partial x^2} + \frac{h^3}{3!}\frac{\partial^3 \phi}{\partial x^3} + \frac{h^4}{4!}\frac{\partial^4 \phi}{\partial x^4} + \cdots\right)$$

$$- \frac{1}{h^2}\left(\phi_{i,j} - h\frac{\partial \phi}{\partial x} + \frac{h^2}{2!}\frac{\partial^2 \phi}{\partial x^2} - \frac{h^3}{3!}\frac{\partial^3 \phi}{\partial x^3} + \frac{h^4}{4!}\frac{\partial^4 \phi}{\partial x^4} + \cdots\right)$$

$$= \frac{h^2}{12}\frac{\partial^4 \phi}{\partial x^4}$$

where all expressions are again evaluated at (i,j).

2.3 Expanding the right-hand side by Taylor gives:

$$A\left(\phi_{i,j} - 2h\frac{\partial \phi}{\partial x} + \frac{4h^2}{2!}\frac{\partial^2 \phi}{\partial x^2} - \frac{8h^3}{3!}\frac{\partial^3 \phi}{\partial x^3} + \frac{16h^4}{4!}\frac{\partial^4 \phi}{\partial x^4} + \cdots\right)$$

$$+ B\left(\phi_{i,j} - h\frac{\partial \phi}{\partial x} + \frac{h^2}{2!}\frac{\partial^2 \phi}{\partial x^2} - \frac{h^3}{3!}\frac{\partial^3 \phi}{\partial x^3} + \frac{h^4}{4!}\frac{\partial^4 \phi}{\partial x^4} + \cdots\right)$$

$$+ C\left(\phi_{i,j} + h\frac{\partial \phi}{\partial x} + \frac{h^2}{2!}\frac{\partial^2 \phi}{\partial x^2} + \frac{h^3}{3!}\frac{\partial^3 \phi}{\partial x^3} + \frac{h^4}{4!}\frac{\partial^4 \phi}{\partial x^4} + \cdots\right)$$

$$+ D\left(\phi_{i,j} + 2h\frac{\partial \phi}{\partial x} + \frac{4h^2}{2!}\frac{\partial^2 \phi}{\partial x^2} + \frac{8h^3}{3!}\frac{\partial^3 \phi}{\partial x^3} + \frac{16h^4}{4!}\frac{\partial^4 \phi}{\partial x^4} + \cdots\right)$$

$$= \frac{\partial^2 \phi}{\partial x^2} - E$$

evaluated at (i,j).

Equating coefficients then gives:

$$A + B + C + D = 0$$
$$-2A - B + C + 2D = 0$$
$$2A + B/2 + C/2 + 2D = 1/h^2$$
$$-\frac{4}{3}A - \frac{B}{6} + \frac{C}{6} + \frac{4}{3}D = 0$$

which solve easily to give $A = D = 1/3h^2$ and $B = C = -1/3h^2$ with $E = 17h^2/36$.

2.4 (i) This example requires expanding ϕ at points like $(i+1, j+1)$ which requires a Taylor expansion in two variables. This is easy to achieve by expanding first in one variable, and then expanding the leading terms of the first expansion in the second variable, simply re-using the standard one-dimensional Taylor series. Hence the expansion which will be used repeatedly in examples of this type is:

$$
\begin{aligned}
\phi_{i+1,j+1} = {} & \phi_{i,j} + k\frac{\partial\phi}{\partial y} + \frac{k^2}{2!}\frac{\partial^2\phi}{\partial y^2} + \frac{k^3}{3!}\frac{\partial^3\phi}{\partial y^3} + \frac{k^4}{4!}\frac{\partial^4\phi}{\partial y^4} + \cdots \\
= {} & h\frac{\partial\phi}{\partial x} + hk\frac{\partial^2\phi}{\partial x\partial y} + h\frac{k^2}{2!}\frac{\partial^3\phi}{\partial x\partial y^2} + h\frac{k^3}{3!}\frac{\partial^4\phi}{\partial x\partial y^3} + \cdots \\
& + \frac{h^2}{2!}\frac{\partial^2\phi}{\partial x^2} + \frac{h^2}{2!}k\frac{\partial^3\phi}{\partial x^2\partial y} + \frac{h^2}{2!}\frac{k^2}{2!}\frac{\partial^4\phi}{\partial x^2\partial y^2} + \cdots \\
& + \frac{h^3}{3!}\frac{\partial^3\phi}{\partial x^3} + \frac{h^3}{3!}k\frac{\partial^4\phi}{\partial x^3\partial y} + \cdots \\
& + \frac{h^4}{4!}\frac{\partial^4\phi}{\partial x^4} + \cdots
\end{aligned}
$$

where each term in the first expansion is re-expanded down the column. The expansion for $\phi_{i+1,j-1}$ will be the same as for $\phi_{i+1,j+1}$ with negative signs down the odd numbered columns as k changes sign. Similar variations then occur for $\phi_{i-1,j+1}$ and $\phi_{i-1,j-1}$. To get a cross-derivative clearly needs points like these, as expansions of $\phi_{i,j+1}$ will only ever involve partial derivatives of y. The obvious approximation can be generated as:

$$
\begin{aligned}
\frac{\partial^2\phi}{\partial x\partial y} &= \frac{\frac{\phi_{i+1,j+1}-\phi_{i-1,j+1}}{2h} - \frac{\phi_{i+1,j-1}-\phi_{i-1,j-1}}{2h}}{2k} \\
&= (\phi_{i+1,j+1} - \phi_{i-1,j+1} - \phi_{i+1,j-1} + \phi_{i-1,j-1})/4hk
\end{aligned}
$$

Expanding these four terms using the above Taylor expansion leads to considerable cancellation, and leaves an error term:

$$E = \left(\frac{hk^3}{3!}\frac{\partial^4\phi}{\partial x^3\partial y} + \frac{kh^3}{3!}\frac{\partial^4\phi}{\partial x\partial y^3}\right)/hk.$$

(ii) The approximation here is the normal:

$$\frac{\partial^2 \phi}{\partial x^2} + x\frac{\partial^2 \phi}{\partial y^2} = \frac{\phi_{i+1,j} - 2\phi_{i,j} + \phi_{i-1,j}}{h^2} + x_i\frac{\phi_{i,j+1} - 2\phi_{i,j} + \phi_{i,j-1}}{k^2}$$

with the errors as found in Exercise 2.2.

2.5 Expand as suggested to yield

$$\phi_1 = \phi_0 + h_1\frac{\partial \phi}{\partial x} + \frac{h_1^2}{2!}\frac{\partial^2 \phi}{\partial x^2} + \frac{h_1^3}{3!}\frac{\partial^3 \phi}{\partial x^3} + \cdots$$

and

$$\phi_2 = \phi_0 + h_2\frac{\partial \phi}{\partial x} + \frac{h_2^2}{2!}\frac{\partial^2 \phi}{\partial x^2} + \frac{h_2^3}{3!}\frac{\partial^3 \phi}{\partial x^3} + \cdots$$

Take the terms up to and including the h^2 terms and solve for $\partial \phi/\partial x$ and $\partial^2 \phi/\partial x^2$ to give:

$$\frac{\partial \phi}{\partial x} = \frac{h_1^2\phi_2 + (h_2^2 - h_1^2)\phi_0 - h_2^2\phi_1}{h_1 h_2^2 - h_1^2 h_2}$$

and

$$\frac{\partial^2 \phi}{\partial x^2} = \frac{h_1\phi_2 - (h_1 - h_2)\phi_0 - h_2\phi_1}{\frac{h_1 h_2^2 - h_1^2 h_2}{2}}$$

which correctly reduce to the standard forms when $h_2 = -h_1$.

In the next group of exercises the grid is numbered with (1,1) at $t = 0$, $x = 0$, and the line $t = 0$ running (2,1), (3,1) …. The second index increases as the t step moves forward.

2.6 The explicit method gives

$$\phi_{i,j+1} = \phi_{i,j} + 0.25(\phi_{i-1,j} - 2\phi_{i,j} + \phi_{i+1,j})$$

with $x = 0$ forcing $\phi_{1,j} = 0$ and $x = 1$ giving $\phi_{5,j} = 10$. On $t = 0$, $\phi_{i,1} = 2.0$. Iterations yield the results in Table A.3.

Table A.3.

$t\backslash x$	4.0	8.0	12.0
40.0	1.5	2.0	4.0
80.0	1.25	2.375	5.0
120.0	1.2188	2.75	5.5938
160.0	1.2969	3.0781	5.9844

2.7 Write

$$\frac{\partial}{\partial x}\left((1 + x^2)\frac{\partial \phi}{\partial x}\right) = 2x\frac{\partial \phi}{\partial x} + (1 + x^2)\frac{\partial^2 \phi}{\partial x^2}$$

and then the finite different scheme is:

$$\frac{\phi_{i,j+1} - \phi_{i,j}}{k} = 2x_i \left(\frac{\phi_{i+1,j} - \phi_{i-1,j}}{2h} \right)$$

$$+(1 + x_i^2) \left(\frac{\phi_{i-1,j} - 2\phi_{i,j} + \phi_{i+1,j}}{h^2} \right)$$

with $x_i = -1 + (i-1)h$. The boundary conditions give $\phi_{1,j} = \phi_{6,j} = 0$ at $x = -1$ and $x = 1$, and $\phi_{i,1} = 1000(1-x)$ for $i = 2,3$ with the corresponding symmetric values for $i = 4,5$. Hence iterating and tabulating just columns 2 and 3 which by symmetry will be the same as columns 5 and 4 respectively, the results in Table A.4 are obtained.

Table A.4.

$t \backslash x$	-0.6	-0.4
0.04	448.0	704.0
0.08	424.96	642.56
0.12	393.01	590.34

2.8 The choice of $k = rh^2 = 0.005$, and $\phi_{1,i} = \phi_{11,i} = 0$ on $x = 0$ and $x = 1$ are the first conditions to fix. On $t = 0$, $\phi_{i,1} = 2x_i = \phi_{12-i,1}$, for $i = 2,6$. The boundary conditions are symmetric about $x = 1/2$ and the first step of the conventional finite difference scheme then gives the values $\phi_{2,2} = 0.2$, $\phi_{2,3} = 0.4$, $\phi_{2,4} = 0.6$, $\phi_{2,5} = 0.8$ and $\phi_{2,6} = 0.8$, with the correponding symmetric values for $\phi_{2,i}$ with $i = 10(-1)6$.

2.9 The finite difference scheme with the $\partial\phi/\partial x$ term is now:

$$\frac{\phi_{i,j+1} - \phi_{i,j}}{k} = \left(\frac{\phi_{i+1,j} - \phi_{i-1,j}}{2h} \right) + \left(\frac{\phi_{i-1,j} - 2\phi_{i,j} + \phi_{i+1,j}}{h^2} \right).$$

Choosing $r = 1/4$ gives $k = 0.0025$ which requires 8 steps to reach $t = 0.02$. There will not be symmetry here because $\partial\phi/\partial x$ will change sign for $0 < x < 1/2$ against the range $1/2 < x < 1$. (Note this didn't happen in Exercise 2.7 because the first derivative term was multiplied by an x which also changed sign on the interval $[-1,1]$ to preserve the symmetry.) The values are then as given in Table A.5.

2.10 There are two modifications to the previous example. The first derivative term has a factor $-1/x$, and the $x = 0$ boundary condition is a derivative condition which forces $u_{1,j} = u_{2,j}$. The finite difference scheme for $i = 3$ to 10 is:

$$\frac{u_{i,j+1} - u_{i,j}}{k} = -\left(\frac{u_{i+1,j} - u_{i-1,j}}{2hx_i} \right) + \left(\frac{u_{i-1,j} - 2u_{i,j} + u_{i+1,j}}{h^2} \right)$$

Table A.5.

$t \backslash x$	0.1	0.2	0.3	0.4	0.5
	0.6	0.7	0.8	0.9	
0.0025	0.1025	0.2025	0.3025	0.5338	0.7
	0.5163	0.2975	0.1975	0.0975	
0.005	0.1044	0.205	0.3395	0.5225	0.6123
	0.5025	0.3232	0.195	0.09566	
0.0075	0.1060	0.2164	0.3556	0.5026	0.5621
	0.4815	0.3321	0.1994	0.09414	
0.01	0.1098	0.2267	0.3611	0.4833	0.5268
	0.4614	0.3328	0.2033	0.09442	
0.0125	0.1144	0.2342	0.3613	0.4657	0.4993
	0.4432	0.3293	0.2055	0.09549	
0.015	0.1187	0.2391	0.3585	0.4497	0.4766
	0.4266	0.3239	0.2060	0.09654	
0.0175	0.1221	0.2419	0.3541	0.4351	0.4571
	0.4115	0.3173	0.2053	0.9720	
0.2	0.1245	0.2429	0.3487	0.4216	0.4399
	0.3976	0.3103	0.2035	0.9735	

with $x_i = (i-1)h$. The finite difference scheme at $i = 2$ will be the above scheme with $u_{1,j} = u_{2,j}$ to give:

$$\frac{u_{2,j+1} - u_{2,j}}{k} = -\left(\frac{u_{3,j} - u_{2,j}}{2hx_2}\right) + \left(\frac{u_{2,j} - 2u_{2,j} + u_{3,j}}{h^2}\right).$$

The resulting value at (0,0.001) and (0.1,0.001) is $u_{1,2} = u_{2,2} = 0.9885$, and at (0.9,0.001) is $u_{10,2} = 0.19$.

2.11 This example has a mixed condition at $x = 1$ and an extra ϕ term. The finite difference scheme at points $i = 2, 9$ is

$$\frac{\phi_{i,j+1} - \phi_{i,j}}{k} = +\left(\frac{\phi_{i-1,j} - 2\phi_{i,j} + \phi_{i+1,j}}{h^2}\right) - \phi_{i,j}$$

and when $i = 10$ the boundary condition gives

$$\frac{\phi_{11,j} - \phi_{10,j}}{h} + \phi_{10,j} = 1$$

to give the iterative equations:

$$\frac{\phi_{10,j+1} - \phi_{10,j}}{k} = \left(\frac{\phi_{9,j} - 2\phi_{10,j} + \phi_{11,j}}{h^2}\right) - \phi_{10,j}$$

with
$$\phi_{11,j} = (1 - \phi_{10,j})h + \phi_{10,j}.$$

The first few steps then have the values given in Table A.6.

Table A.6.

| $t\backslash x$ | 0.1 | 0.2 | 0.3 | 0.4 | 0.5 |
	0.6	0.7	0.8	0.9	
0.001	0.1120	0.04196	0.09191	0.1618	0.2518
	0.3616	0.4915	0.6414	0.7941	
0.002	0.1937	0.05392	0.09382	0.1637	0.2535
	0.3633	0.4930	0.6410	0.7801	
0.003	0.2601	0.07183	0.09672	0.1655	0.2552
	0.3649	0.4943	0.6395	0.7676	
0.004	0.3150	0.09308	0.1010	0.1674	0.2570
	0.3665	0.4954	0.6371	0.7563	

2.12 Use the expansion in Exercise 2.4(i) to complete expansions of each finite difference term. Tedious but straightforward manipulation results in massive cancellation and the remaining terms are:

$$\frac{\partial \phi}{\partial t} + \frac{k}{2!}\frac{\partial^2 \phi}{\partial t^2} - \frac{\partial^2 \phi}{\partial x^2} - \frac{h^2}{4!}\frac{\partial^4 \phi}{\partial x^4} - \frac{k}{2!}\frac{\partial^3 \phi}{\partial x^2 \partial t} - \frac{k^2}{2!}\frac{\partial^4 \phi}{\partial x^2 \partial t^2}$$

to yield the required partial derivative terms and the truncation error.

2.13 Again expand each term (no need for a double expansion here, so this is not so tedious) to yield:

$$\phi_{i+2,j} - \phi_{i+1,j} - \phi_{i-1,j} + \phi_{i-2,j} = 3h^2\frac{\partial^2 \phi}{\partial x^2} + \frac{5h^4}{4}\frac{\partial^4 \phi}{\partial x^4}.$$

2.14 The expansion now yields:

$$\phi_{i+1,j} - \phi_{i,j+1} - \phi_{i,j-1} + \phi_{i-1,j} = h^2\frac{\partial^2 \phi}{\partial x^2} - k^2\frac{\partial^2 \phi}{\partial y^2} + \frac{2h^4}{4!}\frac{\partial^4 \phi}{\partial x^4} + \frac{2k^4}{4!}\frac{\partial^4 \phi}{\partial y^4}.$$

After division by h^2 the first term yields the expected second derivative, but the second term then has the factor k^2/h^2. Hence the finite difference approximation approximates different derivatives depending on the ratio of k to h. This leads to the topic of consistency in Section 2.4.

2.15 The terms now expand to yield:

$$\phi_{i,j+2} + 2\phi_{i,j+1} - 2\phi_{i,j-1} + \phi_{i,j-2} = 8k\frac{\partial \phi}{\partial y} + \frac{20k^3}{3!}\frac{\partial^3 \phi}{\partial y^3}.$$

2.16 From Exercise 2.14 the method will approximate:

$$\frac{\partial \phi}{\partial t} = \frac{\partial^2 \phi}{\partial x^2} - \frac{k^2}{h^2}\frac{\partial^2 \phi}{\partial y^2}$$

and so for $k \ll h$, we have an approximation to the heat equation, but for $k = h$ the equation is a hyperbolic wave equation.

2.17 Here each term expands in the usual way and yields coefficients of the partial derivatives which are entirely in terms of either h or k causing no consistency problems.

2.18 The A matrix of equation (2.6.5) is now $(I + rT)^{-1}$ to yield the eigenvalues:

$$\lambda_s = \frac{1}{1 + 4r\sin^2\frac{s\pi}{2N}}$$

which are all less than unity to establish stability for all r.

2.19 The A matrix is now:

$$A = \begin{bmatrix} 1 - 2r & r(1+h) & & & \\ r(1-h) & 1-2r & r(1+h) & & \\ & \ddots & \ddots & \ddots & \\ & & r(1-h) & 1-2r & r(1+h) \\ & & & r(1-h) & 1-2r \end{bmatrix}.$$

Hence for $1 - 2r > 0$ we get

$$\|A\|_\infty = r(1-h) + 1 - 2r + r(1+h) = 1$$

which will ensure stability for $r < 1/2$, whereas for $1 - 2r < 0$

$$\|A\|_\infty = r(1-h) + 2r - 1 + r(1+h) = 4r - 1$$

which will give no satisfactory bound.

2.20 The A matrix is now:

$$A = \begin{bmatrix} 1 - 2r & r(1-h/x_1) & & & \\ r(1+h/x_2) & 1-2r & r(1-h/x_2) & & \\ & \ddots & \ddots & \ddots & \\ & & r(1+h/x_{n-1}) & 1-2r & r(1-h/x_{n-1}) \\ & & & r(1+h/x_n) & 1-2r \end{bmatrix}.$$

Hence for $1 - 2r > 0$ we get

$$\|A\|_\infty = r(1+h/x_i) + 1 - 2r + r(1-h/x_i) = 1.$$

Hence again stability for $r < 1/2$.

2.21 The A matrix is now:

$$
A = \begin{bmatrix}
1 - 2r + k & r & & & & \\
r & 1 - 2r + k & r & & & \\
& \ddots & \ddots & \ddots & & \\
& & r & 1 - 2r + k & r & \\
& & & r & 1 - 2r + k
\end{bmatrix}.
$$

2.22 All four exercises here use the substitution (2.6.19). The term $e^{\alpha jh}e^{-i\lambda ih}$ is then cancelled throughout. The term $e^{\alpha k}$ is then isolated as it is clearly this term which has to be less than unity in modulus for stability. Inequalities then follow to yield the result. Hence for Exercise 2.18:

$$
e^{\alpha k} - 1 = re^{\alpha k}e^{i\lambda h} - 2re^{\alpha k} + re^{\alpha k}e^{-i\lambda h}.
$$

Therefore

$$
e^{\alpha k} = \frac{1}{2r[1 - \cos\lambda h] + 1} = \frac{1}{1 + 4r\sin^2\frac{\lambda h}{2}} < 1.
$$

The inequality holds for all r, λ and h to confirm unconditional stability.

For Exercise 2.19:

$$
e^{\alpha k} - 1 = re^{i\lambda h} - 2r + re^{-i\lambda h} + \frac{hr}{2}e^{i\lambda h} - \frac{hr}{2}e^{-i\lambda h}
$$

to give:

$$
e^{\alpha k} = 1 - 2r + 2r\cos\lambda h + ihr\sin\lambda h.
$$

Hence

$$
|e^{\alpha k}|^2 = \left(1 - 4r\sin^2\frac{\lambda h}{2}\right)^2 + r^2h^2\sin^2\lambda h < 1.
$$

Cancelling an r gives

$$
r < \frac{8\sin^2\frac{\lambda h}{2}}{16\sin^4\frac{\lambda h}{2} + h^2\sin^2\lambda h}
$$

$$
= \frac{8}{16\sin^2\frac{\lambda h}{2} + h^2\frac{\sin^2\lambda h}{\sin^2\frac{\lambda h}{2}}}
$$

and the minimum value of the right-hand side will occur when $h = 0$ so that as long as $r < 1/2$ the inequality will hold for all λh.

For Exercise 2.21:

$$
e^{\alpha k} - 1 = re^{i\lambda h} - 2r + re^{-i\lambda h} - k.
$$

Hence

$$r < \frac{1-k}{2\sin^2\frac{\lambda h}{2}} < \frac{1}{2\sin^2\frac{\lambda h}{2}}.$$

If $r < 1/2$ then this inequality holds for all h and λ.

2.23 All this group of examples use variations on the set of linear equations for the implicit method defined by:

$$-\phi_{i-1,j+1}\left[\frac{kD}{2h^2}\right] + \phi_{i,j+1}\left[\frac{2Dk}{2h^2}+1\right] - \phi_{i+1,j+1}\left[\frac{kD}{2h^2}\right]$$

$$= \phi_{i-1,j}\left[\frac{kD}{2h^2}\right] + -\phi_{i,j}\left[\frac{2Dk}{2h^2}-1\right] - \phi_{i+1,j}\left[\frac{kD}{2h^2}\right]$$

which are set up for the four unknowns of the second row of the grid. With the point $(1,1)$ in the bottom left of the grid, the boundary conditions give $\phi_{1,i} = 0$ along $x = 0$, $\phi_{6,i} = 10$ along $x = 1$, and $\phi_{i,1} = 2$ for $i = 2,5$ at $t = 0$. The elements of the tridiagonal matrix A say, appear as coefficients in the formula above on the left-hand side. We will need to solve $Ax = b$ with the b vector made up of the right-hand side for each element. In addition at each end the elements $\phi_{1,2}$ and $\phi_{6,2}$ will be on the boundary, and will need to be added into b_1 and b_4 respectively. The solution vector is then $\phi_{i,2}$ for $i = 2,5$, and gives the results in Table A.7.

Table A.7.

$t\backslash x$	4.0	8.0	12.0	16.0
16.0	1.8179	1.9932	2.0328	2.7288
32.0	1.6679	1.9780	2.1191	3.3304
48.0	1.5437	1.9624	2.2385	3.8321
64.0	1.4409	1.9511	2.3770	4.2542

The same trends appear as in the explicit solution, remembering the different time steps being used. With such a crude grid, no great accuracy can be expected. This will be true of all the examples in this section.

2.24 This exercise has the same equations as the previous one but now has 9 unknowns. The $x = 0$ and $x = 1$ conditions are now both zero, and $\phi_{i,1} = 2x = \phi_{12-i,1}$ for $i = 2,5$ with $\phi_{6,1} = 1$ along $t = 0$. Hence $\phi_{2,2}$ to $\phi_{6,2}$ are 0.1644, 0.3147, 0.4308, 0.4792 and 0.3994 with the corresponding symmetric values to $\phi_{10,2}$.

2.25 The equations from Exercise 2.23 need to have the first derivative term

$$\frac{\partial\phi}{\partial x} = \frac{1}{2}\left\{\frac{\phi_{i+1,j+1}-\phi_{i-1,j+1}}{2h} + \frac{\phi_{i+1,j}-\phi_{i-1,j}}{2h}\right\}$$

added to give:

$$-\phi_{i-1,j+1}\left[\frac{k}{2h^2} - \frac{k}{4h}\right] + \phi_{i,j+1}\left[\frac{2k}{2h^2} + 1\right] - \phi_{i+1,j+1}\left[\frac{k}{2h^2} + \frac{k}{4h}\right]$$

$$= \phi_{i-1,j}\left[\frac{k}{2h^2} - \frac{k}{4h}\right] + -\phi_{i,j}\left[\frac{2k}{2h^2} - 1\right] - \phi_{i+1,j}\left[\frac{k}{2h^2} + \frac{k}{4h}\right]$$

where the Crank–Nicolson philosophy of averaging is used in the first derivative approximation. The other minor change from Exercise 2.24 is in the $t = 0$ boundary condition which is just x not $2x$. The same algorithm then gives for $\phi_{2,2}$ to $\phi_{10,2}$ (the first derivative knocks out the symmetry) 0.09736, 0.1917, 0.2766, 0.3352, 0.3213, 0.3284, 0.2714, 0.1887, 0.09606.

2.26 The finite difference scheme is

$$\phi_{i,j+1} - \phi_{i,j} = \frac{k}{2h^2}(\phi_{i+1,j+1} - 2\phi_{i,j+1} + \phi_{i-1,j+1})$$

$$+ \frac{k}{2h^2}(\phi_{i+1,j} - 2\phi_{i,j} + \phi_{i-1,j}) - k\phi_{i,j}.$$

Hence the iteration matrices are:

$$A = [2I - rT]^{-1}[2(1-k)I + rT]$$

with T as in (2.7.7). Hence the eigenvalues for stability are:

$$\lambda_s = \frac{2(1-k) - 4r\sin^2\frac{s\pi}{2n}}{2 + 4r\sin^2\frac{s\pi}{2n}}$$

which are less than unity for all r.

2.27 The finite difference scheme is:

$$\frac{u_{i,j+1} - u_{i,j}}{k} = \frac{k}{h^2}(u_{i-1,j} - 2u_{i,j} + u_{i+1,j}) + \frac{k}{2r_ih}(u_{i+1,j} - u_{i-1,j})$$

with $r_i = 0.5 + (i-1)h$. The boundary conditions give $u_{1,j} = 100$ at $r = 0.5$ and $u_{7,i} = 0$ at $r = 2$. At $t = 0$, $u_{i,1} = 15$ for $i = 2,6$. Conventional explicit iteration then gives the results in Table A.8.

Table A.8.

$t\backslash x$	0.75	1.0	1.25	1.5	1.75
0.03125	50.417	15.0	15.0	15.0	6.964
0.0625	50.417	30.495	15.0	10.647	6.964
0.09375	59.455	30.494	19.579	10.647	4.943
0.125	59.455	37.025	19.579	11.651	4.943

These results will not be at all accurate with such a crude step size. By the nature of the explicit method, a large initial discontinuity at $r = 0.5$ in particular will move into the solution space only at every other step as can be seen by the $x = 0.75$ column. A very fine step size is required in the x direction to resolve this difficulty, with the knock-on effect of an even smaller k value to keep $r < 1/2$.

2.28 The finite difference scheme is:

$$\frac{u_{i,j+1} - u_{i,j}}{k}$$

$$= \frac{k}{2h^2}(u_{i-1,j+1} - 2u_{i,j+1} + u_{i+1,j+1})$$

$$+ \frac{k}{2h^2}(u_{i-1,j} - 2u_{i,j} + u_{i+1,j}) + \frac{k}{4r_ih}(u_{i+1,j+1} - u_{i-1,j+1})$$

$$+ \frac{k}{4r_ih}(u_{i+1,j} - u_{i-1,j})$$

The solution now follows closely the problems in Section 2.7. The tridiagonal equations are set up for each row, the coefficients of the terms $u_{i-1,j+1}$, $u_{i,j+1}$ and $u_{i+1,j+1}$ appearing as matrix elements and all the other terms appearing in the right-hand side vector. The solution is then as given in Table A.9.

Table A.9.

$t\backslash x$	0.75	1.0	1.25	1.5	1.75
0.0625	51.426	25.846	17.749	13.962	7.222
0.125	56.977	37.304	23.031	13.542	6.952
0.1875	61.340	42.206	27.488	16.346	7.594
0.25	63.506	45.744	30.781	18.107	8.806

2.29 The boundary condition at $r = 2$ now becomes a radiative derivative condition which gives the added equation

$$\frac{u_{7,j} - u_{6,j}}{h} = 1.$$

Hence the iteration by the explicit method is as exercise 2.27 with

$$\frac{u_{6,j+1} - u_{6,j}}{k} = \frac{1}{h^2}(u_{5,j} - 2u_{6,j} + u_{7,j}) + \frac{k}{2r_6h}(u_{7,j} - u_{5,j})$$

with $u_{7,j}$ given by the previous equation. The result is then as given in Table A.10.

The effect of the changed right-hand boundary condition propagates into the solution as the time step proceeds.

Table A.10.

$t \backslash x$	0.75	1.0	1.25	1.5	1.75
0.03125	50.417	15.0	15.0	15.0	15.134
0.0625	50.417	30.495	15.0	15.073	15.206
0.09375	59.455	30.494	22.013	15.111	15.278
0.125	59.455	38.393	22.034	18.365	15.334

The same problem solved implicitly will involve adding the derivative boundary condition as another equation. The new values are in Table A.11.

Table A.11.

$t \backslash x$	0.75	1.0	1.25	1.5	1.75
0.0625	51.439	25.906	18.007	15.034	11.594
0.125	57.057	37.607	23.969	16.285	11.024
0.1875	61.571	42.888	29.023	19.068	11.908
0.25	63.916	46.753	32.661	21.882	13.529

Now the implicit nature of the method (or the use of the information transmitting characteristics t = constant) allows the change in the right-hand boundary condition to be felt throughout the range. The value of $u_{7,j}$ can be found from the derivative condition and has not been tabulated here.

2.30 For spherical problems, the change from the algorithms in the cylindrical cases is that the first derivative term has the factor $2/r$ not $1/r$. The r grid has been made finer to just 0.1 and the inner limit has changed from 0.5 to 0.1. Hence with these changes the explicit code gives the results in Table A.12.

Table A.12.

$t \backslash x$	0.2	0.3	0.4	0.5
	0.6	0.7	0.8	0.9
0.005	36.25	15.00	15.00	15.00
	15.00	15.00	15.00	6.667
0.01	36.25	22.083	15.000	15.00
	15.00	15.00	10.313	6.667
0.015	41.562	22.083	17.656	15.00
	15.00	12.321	10.313	4.583
0.02	41.562	25.625	17.656	16.063
	13.438	12.321	7.969	4.583

The same erratic transmission of the boundary conditions into the solution occurs as is observed in the cylindrical case.

2.31 Following the lines of the cylindrical case with the same changes as in the previous exercise gives the results in Table A.13.

Table A.13.

| $t\backslash x$ | 0.2 | 0.3 | 0.4 | 0.5 |
	0.6	0.7	0.8	0.9
0.01	20.231	16.845	15.610	15.176
	14.966	14.590	13.130	6.981
0.02	21.113	18.371	16.552	15.522
	14.724	13.343	10.163	5.741
0.03	21.786	19.171	17.238	15.745
	14.226	12.044	8.959	4.747
0.04	22.151	19.698	17.623	15.749
	13.732	11.259	8.062	4.230

As expected the step function boundary discontinuity at $r = 0.1$ will be the largest source of error.

2.32 Again the changes to Exercise 2.30 are applied to the solution of Exercise 2.29 to employ spherical polars. The explicit solution gives the results in Table A.14.

Table A.14.

| $t\backslash x$ | 0.2 | 0.3 | 0.4 | 0.5 |
	0.6	0.7	0.8	0.9
0.005	36.25	15.00	15.00	15.00
	15.00	15.00	15.00	15.056
0.01	36.25	22.083	15.00	15.00
	15.00	15.00	15.031	15.086
0.015	41.562	22.083	17.656	15.00
	15.00	15.018	15.049	15.117
0.02	41.562	25.625	17.656	16.063
	15.010	15.028	15.074	15.142

and for the implicit version, the results are in Table A.15.

Table A.15.

$t\backslash x$	0.2	0.3	0.4	0.5
	0.6	0.7	0.8	0.9
0.01	20.231	16.845	15.612	15.187
	15.019	14.830	14.179	11.466
0.02	21.114	18.376	16.572	15.602
	15.012	14.268	12.642	9.875
0.03	21.790	19.193	17.318	15.995
	14.907	13.584	11.648	8.816
0.04	22.168	19.763	17.816	16.231
	14.752	13.072	10.898	8.092

Chapter 3

3.1 The analytic solution is attempted first. The characteristic and the equation along the characteristic satisfies

$$\frac{dx}{x^2 u} = \frac{dy}{e^{-y}} = -\frac{du}{u^2}$$

which gives easily

$$\frac{1}{u} = e^y - A$$

from the second equation. The first then gives

$$\frac{e^y}{e^y - A}\frac{dy}{dx} = \frac{1}{x^2}.$$

Putting $z = e^y$ deals with the y integral to leave

$$\ln(e^y - A) = B - \frac{1}{x}.$$

But $u = 1$ when $y = 0$ to give $A = 0$, and we require the characteristic through (1,0) which gives $B = 1$. Rearranging gives

$$u = e^{1/x - 1}$$

and

$$y = \ln(1/u)$$

from which numerically, $u = 0.9131$ at $(1.1, y)$ and $y = 0.09091$.

Numerically the characteristic is

$$x_R^2 u_R(y_P - y_R) = e^{-y_R}(x_P - x_R)$$

or

$$y_P = x_P - 1$$

to give $y_P = 0.1$ which is in good agreement with the analytic value.

Along the characteristics:

$$x_R^2 u_R(u_P - u_R) = -u_R^2(x_P - x_R)$$

to give $u_P = 1 - 0.1 = 0.9$ again in good agreement with analytic values. Using these values at P to form averages gives

$$1.05^2 * 0.95(y_P) = e^{-0.05}(0.1)$$

and the corrected value $y_P = 0.09082$ which is very good agreement. Continuing,

$$u_p = \frac{-0.95^2(0.1)}{1.05^2 * 0.95} + 1 = 0.9138.$$

3.2 Analytically this problem leads to several blind alleys! The characteristic equation is

$$dx = \frac{\sqrt{u}\,dy}{x} = \frac{du}{2x}$$

which solves easily to give $x^2 = u + A$ and $(y - B)^2 = u$. Consider the characteristic through $(2, 5)$. Suppose this intersects $x = 0$, then as $u = 0$ on this curve we are left with $u = x^2$ and $u = (y - B)^2$. The constant B is fixed to go through $(2, 5)$ and there are two solutions $B = 7$ or $B = 3$. These curves are both a pair of straight lines: $x^2 = (y - 7)^2$ or $x^2 = (y - 3)^2$. In both cases the positive gradient lines intersect $y = 0$ in the region of interest which contradicts our supposition. The negative gradient lines intersect both $y = 0$ and $x = 0$ and so fail to satisfy both conditions. Hence we are forced to the other possibility that $u = 0$ on $y = 0$.

Now we get hyperbolae, and the conditions give $B = 0$, and hence $y^2 = x^2 - A$, which leaves $A = -21$ to force the curve through $(2, 5)$. The solution is then $u = 25$ as $y^2 = u$.

A similar plethora of possible solutions arises at $(5, 4)$. Assume an intersection with $y = 0$ to give $B = 0$ then to go through $(5, 4)$ gives $A = 9$. The curve is $x^2 = y^2 + 9$. This intersects $y = 0$ at $x = \pm 3$ at which $u \neq 0$, so this time the hyperbolae fail to work.

Try the alternative: $u = 0$ when $x = 0$ gives $x^2 = u$ and $(y - B)^2 = u$, and so either $B = -1$ or $B = 9$ to get the curve through $(2, 5)$. We are restricted to $x > 0$ and $y > 0$, and a curve which intersects $x = 0$. The alternative which works is $x = y + 1$ with $u = 25$ as the solution.

The next stage is to make $u = x$ along $y = 0$ and seek the characteristic through $(4, 0)$ where $u = 4$. Hence $A = 12$ and $B^2 = 4$. Hence at $x = 4.05$, $u = x^2 - A = 4.4025$. In the region of interest $y = \sqrt{u} + B = 0.982$.

Finally, and straightforwardly the numerical solution gives $4.05 - 4 = 2/4(y_P - y_R)$ and $y_P = 0.1$ in agreement with the above, and

$$\frac{2(y_P - y_R)}{4} = \frac{u_P - 4}{8}$$

to give $u_P = 4.4$.

3.3 Following the routine approach, the characteristic equations are

$$dx = \frac{dy}{3x^2} = \frac{du}{x + y}$$

which solve easily to give

$$x^3 = y + A \qquad \text{and} \qquad u = x^2/2 + x^4/4 - Ax + B.$$

On $y = 0$, $u = x^2$ and the characteristic through $(3, 10)$ is required. These conditions give rapidly that $A = 8$. The characteristic intersects $y = 0$ at $x = 2$, and then $B = 14$ for $u = 4$. Hence $u(3, 19) = 59/4$.

It is feasible here to get a solution at a specific point by characteristics numerically, rather than being led to a point by the characteristics, as the characteristic equation has no u dependence. Hence the characteristic through $(3, 19)$ can be tracked backwards to where it crosses $y = 0$. Then the equation along the characteristic can be tracked forward to $(3, 19)$ to find u. Let x_i and x_{i-1} be two consecutive points on the characteristic. Then the finite difference form on the characteristic equation in the usual way is:

$$3x^2(x_i - x_{i-1}) = y_i - y_{i-1}$$

and again in the usual way take x to be equal to the value at the trailing end of the step. The y coordinate moves from 19 down to 0. Hence moving in y steps of 1, the iteration is:

$$x_{i-1} = x_i - (y_i - y_{i-1})/(3x_i^2)$$

and $y_i = i$. Hence the sequence of x's yield $x_0 = 2.0314$, in agreement with the analytic result. Now use the equation along the characteristic in the forwards direction to find u. Now averaging can be used for coefficients in the traditional way, and the basic iteration is

$$u_{i+1} = u_i + (y_{i+1} - y_i)\frac{x + y}{3x^2}$$

with $x = (x_{i+1} + x_i)/2$ and $y = (y_{i+1} + y_i)/2$. This iteration gives $u = 14.83$.

3.4 In this example the characteristic equations are

$$\frac{dx}{x-y} = \frac{dy}{u} = \frac{du}{x+y}.$$

Hence in finite difference form

$$\frac{x_P - x_R}{x_R - y_R} = \frac{y_P - y_R}{u_R}$$

to give $y_P = 0.1$. Along the characteristic

$$\frac{y_P - y_R}{u_R} = \frac{u_P - u_R}{x_R + y_R}$$

to give $u_P = 1.1$. Now using these values to give averages yields $y_P = 0.105$ and $u_P = 1.11$.

3.5 This is a direct application of (3.2.5) and shows similar effects to the worked example. The crucial characteristics are again those through (0,0) and $x = 1$, $t = 0$. The results are shown in Table A.16.

Table A.16.

$t\backslash x$	0.25	0.5	0.75	1.0	1.25	
	1.50	1.75	2.0	2.25	2.5	
	2.75	3.0	3.25	3.5	3.75	4.0
0.125	−0.1094	−0.2344	−0.2344	−0.3906	−0.1250	
	−0.3750	−0.6250	−0.8750	−1.1250	−1.3750	
	−1.625	−1.875	−2.125	−2.375	−2.625	−2.875
0.250	−0.0059	−0.1875	−0.2588	−0.1016	−0.0615	
	−0.2500	−0.500	−0.750	−1.000	−1.250	
	−1.500	−1.750	−2.000	−2.250	−2.500	
0.375	0.1128	−0.1104	−0.2517	−0.1655	−0.0530	
	−0.1481	−0.375	−0.625	−0.875	−1.125	
	−1.375	−1.625	−1.875	−2.125		
0.5	0.2390	−0.0091	−0.2095	−0.2119	−0.0833	
	−0.0840	−0.2587	−0.5	−0.75	−1.000	
	−1.25	−1.5	−1.75			

These results can be compared with the same problem solved using the Wendroff method in Exercise 3.9. Note that the right-hand side loses an x value at every step.

Hence to compute u at $x = 1$ and $t = 4$ requires 32 time steps and therefore x must range to 9.0.

Table A.17.

$t\backslash x$	0.25	0.5	0.75	1.0	1.25	
	1.50	1.75	2.0	2.25	2.5	
	2.75	3.0	3.25	3.5	3.75	4.0
0.125	0.1016	0.3203	0.6641	1.1328	1.7266	
	2.4453	3.2891	4.2578	5.3516	6.5703	
	7.9141	9.3828	10.977	12.695	14.539	16.508
0.250		0.3960	0.7643	1.2560	1.8713	
	2.6101	3.4725	4.4584	5.5678	6.8008	
	8.1573	9.6374	11.241	12.696	14.819	
0.375			0.8407	1.3413	1.9626	
	2.7044	3.5670	4.5501	5.6539	6.8783	
	8.2234	9.6891	11.275	12.982		
0.5				1.3626	1.9692	
	2.6922	3.5316	4.4873	5.5595	6.7481	
	8.0531	9.4744	11.012			

3.6 This is an extended problem and will use (3.2.17). The required partial derivatives with respect to x and t of a and c are trivial. The direct application of the method gives the results in Table A.17.

In this example there is no condition along $x = 0$, and now the computed points involve the loss of an x value at both ends of the range. These values can be compared with the Wendroff approach in Exercise 3.10.

3.7 This problem is again similar to the previous one. Now the range of x used at $t = 0$ has to be fixed to end up at the point $x = 0.5$, $t = 0.5$. As an x value is lost at every step the x range at $t = 0$ will be $-0.5 < x < 1.5$. These values are shown in Table A.18.

3.8 The problem of extending the Lax–Wendroff formula further into the non-linear regime occurs when a and c are functions of u, and not just x and t. In (3.2.17), $\partial a/\partial x$, $\partial a/\partial t$, $\partial c/\partial x$ and $\partial c/\partial t$ are required and will lead to further u derivatives and hence a non-linear difference equation. However if a is just a function of x and t as before, but just c depends on u, and then in the form that both $\partial c/\partial x$ and $\partial c/\partial t$ are linear in u, then the terms

$$k^2 c_t/2 \qquad \text{and} \qquad -ah^2 c_x/2$$

in (3.2.17) will yield terms involving just $\partial u/\partial x$, which can be differenced directly, and $\partial u/\partial t$ which will need conversion to an x derivative using (3.2.15) and then differencing can be employed. Much beyond this, any further advance would seem to yield diminishing returns as

Table A.18.

$t\backslash x$	-0.4 0.2 0.8	-0.3 0.3 0.9	-0.2 0.4 1.0	-0.1 0.5 1.1	0.0 0.6 1.2	0.1 0.7 1.3	1.4
0.05	0.9818	0.9866	0.9915	0.9964	1.0012	1.0061	
	1.0110	1.0159	1.0207	1.0256	1.0305	1.0354	
	1.0403	1.0451	1.0500	1.0549	1.0597	1.0646	1.0695
0.1		0.9767	0.9862	0.9957	1.0052	1.0147	
	1.0242	1.0337	1.0432	1.0527	1.0622	1.0718	
	1.0813	1.0908	1.1003	1.1098	1.1193	1.1288	
0.15			0.9842	0.9981	1.012	1.0259	
	1.0398	1.0538	1.0677	1.0816	1.0955	1.1094	
	1.1233	1.1373	1.1512	1.1651	1.1790		
0.2				1.0039	1.0220	1.0401	
	1.0582	1.0763	1.0944	1.1125	1.1305	1.1486	
	31.1667	1.1848	1.2029	1.2210			
0.25					1.0352	1.0573	
	1.0794	1.1014	1.1235	1.1455	1.1676	1.1897	
	1.2117	1.2338	1.2559				
0.3						1.0778	
	1.1036	1.1294	1.1553	1.1811	1.2069	1.2327	
	1.2585	1.2844					
0.35							
	1.1311	1.1605	1.1899	1.2193	1.2487	1.2780	
	1.3074						
0.4							
		1.1948	1.2276	1.2603	1.2931	1.3258	
0.45							
			1.2684	1.3043	1.3403		
0.5							
				1.3516			

non-linear difference equations will result in considerable increase in computational work.

3.9 Though the Wendroff method needs conditions along $x = 0$ to get it started, there is no loss of computed values at the endpoints as we have seen for Lax–Wendroff. Hence the x values need only run from 0 to 2.5. This is a direct application of the method and gives the results in Table A.19.

Table A.19.

$t\backslash x$	0.25	0.5	0.75	1.0	1.25
	1.50	1.75	2.0	2.25	2.5
0.125	-0.1042	-0.2361	-0.2338	-0.1096	-0.0468
	-0.4011	-0.6163	-0.8780	-1.1240	-1.3753
0.250	-0.0069	-0.1852	-0.2523	-0.1862	-0.0631
	-0.1595	-0.5534	-0.7245	-1.0111	-1.2455
0.375	0.1273	-0.0972	-0.2368	-0.2354	-0.1288
	-0.0733	-0.3195	-0.6884	-0.8321	-1.149
0.5	0.2508	-0.0113	-0.1800	-0.2554	-0.1932
	-0.0889	-0.1502	-0.4989	-0.7994	-0.9486

As expected from our knowledge of characteristics, it will be along the two characteristics through the discontinuities at $(0,0)$ and $x = 1$, $t = 0$ that the maximum error will appear, and indeed the largest discrepancy between the two methods. However outside these regions good agreement is observed.

3.10 This example shows the problem of using Wendroff's method. An analytic result is needed along $x = 0$ to allow the method to operate. It is probably true that if such a solution can be found the whole problem could be completed analytically, but at least the method is demonstrated. The characteristic equations are

$$dt = \frac{dx}{3t^2} = \frac{du}{t + x}$$

to give

$$t^3 = x + A \quad \text{and} \quad u = t^2/2 + t^4/4 - tA + B.$$

To get the solution at any point $(0, T)$ say, then $A = T^3$ and so $u = t^2/2 + t^4/4 - T^3 t + B$. This characteristic through $(0, T)$ intersects $t = 0$ at $x = -T^3$ and hence as $u = x^2$ on this line, $u = T^6$. Hence the solution along $x = 0$ is $u = T^2/2 - 3T^4/4 + T^6$ for $t = T$. The Wendroff method was formulated in terms of x and y, so here use t as the y variable and leave x as it is. The results are shown in Table A.20.

3.11 As with the Lax–Wendroff method, any dependence of either a or b on u will result in non-linear finite differences, which will require the solution of non-linear equations for each step. Stability analysis will not be straightfoward.

Table A.20.

$t \backslash x$	0.25	0.5	0.75	1.0	1.25	
	1.50	1.75	2.0	2.25	2.5	
	2.75	3.0	3.25	3.5	3.75	4.0
0.125	0.0999	0.3172	0.6595	1.1268	1.7191	
	2.4363	3.2786	4.2459	5.3382	6.5555	
	7.8978	9.3691	10.957	12.675	14.517	16.484
0.250	0.1449	0.3860	0.7523	1.2434	1.8579	
	2.6009	3.4671	4.4583	5.5745	6.8157	
	8.1820	9.6731	11.289	13.031	14.897	16.888
0.375	0.1872	0.4403	8189	1.3222	1.9507	
	2.7041	3.5825	4.5859	5.7143	6.9678	
	8.3462	9.8496	11.478	13.231	15.110	17.113
0.5	0.2181	0.4658	0.8388	1.3370	1.9599	
	2.7081	3.5810	4.5791	5.7022	6.9502	
	8.3233	9.8213	11.444	13.192	15.065	17.063

If c depends on u linearly then there will be an extra term in $u_{i,j}$ to account for and the method should in principle still work.

3.12 The solutions here are applications of the finite difference form along the characteristic given by (3.3.8) and the equations for the first derivatives along these characteristics given by (3.3.13) and (3.3.14). The final solution value follows from (3.3.16). Iterative improvement using averages then employs (3.3.17) to (3.3.20). In this first exercise, the characteristic gradients are $f = 2x$ and $g = -2x$. The characteristics satisfy the finite difference forms

$$y_R = 0.6(x_R - 0.3) \qquad \text{and} \qquad y_R = -0.8(x_R - 0.4)$$

with solution $x_R = 0.35714$ and $y_R = 0.034286$. The equations for p_R and q_R are

$$0.6(p_R - 0.6) - 0.36q_R = 0 \qquad \text{and} \qquad -0.8(p_R - 0.8) - 0.64q_R = 0$$

with solution $p_R = 0.68571$ and $q_R = 0.14286$. Equation (3.3.16) gives $u_R = 0.12918$.

Repeating the procedure using these first R values to form averages along the characteristics gives $x_R = 0.35354$ and $y_R = 0.035180$. The first derivatives are $p_R = 0.69314$ and $q_R = 0.14068$ with the improved u value of 0.12709.

3.13 The R coordinate satisfies

$$y_R = 1.4(x_R - 0.4) \qquad \text{and} \qquad y_R = -1.5(x_R - 0.5)$$

with solution $x_R = 0.45172$ and $y_R = 0.072414$. The equations for p_R and q_R are

$$1.4(p_R - 1.0) - 2.24(q_R - 1.0) = 0$$
$$\text{and} \quad -1.5(p_R - 1.0) - 2.25(q_R - 1.0) = 0$$

with solution $p_R = 1.0$ and $q_R = 1.0$. Equation (3.3.16) gives $u_R = 0.52414$.

Repeating the procedure using averages along the characteristics gives $x_R = 0.45167$, $y_R = 0.073668$, $p_R = 1.0$ and $q_R = 1.0$ with the improved u value of 0.52533.

Computing the values at S using Q and W gives

$$y_S = 1.5(x_S - 0.5) \quad \text{and} \quad y_S = -1.4(x_S - 0.6)$$

with solution $x_S = 0.54828$ and $y_S = 0.072414$. The equations for p_S and q_S are

$$1.5(p_S - 1.0) - 2.25(q_S - 1.0) = 0$$
$$\text{and} \quad -1.4(p_S - 1.0) - 2.24(q_S - 1.0) = 0$$

with solution $p_S = 1.0$ and $q_S = 1.0$. Equation (3.3.16) gives $u_S = 0.62069$.

Again repeating the procedure using averages along the characteristics gives $x_R = 0.54833$, $y_R = 0.073668$, $p_R = 1.0$ and $q_R = 1.0$ with the improved u value of 0.6220.

Finally using R and S to get T gives

$$y_T - 0.073662 = 1.4517(x_T - 0.45167)$$
$$\text{and} \quad y_T - 0.073662 = -1.4517(x_T - 0.54833)$$

with solution $x_R = 0.5000$ and $y_R = 0.14383$. The equations for p_R and q_R are

$$1.4517(p_R - 1.0) - 2.2477(q_R - 1.0) = 0$$

and

$$-1.4517(p_R - 1.0) - 2.2477(q_R - 1.0) = 0$$

with solution $p_R = 1.0$ and $q_R = 1.0$. Equation (3.3.16) gives $u_R = 0.64383$.

Repeating the procedure using averages along the characteristics gives $x_R = 0.5000$, $y_R = 0.14500$, $p_R = 1.0$ and $q_R = 1.0$ with the improved u value of 0.6450.

3.14 In this problem the level of non-linearity has been increased by making the coefficients functions of u and p. This causes no problems for the characteristic method, as all the initial values are known at P and Q to allow the solution to progress. The equations are now

$$y_R = 1.25(x_R - 0.5) \quad \text{and} \quad y_R = -1.2(x_R - 0.6)$$

with solution $x_R = 0.5489$ and $y_R = 0.061224$. The equations for p_R and q_R are

$$1.25(p_R - 1.0) - 1.25(q_R - 1.0) + 0.5(0.061224) = 0$$

and

$$-1.2(p_R - 1.2) - 1.632(q_R - 1.0) + 0.6(0.061224) = 0$$

with solution $p_R = 1.0836$ and $q_R = 1.081$. Equation (3.3.16) gives $u_R = 1.3656$.

Repeating the procedure using averages along the characteristics gives $x_R = 0.54661$, $y_R = 0.060958$, $p_R = 1.0862$ and $q_R = 1.060$ with the improved u value of 1.3628.

3.15 To get a quadratic fit of the form

$$y = \alpha x^2 + \beta x + \gamma$$

will require three conditions to fix the constants. To this end, force y and its two derivatives to be correct at the point P say to give the equations:

$$\begin{aligned} y_P &= \alpha x_P^2 + \beta x_P + \gamma \\ y_P' &= 2\alpha x_P + \beta = f_P \\ y_P'' &= 2\alpha. \end{aligned}$$

The first two equations are the same ones as used for the linear fit, f_P being the gradient of the f characteristic through P. But f is known from the characteristic equation and hence its derivative can also be found to effect the third equation.

Similarly the characteristic through Q can be found using the other characteristic solution g. The intersection will involve the solution of a quadratic and care is needed to choose the correct root.

For highly non-linear problems as in Exercise 3.14, the coefficients depend on u and its derivatives and then the second derivative required for a quadratic fit will involve unknown derivatives of u, and the method will not be feasible.

3.16 The equations are now

$$y_R = 1.00x_R \quad \text{and} \quad y_R = -0.3894(x_R - 0.5)$$

with solution $x_R = 0.14013$ and $y_R = 0.14013$. The equations for p_R and q_R are

$$(2.0)(1.0)p_R - q_R = 0 \quad \text{and} \quad (2.0)(-0.3894)(p_R - 4.0) - 0.60653q_R = 0$$

with solution $p_R = 1.5288$ and $q_R = 3.0576$. Equation (3.3.16) gives $u_R = 0.32135$.

Repeating the procedure using averages along the characteristics gives $x_R = 1.5378$, $y_R = 1.5228$, $p_R = 1.4123$ and $q_R = 2.8413$ with the improved u value of 0.32493.

3.17 For the problems in this section the formulae being applied are (3.4.6) and (3.4.7). The first problem is a more stringent test of the method than the worked example in that the string is displaced initially at the midpoint and then released giving a discontinuity in the derivative at $x = 1/2$ and $t = 0$. The results are shown in Table A.21 for the range $0 < x < 0.5$, the other half of the range being symmetric.

Table A.21.

$t\backslash x$	0.1	0.2	0.3	0.4	0.5
0.05	0.02500	0.0500	0.07500	0.1000	0.1875
0.1	0.02500	0.0500	0.7500	0.0984	1.0312
0.15	0.02500	0.0500	0.07461	0.09219	0.08516
0.2	0.0250	0.04990	0.07246	0.07978	0.07070
0.25	0.02498	0.04922	0.06650	0.06328	0.06079
0.3	0.02477	0.04680	0.05542	0.04696	0.05212
0.35	0.02388	0.04102	0.04007	0.03405	0.04088
0.4	0.02130	0.03072	0.02344	0.02434	0.02621
0.45	0.01576	0.01625	0.00887	0.01488	0.01061
0.5	0.0064	−0.0002	−0.0023	0.0029	−0.0029

The frequency is close to the worked example with a parabolic initial shape.

3.18 Using the same approach as the previous example yields the results shown in Table A.22.

3.19 The equations need modifying to account for the damping term to become

$$u_{i,j+1}\left(1 + \frac{\kappa k}{2}\right) = r^2 u_{i-1,j} + 2(1-r^2)u_{i,j} + r^2 u_{i+1,j} - u_{i,j-1}\left(1 - \frac{\kappa k}{2}\right)$$

Table A.22.

$t\backslash x$	0.1	0.2	0.3	0.4	0.5
0.05	0.03815	0.07257	0.09989	0.1174	0.12347
0.1	0.03675	0.06990	0.09621	0.1131	0.1189
0.15	0.03444	0.06551	0.09017	0.1060	0.11146
0.2	0.03129	0.05953	·0.08193	0.9631	0.10127
0.25	0.02738	0.05208	0.07168	0.08427	0.08860
0.3	0.02280	0.04336	0.05968	0.070157	0.07377
0.35	0.01765	0.03358	0.04621	0.05433	0.05713
0.4	0.01208	0.02298	0.03162	0.03718	0.03909
0.45	0.00621	0.01181	0.01626	0.01911	0.02009
0.5	0.0002	0.0004	0.0005	0.0006	0.0006

for a general point and

$$u_{i,1}\left(1 + \frac{\kappa k}{2}\right) = r^2 f_{i-1} + 2(1 - r^2)f_i + r^2 f_{i+1} - 2kg_i\left(1 - \frac{\kappa k}{2}\right)$$

for the first computed row. The damped equation then gives the results in Table A.23, again with just every other step recorded.

Table A.23.

$t\backslash x$	0.1	0.2	0.3	0.4	0.5
0.1	0.04029	0.07478	0.09960	0.1149	0.1194
0.2	0.03032	0.06051	0.08457	0.09921	1.0396
0.3	0.02023	0.04072	0.06068	0.07438	0.07882
0.4	0.01063	0.02049	0.03112	0.04083	0.04428
0.5	0.0006	0.0011	0.0008	0.0017	0.0021
0.6	−0.0094	−0.0187	−0.0283	−0.0376	−0.0419
0.7	−0.0193	−0.0387	−0.0566	−0.0717	−0.0782
0.8	−0.0292	−0.0573	−0.0817	−0.0971	−0.1014
0.9	−0.0380	−0.0722	−0.0979	−0.1115	−0.1141
1.0	−0.0432	−0.0789	−0.1023	−0.1151	−0.1189
1.1	−0.0414	−0.0740	−0.0963	−0.1097	−0.1140
1.2	−0.0317	−0.0593	−0.0815	−0.0953	−0.0987

The damping is quite marked compared with the original worked example. The central swing in the negative direction reaches to just −0.1189, whereas the damp-free motion returns to the starting displacement of −0.125.

Chapter 4

In this chapter most of the exercises involve setting up quite moderate-sized matrices for Laplace's equation and then solving these numerically. Except for a few problems, the matrices involved have order up to around 100, and hence it is imperative to program the solutions. None of the problems is too large for straight Gaussian elimination, though experimentation with iterative methods may be encouraged after the final section is covered. For the solutions here, a Gauss routine was used with a series of setup routines which fix the matrix elements. These are first all set to zero, and the diagonal elements set to -4.0 for the standard computational molecule. For these solutions the grid system has been numbered from the bottom left-hand grid point, row by row to the top right. Deviations from this are outlined in the solutions. The four unit matrix elements from the molecule Figure 4.1 can also be coded in using loops. In a rectangular grid the element N will be element $i + r$ where i is the current element and r is the number of elements in a row. Similarly the element S is element $i - r$. In general the element E is $i + 1$ and W is $i - 1$, and in all cases if the required element is on the boundary, then the right hand side vector, say $\mathbf{b}$, is updated accordingly.

Hence there is a programmed record of the matrix elements used, and variations on the problem can be easily encompassed. For irregular boundaries, the relevant grid lengths are computed (with care to remember that the actual grid length is θh) and the special values added to the matrix using (4.2.5) and (4.2.6). With derivative boundary conditions, extra equations will arise whose elements are added to the matrix.

Many of the problems can take advantage of symmetry so that just under half the equations can be eliminated trivially using the relevant relations $x_i = x_j$. If the symmetry is ignored, the resulting symmetric values can act as a check. The grids here are quite crude so the accuracy obtained will be quite low.

Care is needed not to forget the grid size values h and k. With any non-homogeneous term these lengths will appear explicitly and not cancel. So watch out for Poisson's equation, and derivative boundary conditions such as $\partial\phi/\partial x = 1$.

4.1 This is a standard square grid with 3 rows of 3 points. The grid elements are as described above and the right-hand side will then be $-2h^2$ after multiplying through by h^2. Symmetry will reduce the unknowns to just 3, that is with the numbering as above, $\phi_1 = \phi_3 = \phi_7 = \phi_9$ at the corners, and $\phi_2 = \phi_4 = \phi_6 = \phi_8$ at the edge midpoints. The solution is $\phi_1 = 11/32$, $\phi_2 = 7/16$ and at the centre $\phi_5 = 9/16$.

4.2 This is another standard square grid with 4 rows of 4 points. A typical row has $A_{8,4} = 1$, $A_{8,7} = 1$, $A_{8,8} = -4$, $A_{8,9} = 0$ (as this element is on the boundary and yields $-\frac{3}{5}$ to b_8 from conditions (iv) which imply $\phi = 2y$) and $A_{8,12} = 1.0$. The solution now is as given in Table A.24.

Table A.24.

n	ϕ_n	n	ϕ_n
1	0.14361	9	0.54006
2	0.21327	10	0.68921
3	0.26855	11	0.83639
4	0.32079	12	0.99751
5	0.32115	13	0.78988
6	0.44094	14	0.93945
7	0.54012	15	1.11873
8	0.61461	16	1.33906

4.3 Use the same setup routine as in Exercise 4.2, but add elements ϕ_{17}, ϕ_{18}, ϕ_{19} and ϕ_{20} on the $\phi = 0$ boundary at $y = 1/5$ to $y = 4/5$, and similarly on $\phi = 1$ add ϕ_{21} to ϕ_{24}. The four derivative boundary equations for the latter four points are $\phi_4 = \phi_{21}$ up to $\phi_{16} = \phi_{24}$ which just give 1 and -1 in the matrix elements, and on $\phi = 0$ the first edge element gives

$$\frac{\phi_1 - \phi_{17}}{1/5} = 1$$

and the last gives

$$\frac{\phi_{13} - \phi_{20}}{1/5} = 1$$

where the h value is needed explicitly, and there is now a contribution of $-1/5$ to b_{17} to b_{20}. The b elements will also change and there will be extra 1's for elements 1, 5, 9, 13 and 4, 8, 12, 16 which now have unknown boundary elements for their west and east neighbours respectively, rather than known boundary elements. These equations solve to give the results in Table A.25

Table A.25.

n	ϕ_n	n	ϕ_n
1	0.029339	9	0.38778
2	0.11960	10	0.51604
3	0.16090	11	0.57943
4	0.17550	12	0.57277
5	0.16842	13	0.67887
6	0.28815	14	0.80882
7	0.34852	15	0.88039
8	0.36560	16	0.77329

4.4 There are only three equations to demonstrate the use of a fixed grid on a non-standard boundary. With the usual numbering these are:

$$-4\phi_1 + \phi_2 + \phi_3 = -0.25$$
$$\phi_1 - 4\phi_2 = -0.75$$
$$\phi_1 - 4\phi_3 = -1.75$$

with solution $\phi_1 = 0.25$, $\phi_2 = 0.25$ and $\phi_3 = 0.5$.

4.5 The inverted L-shape has its nodes numbered as usual from the bottom left with points 1 and 2, 3 and 4, and 5 and 6 making up the rows for $1/2 < x < 1$ and $y < 1/2$. From then on there is full width to give rows with grid numbers 7 to 11, and 12 to 16. The matrix is then conventional but the narrow bottom rows mean that the north and south unit elements are only 2 elements from the diagonal whereas for elements 5 onwards these are the full 5 elements from the diagonal. As a guide row 3 has elements $a_{1,1} = 1.0$, $a_{3,3} = -4.0$, $a_{3,4} = 1$ and $a_{3,5} = 1.0$, with the east element being a zero boundary point. For element 11 the row is $a_{11,6} = 1.0$, $a_{11,10} = 1.0$, $a_{11,11} = -4$, and $a_{11,16} = 1.0$. The west element is a boundary point giving $b_{11} = -1$. The solution is given in Table A.26.

Table A.26.

n	ϕ_n	n	ϕ_n
1	0.34447	5	0.43227
2	0.67659	6	0.73623
3	0.36799	10	0.62486
4	0.69520	11	0.81746
		16	0.90873

The other points follow by symmetry.

4.6 The diagonal just adds three points to the system namely x_{17} at $\{1/3, 1/2\}$, x_{18} at $\{1/2, 1/2\}$ (the original indented corner point) and x_{19} at $\{1/2, 1/3\}$. The modified table is in Table A.27.

Again symmetry gives the other values.

4.7 This is just a larger version of Exercise 4.5 but with a total of 85 grid points, that is from elements 1 to 5 in the bottom narrow row, rising to elements 21 to 25 for $y = 5/12$ and $1/2 < x < 1$. The top five rows each have 11 elements to leave the final row with elements 75 to 85. There is again symmetry from the indented corner to the top right-hand corner. The solution below is just for those elements corresponding to elements in Exercise 4.5.

Table A.27.

n	ϕ_n	n	ϕ_n
1	0.382433	6	0.81139
2	0.70026	10	0.74681
3	0.49614	11	0.87377
4	0.75195	16	0.93689
5	0.61986	17	0.23030
		18	0.42508

Table A.28.

n	ϕ_n	n	ϕ_n
7	0.34535	27	0.45486
9	0.67762	29	0.74554
17	0.37296	49	0.63480
19	0.69914	51	0.82227
		73	0.91080

Hence we see that around two figures were obtained in the course grid calculation.

4.8 In moving from Exercises 4.6 to 4.8 add points 86 to 95 to fill in the corner. These have been counted row-wise downwards with $x = 1/4$, $y = 1/2$ being point 86, $\{1/2, 1/2\}$ is 89 to make up the first row, then 90 to 92 for the row below, down to the single row element 95 at $\{1/2, 1/4\}$.

The equivalent table to Table A.27 is Table A.29.

Table A.29.

n	ϕ_n	n	ϕ_n
7	0.39308	29	0.81333
9	0.70449	49	0.74948
17	0.50123	51	0.87503
19	0.75468	73	0.93750
27	0.62403	87	0.23896
		89	0.43264

Again the comparison with Exercise 4.6 shows that the course grid was achieving a little under two-figure accuracy.

4.9 The final problem in this group has a locally fine grid above $y = 0$ where the most rapid changes in ϕ are expected. Numbering according to the usual style gives the bottom row with elements 1 to 7, and the second row up from 8 to 14 to complete the fine grid. Along $y = 1/2$

are the elements 15, 16 and 17 and then along $y = 3/4$ are 18, 19 and
20. Points 8, 10, 12 and 14 are the ones to need interpolation having
no north neighbours. The rule given in the exercise will be used for
each of these points to set the matrix elements. Hence $a_{10,10} = 1.0$,
$a_{10,9} = -0.5$ and $a_{10,11} = -0.5$ are the settings for the interpolation
given in the exercise. Elements 14 and 8 will differ from this in being
at the ends of the line and therefore contributing to b, though the
contribution from the boundary conditions is zero.

Hence the resulting values, accounting for symmetry, are given in Table
A.30.

Table A.30.

n	ϕ_n	n	ϕ_n
1	0.32199	10	-0.21035
2	0.50032	11	0.27997
3	0.45999	15	0.096914
4	0.54999	16	0.13129
8	-0.21234	18	0.037068
9	0.21930	19	0.051356

4.10 The same routine can be followed as in the previous section for most
points on the grids in this section with any curved boundary points
using (4.2.5) and (4.2.6) as relevant. Hence in the solutions here the
required θ value and any special details will be reported but routine
application of the Figure 4.1 computational molecule will be assumed.
Again results will be quoted taking symmetry into account. In Exercise
4.10 the only special points are 2 and 7 with $\theta = 1/2$ in the y direction.
The derivative condition gives $x_4 = x_5$ as an extra equation. This is a
non-homogeneous problem so care is needed to remember to multiply
through by h^2 in fixing the b values. Hence for point 2 the matrix
elements are: $a_{2,2} = -2.0 - 4.0$ (the -2.0 from the conventional x
direction and the -4.0 from $-2/\theta$), $a_{2,1} = 1.0$ (conventionally), $a_{2,4} =$
$4/3$ (being $2/(1+\theta)$). The contribution to b_2 is $-0.5 - \frac{8}{3}\frac{3}{4} + 4\frac{1}{4}\frac{1}{16^2}$. At
$(3/4, 1/4)$, $\phi = 2y = 1/2$ and the finite difference weight is 1.0 to give
the -0.5. At $(1/2, 3/8)$, $\phi = 3/4$ and the curved boundary weight is
$2/\theta/(1+\theta) = 8/3$, and finally the non-homogeneous part gives $4x^2y^2h^2$
with $x = 1/2$ and $y = 1/4$, remembering that the factor h^2 has been
multiplied through. Then $\phi_1 = 0.91273$, $\phi_2 = 0.74574$, $\phi_3 = 0.90617$
and $\phi_4 = 0.79922 = \phi_5$, with symmetry giving the other points.

4.11 The grid numbering is again from the bottom left and row by row.
The $y = 1/3$ line has therefore numbers 1 to 5 and the $y = 2/3$ line
numbers 6 to 10. There are only two θ values to find, the values at
1, 5, 7 and 9 being t say and at 6 and 10 being s. Hence the circle is

$x^2 + y^2 = 1$, so $t = \left(\sqrt{(8/9)} - 2/3\right) * 3$, remembering that the actual length of the arm of the grid is $h\theta$, hence the multiplying 3. Similarly $s = \left(\sqrt{(5/9)} - 2/3\right) * 3$. On the boundary, $\phi = \cos^2 \theta$, to give 8/9 on $y = 1/3$, 4/9 on $y = 2/3$, 4/9 on $x = 2/3$ and 1/9 on $x = 1/3$. The non-standard equations are then at points 5, 9 and 10 and are:

$$\phi_{10} - 2\phi_5 - \frac{2}{t}\phi_5 + \frac{2}{1+t}\phi_4 = -1 - \frac{8}{9}\frac{2}{t(1+t)}$$

$$\phi_{10} - 2\phi_9 + \phi_8 - \frac{2}{t}\phi_9 + \frac{2}{(1+t)}\phi_4 = -\frac{2}{t(1+t)}\frac{1}{9}$$

$$-\frac{2}{s}\phi_{10} + \frac{2}{1+s}\phi_9 - \frac{2}{s}\phi_{10} + \frac{2}{1+s}\phi_5 = -\frac{2}{s(1+s)}\left(\frac{4}{9} + \frac{5}{9}\right).$$

The other equations have the standard weights and the solution yields $\phi_3 = 0.71259$, $\phi_4 = 0.73105$, $\phi_5 = 0.79143$, $\phi_8 = 0.38824$, $\phi_9 = 0.42019$ and $\phi_{10} = 0.52021$.

4.12 Equation (4.2.7) is now being solved with $\phi = 1$ on $\theta = 0$, $r = 0$ and $\theta = \pi$. Letting the r step be $h = 1/3$ and the θ step be $k = \pi/4$, will mean that the h and k factors cannot be cancelled through and will appear in each equation. The grid is regular so there are only standard finite difference approximations. The row $\theta = \pi/4$ has grid points 1 and 2, the row $\theta = \pi/2$ has points 3 and 4 and the row $\theta = 3\pi/4$ has points 5 and 6. In this way points 3 and 4 will coincide with 3 and 8 of the previous exercise and allow easy comparison. A typical row at point 3 is

$$-\phi_3(2/h^2 + 9 * 2/k^2) + \phi_4(1/h^2 + 3/2h) + \phi_1(9/k^2) + \phi_5(9/k^2)$$
$$= -1/h^2 + 3/2h$$

where the $1/r^2$ gives the factors of 9, and the boundary value is $\phi = 1$. These equations solve to give $\phi_1 = 0.80125$, $\phi_2 = 0.63353$, $\phi_3 = 0.69640$ and $\phi_4 = 0.36854$, and values of ϕ_3 and ϕ_4 here compare well with the corresponding ϕ_3 and ϕ_8 from the previous exercise.

4.13 This problem is very similar to the worked example in the text and the one non-standard point is 3. The θ value at this point is $t = (\cos(\pi/12) - 3/4) * 4$. The equations for points 3 and 4 for example are:

$$\phi_3(-2/th^2 - 2/k^2 * 16/9 - (1+t)/t/h * 4/3)$$
$$+\phi_2(2/h^2/(1+t) - t/(1+t)/h * 4/3) + \phi_6(1/k^2 * 16/9)$$
$$= -2/h^2/t/(1+t) * 1/4 - 1/h/t/(1+t) * 4/3 * 1/4$$
$$\phi_4(-2/h^2 - 2/k^2 * 16)$$
$$+\phi_5(1/h^2 + 1/2/h * 4) + \phi_7/k^2 * 16 + \phi_1/k^2 * 16$$
$$= 0$$

where the terms have not been simplified so show the $1/r$ and $1/r^2$ terms. Also for point 3, $\phi = 1/4$ on the boundary. The solution then is given in Table A.31

Table A.31.

n	ϕ_n	n	ϕ_n
1	0.073764	5	0.27849
2	0.14260	6	0.39972
3	0.20344	7	0.20316
4	0.14298	8	0.40071
		9	0.58615

4.14 The three points, 1, 2 and 3, lie on $y = 1/2$ at $x = 1/2$, $x = 1$ and $x = 3/2$. The curved boundary formula requires the values of t_1 at point 1, t_2 at 2, t_3 at 3 all in the y direction and s_1 at 3 in the x direction. The ellipse has $y = \sqrt{1 - x^2/4}$ and gives $t_1 = \left(\sqrt{15/16} - 0.5\right) * 2$, $t_2 = \left(\sqrt{3/4} - 0.5\right) * 2$ and $t_3 = \left(\sqrt{7/16} - 0.5\right) * 2$. Similarly $x = 2\sqrt{1 - y^2}$ and $s_1 = \left(\sqrt{3} - 1.5\right) * 2$. At points 1, 2 and 3 the equations are then

$$\frac{1/2 - 2\phi_1 + \phi_2}{1/4} + \frac{1/2 t_1 - (1 + t_1)\phi_1 + 1}{1/4 t_1 (1 + t_1)/2} = 0$$

$$\frac{\phi_1 - 2\phi_2 + \phi_3}{1/4} + \frac{t_2 - (1 + t_2)\phi_2 + 1}{1/4 t_2 (1 + t_2)/2} = 0$$

$$\frac{s_1 \phi_2 - (1 + s_1)\phi_3 + 1}{1/2 s_1 (1 + s_1)} + \frac{3/2 t_3 - \phi_3 (1 + t_3) + 1}{t_3 (1 + t_3)/2} = 0$$

with solution $\phi_1 = 0.63732$, $\phi_2 = 0.77467$ and $\phi_3 = 0.93471$. This solution can be expected to be quite crude with so few points.

4.15 Numbering in the usual way will give points 1 to 7 along $y = 1/4$, 8 to 13 along $y = 1/2$ and 14 to 18 along $y = 3/4$. There are now vertical non-regular intersections t_1 to t_7 at points 14 to 18 in the top row and 13 and 7 in the lower two rows respectively. There are also three horizontal intersections at 18, 13 and 7, namely s_1, s_2 and s_3. By the same geometry as in the previous exercise these have values:

$$t_1 = \left(\sqrt{63/64} - 0.75\right) * 4, \qquad t_2 = \left(\sqrt{15/16} - 0.75\right) * 4$$

$$t_3 = \left(\sqrt{55/64} - 0.75\right) * 4, \qquad t_4 = \left(\sqrt{3/4} - 0.75\right) * 4$$

$$t_5 = \left(\sqrt{39/64} - 0.75\right) * 4, \qquad t_6 = \left(\sqrt{7/16} - 0.5\right) * 4$$

$$t_7 = \left(\sqrt{15/64} - 0.25\right) * 4, \qquad s_1 = \left(\sqrt{7/4} - 1.25\right) * 4$$

$$s_2 = \left(\sqrt{3} - 1.5\right) * 4, \qquad s_3 = \left(\sqrt{15/4} - 1.75\right) * 4.$$

The now conventional curved boundary formulae are used at these points with standard representations at the others to yield the solution in Table A.32.

Table A.32.

n	ϕ_n	n	ϕ_n
1	0.25587	10	0.38465
2	0.31047	11	0.44606
3	0.38959	12	0.56681
4	0.48822	13	0.86561
5	0.61725	14	0.50561
6	0.78897	15	0.66298
7	0.92301	16	0.74614
8	0.33801	17	0.83279
9	0.34641	18	0.95555

The three results which correspond to those in Exercise 4.14 are x_9, x_{11} and x_{13} and these show the crudity of the previous calculation.

4.16 On $y = 1/3$ are the row of points 1 to 5, on $y = 2/3$ the points 6, 7, 8, 9 and 10. These form the basic crude grid. The fine grid has points 13 to 17 on $y = 5/6$ with $-1/3 < x < 1/3$ and the intercepts with $y = 2/3$ by $x = \pm 1/6$ are points 11 and 12. Hence point 11 is half way between 7 and 8 of the course grid. Let s be the intercept at both points 5 and 10 in both directions, and t be the vertical intercept from points 9 and 7 to the boundary, and the horizontal intercepts at points 1 and 5. On the fine grid t_1 and t_2 are the intercepts at the points 13 and 14 respectively, with identical values at 16 and 15. These four values are:

$$t = \left(\sqrt{8/9} - 2/3\right) * 3$$
$$s = \left(\sqrt{5/9} - 2/3\right) * 3$$
$$t_1 = \left(\sqrt{8/9} - 5/6\right) * 6$$
$$t_2 = \left(\sqrt{35/36} - 5/6\right) * 6.$$

Note the $1/h$ factor altering with the grid from 3 to 6. The main equations are now set up as for the previous exercises with the additional interpolation formulae:

$$\phi_{13} = \phi_7 * t_1/2$$
$$\phi_{11} = \phi_7/2 + \phi_8/2$$

with symmetric equations for ϕ_{17} and ϕ_{12}. The solution is given in Table A.33.

Table A.33.

n	ϕ_n	n	ϕ_n
1	0.0090191	8	0.019027
2	0.031059	11	0.12119
3	0.063096	13	0.017118
6	0.058385	14	0.11978
7	0.052122	15	0.35476

Here we see the influence of the point boundary value spreading through the solution space.

4.17 The further fine grid involves the line $y = 11/12$ running from $x = -1/6$ to $x = 1/6$. The intercepts at steps of $1/12$ are points 20 to 24. Points 18 and 19 are the interpolatory points at $(-1/12, 5/6)$ and $(1/12, 5/6)$. All the other points are as in Exercise 4.16.

The intercepts t, s, t_1 and t_2 are still as before with the additional ratios for the vertical intercepts t_3 and t_4 at points 20 and 21. These values are:

$$t_3 = \left(\sqrt{35/36} - 11/12\right) * 12$$
$$t_4 = \left(\sqrt{143/144} - 11/12\right) * 12.$$

Interpolation formulae are used at points 11, 12, 13 and 17 as before and additionally at 18, 19, 20 and 24 on the very fine grid. The solution is then as given in Table A.34.

Table A.34.

n	ϕ_n	n	ϕ_n
1	0.0052504	11	0.070554
2	0.018081	13	0.0099656
3	0.036732	14	0.069733
6	0.033989	15	0.20810
7	0.030343	20	0.029014
8	0.11076	21	0.13002
		22	0.36054

The points 1 to 17 correspond to the same numbers in the previous exercise, and we see the courser grid over-exaggerates the propagation of the point boundary condition into the surrounding space. Problems of this type are notoriously hard to get high accuracy.

4.18 Again the point numbering is from the bottom left with the row $y = 0.2$ containing points 1 to 7, $y = 0.4$ points 8 to 14, and $y = 0.6$ points 15 to 17 and 18 to 20. Between these two groups is the boundary point at the base of the dish. Points 21 and 22 are on $y = 0.8$ on the left-hand side of the dish and points 23 and 24 to the right. There is only one intercept to be calculated as the vertical arm at 17 is the same length as the horizontal arm at 22, namely $s = 2 - \sqrt{3}$ multiplied by the step of $1/5$. The solution is then similar to previous problems with values as given in Table A.35.

Table A.35.

n	ϕ_n	n	ϕ_n
1	0.20709	10	0.40838
2	0.20839	11	0.24362
3	0.19362	15	0.64001
4	0.15771	16	0.69457
8	0.41998	17	0.76346
9	0.43283	21	0.84550
		22	0.94198

4.19 The single-element first block gives $x_1 = 0.4$ for the first iteration. For the second (2×2) block, this x_1 value is added to the right-hand side (because the coefficient is -1 in both cases) and the resulting values are $x_2 = 1.92$ and $x_3 = -0.36$. For the second iteration the first block will be:

$$5x_1 = 2 - 2x_2 + x_3$$

where the existing values are used on the right-hand side to give -0.44. The (2×2) block gives $x_2 = 1.668$ and $x_3 = -0.444$. These values are already looking like the correct answers. After 14 iterations a full 12 digits is obtained, namely

$$x_1 = -0.363636363636$$
$$x_2 = 1.69090909091$$
$$x_3 = -0.436363636364.$$

By reducing the diagonal dominance the convergence to the same 12-figure accuracy takes 14 iterations when $a_{33} = 3.0$, 12 iterations when $a_{33} = 2.0$, 20 iterations when $a_{33} = 1.0$ and 25 iterations when $a_{33} = 0.0$.

4.20 Applying straight Gauss–Seidel to this problem gives

$$x_1 = 0.55$$
$$x_2 = 0.76$$
$$x_3 = 0.76$$
$$x_4 = 0.83$$

in 22 iterations. Block iteration in (2×2) blocks yields the same result in 15 iterations. A (2×2) Gauss elimination uses only 6 multiplicative operations giving a gain over the Seidel process overall.

4.21 For this problem the two choices of blocks make little difference to the number of iterations. Choosing two (3×3) blocks gives

$$x_1 = 7.11156186612, \qquad x_2 = 14.2677484787$$
$$x_3 = 4.06693711967, \qquad x_4 = 4.15618661258$$
$$x_5 = -3.56997971602, \qquad x_6 = 1.85192697769$$

in 35 iterations, and the choice of three (2×2) blocks gives the same result in 36 iterations.

4.22 The judgement with this problem is where to fix the blocks. There are several rows without diagonal dominance which could slow up or fail with single-element iteration. So rows 1, 2, 3, 4, 5 and 11 look troublesome. The banded structure is quite tight except for element $a_{12,5}$. The suggestion is to block the early rows into a (5×5) block. Then maybe a (2×2) block for rows 6 and 7, ending with another (5×5). Hence the blocks are quite small and in fact only 13 non-zero elements lie outside a block out of 60 non-zero elements. The result is 12 figures after 93 iterations with the result:

$$x_1 = 0.301757909087, \qquad x_2 = 0.0312849271306$$
$$x_3 = -0.145544489171, \qquad x_4 = 0.176961471424$$
$$x_5 = 0.125166155776, \qquad x_6 = -0.00514960631633$$
$$x_7 = 0.124840456038, \qquad x_8 = 0.0707828133429$$
$$x_9 = 0.0296080625720, \qquad x_{10} = 0.0487457647817$$
$$x_{11} = 0.237353858371, \qquad x_{12} = -0.0127473880690.$$

Chapter 5

5.1 Let $u = \alpha_0 + \alpha_1 x + \alpha_2 x^2 + \cdots$, then $u' = \alpha_1 + 2x\alpha_2 + 3x^2\alpha_3 + \cdots$ and $u'' = 2\alpha_2 + 6x\alpha_3 + \cdots$. Hence the initial conditions give $\alpha_0 = 0$ and $\alpha_1 = 2$ and the starting point for the collocation is the series $u = 2x + \alpha_2 x^2 + \alpha_3 x^3 + \cdots$. The error E is

$$E = 2\alpha_2 + 6\alpha_3 x + 2x + \alpha_2 x^2 + \alpha_3 x^2 + \cdots - x$$

and the coefficients of the α's can be easily found at the collocation points $1/2$ and 1 to give the equations

$$\begin{bmatrix} 2.25 & 3.125 \\ 3.0 & 7.0 \end{bmatrix} \begin{bmatrix} \alpha_2 \\ \alpha_3 \end{bmatrix} = \begin{bmatrix} -0.5 \\ -1.0 \end{bmatrix}$$

with solution $\alpha_2 = -0.058824$ and $\alpha_3 = -0.11765$.

5.2 There are two ways to tackle a boundary problem like this in terms of the starting point. For a hand calculation use $u = x(1 - x)(\alpha_1 + x\alpha_2 + \cdots)$ which fixes the boundary conditions a priori. The equivalent which is probably easier to programme for a general case is to just use the series as in the previous exercise, then fix α_0 to force the condition $u(0) = 0$ to give $\alpha_0 = 0$, and let the condition $u(1) = 0$ give an extra linear equation for solution. This latter approach will be used on the next examples where higher order collocation is required and hand calculations will become tedious. Using the first approach here gives

$$E = 2(\alpha_2 - \alpha_1) - 6\alpha_2 x + 2\alpha_1 x + 2x^2(\alpha_2 - \alpha_1) - 2\alpha_2 x^3 + x^2.$$

Putting $x = 1/3$ and $x = 2/3$ gives linear equations:

$$\begin{bmatrix} -1.5556 & 0.14815 \\ -1.5556 & -1.7037 \end{bmatrix} \begin{bmatrix} \alpha_2 \\ \alpha_3 \end{bmatrix} = \begin{bmatrix} -1/9 \\ -4/9 \end{bmatrix}$$

with solution $\alpha_2 = -0.08857$ and $\alpha_3 = -0.18$.

Extending the process by one term gives

$$\begin{aligned} E = \; & 2(\alpha_2 - \alpha_1) + 6x(\alpha_3 - \alpha_2) - 12x^2\alpha_3 + 2\alpha_1 x \\ & + 2x^2(\alpha_2 - \alpha_1) + x^3(\alpha_3 - \alpha_2) - x^4\alpha_3 + x^2 \end{aligned}$$

and with $x = 1/4$, $x = 1/2$ and $x = 3/4$ the linear equations give $\alpha_1 = 0.11747$, $\alpha_2 = 0.12493$ and $\alpha_3 = 0.068603$.

5.3 For the next set of examples, higher order collocations are achieved using some programming aid. The routine coef has been set up with parameters *prime*, *alsub* and *xval*. The routine delivers the value of the coefficient of α_{alsub} at $x = xval$ in the expansion of the *prime*-th derivative of y. Given an ordinary differential equation, the collocation matrix elements are immediate. Hence for this exercise they are:

$$a_{i,j} = \text{coef}(2, j, x_i) + 4\text{coef}(1, j, x_i) + 3\text{coef}(0, j, x_i)$$

that is $y'' + 4y' + 3y$. The x values then range across the collocation points with i, such as $x_i = i/(k+1)$ where k is the number of collocation points in total. In my solutions I range k from 2 up to 5 to give a meaningful comparison with the analytic values. The right-hand side vectors are just the equation right-hand sides and any predetermined coefficients from initial conditions. The boundary conditions give extra equations which can be set up before a call to an equation solver. In this way a sequence of collocations can be run off for each example. For this example, the initial condition gives $\alpha_0 = 0$, and the right-hand side vectors are

$$b_i = 10\exp(-2x_i).$$

The boundary condition gives the extra equation

$$\alpha_1 + \alpha_2 + \alpha_3 + \cdots = 1.$$

The results are shown in Table A.36. For each collocation with k α's and k running from 2 to 5, the α's have been found and the resulting approximate solution has been tabulated across the range of x to allow comparison with the analytic solution and the increasingly accurate (hopefully) computed solution.

Table A.36.

$x \backslash k$	2	3	4	5	Analytic solution
0.0	0.00	0.00	0.00	0.00	0.00
0.1	0.20978	0.18125	0.18062	0.17776	0.17762
0.2	0.39655	0.35831	0.36249	0.35927	0.35940
0.3	0.55979	0.52148	0.53032	0.52702	0.52730
0.4	0.69897	0.66362	0.67488	0.67146	0.67182
0.5	0.81358	0.78018	0.79167	0.78853	0.78894
0.6	0.99311	0.86918	0.88094	0.87769	0.87814
0.7	0.96703	0.93121	0.94355	0.94051	0.94095
0.8	1.0048	0.96943	0.98268	0.97960	0.98005
0.9	1.016	0.98959	1.001	0.99818	0.99863
1.0	1.00	1.00	1.00	1.00	1.00
α_0	0.0	0.0	0.0	0.0	
α_1	2.2112	1.7947	1.7309	1.6749	
α_2	−1.1249	0.39344	1.1634	1.5476	
α_3	−0.086253	−2.2605	−4.4600	−5.7698	
α_4		1.0723	3.6151	5.9231	
α_5			−1.0495	−3.0507	
α_6				0.675	

Table A.36 shows increasing accuracy as the number of collocation points is increased.

5.4 Following the same lines as the previous exercise, $a_{i,j} = \mathrm{coef}(2, j, x_i) + 4\mathrm{coef}(1, j, x_i) + 4\mathrm{coef}(0, j, x_i)$ and $b_i = \cos(x_i)$. The corresponding results are shown in Table A.37.

5.5 Each problem now introduces an extra complication. Here the equation is third order. The boundary conditions at $x = 0$ give the series $y = 1 + \alpha_2 x^2 + \alpha_3 x^3 + \cdots$, and the matrix coefficients are in the usual way

$$\begin{aligned}
a_{i,j} &= \mathrm{coef}(3, j+1, x_i) - 6\mathrm{coef}(2, j+1, x_i) \\
&\quad + 12\mathrm{coef}(1, j+1, x_i) - 8\mathrm{coef}(0, j+1, x_i)
\end{aligned}$$

Table A.37.

$x \backslash k$	2	3	4	5	Analytic solution
0.0	0.00	0.00	0.00	0.00	0.00
0.1	0.45689	0.52998	0.52879	0.53121	0.53124
0.2	0.78202	0.88333	0.87532	0.87817	0.87805
0.3	0.99506	1.1009	1.0869	1.0899	1.0897
0.4	1.1157	1.2177	1.2008	1.2039	1.2036
0.5	1.1635	1.2625	1.2452	1.2482	1.2480
0.6	1.1583	1.2583	1.2413	1.2442	1.2439
0.7	1.1197	1.2220	1.2047	1.2073	1.2070
0.8	1.0673	1.1646	1.1466	1.1492	1.1489
0.9	1.0209	1.0909	1.0758	1.0782	1.0779
1.0	1.00	1.00	1.00	1.00	1.00
α_0	0.0	0.0	0.0	0.0	
α_1	5.2933	6.3342	6.3857	6.4313	
α_2	-7.5716	-11.150	-12.014	-12.313	
α_3	3.2783	8.3118	10.901	11.906	
α_4		-2.4958	-5.5446	-7.3162	
α_5			1.2720	2.8139	
α_6				-0.52215	

where we see the parameter $j+1$ as the series starts at α_2. The vector $b_i = 8$, from the $8y$ term, as the series starts $1 + \cdots$ to give a right-hand side. The final equation is

$$\alpha_2 + \alpha_3 + \cdots = e^2 - 1$$

to force the boundary condition $y = e^2$ at $x = 1$. The first term in the y expansion gives the extra 1. The results are shown in Table A.38.

As the order of the approximation increases the solution converges well to the analytic values.

5.6 In this example the coefficients themselves become functions of x, and hence in the program the coefficients are now represented by short functions rather than real variables. Also both boundary conditions are used to give extra equations. With these small changes the coefficient matrix is

$$
\begin{aligned}
a_{i,j} \quad = \quad & x_i^2 \text{coef}(2, j-1, x_i) + 6x_i \text{coef}(1, j-1, x_i) \\
& + 6\text{coef}(0, j-1, x_i)
\end{aligned}
$$

with the series starting at α_0, necessitating the parameter $j-1$. The vector $b_i = 1/x_i^2$ with no contribution from the y's. The boundary

Table A.38.

$x \backslash k$	2	3	4	5	Analytic solution
0.0	1.00	1.00	1.00	1.00	1.00
0.1	1.0467	0.98105	1.0048	1.0011	1.0016
0.2	1.1678	0.95214	1.0235	1.0132	1.0144
0.3	1.3458	0.94764	1.0721	1.0547	1.0568
0.4	1.5806	1.0015	1.1790	1.1543	1.1573
0.5	1.9061	1.1589	1.3872	1.3552	1.3591
0.6	2.3700	1.4886	1.7598	1.7219	1.7265
0.7	3.0445	2.0946	2.3878	2.3471	2.3520
0.8	4.0243	3.1279	3.4015	3.3634	3.3681
0.9	5.4261	4.7990	4.9843	4.9576	4.9607
1.0	7.3891	7.3891	7.3891	7.3891	7.3891
α_0	1.0	1.0	1.0	1.0	
α_1	0.0	0.0	0.0	0.0	
α_2	5.5067	−2.7368	0.47975	0.080839	
α_3	−8.4119	9.3325	−0.65939	1.7736	
α_4	9.2943	−10.145	6.6319	0.60571	
α_5		9.9382	−4.0855	4.1233	
α_6			4.0223	−1.6259	
α_7				1.5933	

conditions give the extra equations

$$\alpha_0 + \alpha_1 + \alpha_2 + \cdots = 1$$
$$\alpha_0 + 2\alpha_1 + 4\alpha_2 + \cdots = 2.$$

The results are shown in the usual way in Table A.39.

5.7 For this example

$$a_{i,j} = (1 - x_i^2)\text{coef}(2, j, x_i) - x_i\text{coef}(1, j, x_i)$$

with $b_i = 0$. The boundary condition at $x = 1$ give $\alpha_0 + \alpha_1 + \cdots = \pi/2$, to give the results in Table A.40.

This solution has a singular derivative at $x = 1$ and the numerical solution is less accurate in this regime than at the other end of the range.

5.8 The examples in this section were solved by extending the codes of the previous section. The routine coef is again used to denote the terms in the expansion which are coefficients of α_i. Now the linear equations have a matrix of coefficients which is an integral for each element. For simple examples with two or three unknowns these integrals can be

Table A.39.

$x \backslash k$	2	3	4	5	Analytic solution
1.0	1.00	1.00	1.00	1.00	1.00
1.1	1.6065	1.80865	1.8137	1.8472	1.8529
1.2	2.0068	2.2743	2.2452	2.2813	2.2810
1.3	2.2361	2.4995	2.4380	2.4728	2.4694
1.4	2.2394	2.5677	2.4905	2.5255	2.5206
1.5	2.3219	2.5436	2.4646	2.4994	2.4933
1.6	2.2486	2.4734	2.3962	2.4286	2.4219
1.7	2.1447	2.3845	2.3035	2.3335	2.3368
1.8	2.0452	2.2857	2.1969	2.2273	2.2203
1.9	1.9853	2.1671	2.0882	2.1172	2.1099
2.0	2.000	2.0000	2.000	2.000	2.000
α_0	-24.124	-61.790	-110.95	-190.15	
α_1	48.885	152.68	328.07	656.45	
α_2	-29.610	-134.96	-381.08	-944.05	
α_3	5.8495	52.862	222.88	733.95	
α_4		-7.7885	-65.725	-324.82	
α_5			7.8033	77.356	
α_6				-7.7239	

Table A.40.

$x \backslash k$	2	3	4	5	Analytic solution
0.0	0.00	0.00	0.00	0.00	0.00
0.1	0.12933	0.10993	0.11917	0.10913	0.10017
0.2	0.25272	0.22777	0.23419	0.22362	0.20136
0.3	0.37472	0.34990	0.35131	0.34029	0.30469
0.4	0.49986	0.47534	0.47316	0.46063	0.41152
0.5	0.63268	0.60570	0.60103	0.58715	0.52360
0.6	0.77772	0.74522	0.73714	0.72249	0.64350
0.7	0.93951	0.90072	0.88701	0.87096	0.77540
0.8	1.1226	1.0817	1.0616	1.0428	0.92730
0.9	1.3315	1.3001	1.2799	1.2610	1.1198
1.0	1.5708	1.5708	1.5708	1.5708	1.5708
α_0	0.0	0.0	0.0	0.0	
α_1	1.3381	1.0412	1.2470	1.0308	
α_2	-0.5236	0.69527	-0.78231	0.8826	
α_3	0.75631	-1.2540	2.6186	-3.5498	
α_4		1.0882	-3.4226	7.9961	
α_5			1.9101	-8.4095	
α_6				3.6151	

found analytically, but to obtain more accurate solutions this process is time-wasting. In addition to the equation-solving routine used in the previous section, an integrator has also been employed. A simple Gauss–Legendre routine with 16 points is more than sufficient for all these examples. For details of such routines see Evans (1993). The routine is:

$$\text{integrate}(f, a, b)$$

where f is the function being integrated from a to b. For readers wishing to remain in analytic mode, this operation will need to be completed analytically and the above notation enables the required calculations to be succinctly written down. From equations (5.3.4) and (5.3.5) the required coefficients are just the integrals of the product of two calls to coef with right-hand sides which will depend on the right-hand sides of the differential equation and any constant terms in the α expansion.

Hence for this first example

$$a_{i,j} = \text{integrate}(f_{i,j}, 0, 1)$$

where

$$f_{i,j} = c_i * c_j$$

and

$$c_i = \text{coef}(2, i, x) + \text{coef}(0, i, x).$$

The term c_i is the coefficient of α_i in the error term E exemplified by (5.3.3) in the worked example, and here corresponds to $u'' + u$ to give the parameters 2 and 0 for the level of derivative as in the previous section.

The right-hand side vector b_i is given by

$$b_i = \text{integrate}(g_i, 0, 1)$$

with

$$g_i = (-x) * c_i.$$

A $(-x)$ comes from the right-hand side of the differential equation, and a $2x$ from the first term of the expansion of u namely

$$u = 2x + \alpha_2 x^2 + \alpha_3 x^3 + \cdots$$

so forcing the correct boundary conditions. The linear equations are

$$\begin{bmatrix} 5.5333 & 8.166 \\ 8.166 & 14.5429 \end{bmatrix} \begin{bmatrix} \alpha_2 \\ \alpha_3 \end{bmatrix} = \begin{bmatrix} -1.25 \\ -2.2 \end{bmatrix}$$

with solution $\alpha_2 = -0.01538$ and $\alpha_3 = -0.1426$. The analytic solution is $u = x + \sin x$ which expands for small x as $2x - x^3/6 + \cdots$ which compares well with the above series.

5.9 The expansion is now

$$u = \alpha_1 x + \alpha_2 x^2 + \cdots$$

so that $u(0) = 0$. With the same notation as above the equations are as in Exercise 5.8, but with

$$c_i = \text{coef}(2, i, x) + 2\text{coef}(0, i, x)$$

and

$$g_i = (-x * x) * c_i.$$

In this case the contribution is entirely from the right-hand side of the differential equation. The second boundary condition gives the additional equation

$$\alpha_1 + \alpha_2 + \cdots = 1.$$

The solution for the quadratic and cubic approximations is shown in Table A.41.

Table A.41.

$x \backslash k$	2	3	Analytic solution
0.0	0.00	0.00	0.00
0.1	0.02700	0.012155	0.011118
0.2	0.04800	0.024302	0.021897
0.3	0.06300	0.035431	0.031824
0.4	0.07200	0.044532	0.040199
0.5	0.07500	0.050595	0.046158
0.6	0.07200	0.052611	0.048683
0.7	0.06300	0.049569	0.046626
0.8	0.04800	0.040460	0.038728
0.9	0.02700	0.024273	0.023652
1.0	0.00	0.00	0.00
α_0	0.0	0.0	
α_1	0.3	0.11823	
α_2	-0.3	0.050082	
α_3		-0.16831	

5.10 In this example the routine coef has not been used, the c_i's being coded directly after the following short analysis. It is required to start with

$$u = \alpha_1 + \alpha_2 x + (1 - x)(2 - x)(\alpha_3 + \alpha_4 x)$$

and the boundary conditions fix $\alpha_1 = -1$ and $\alpha_2 = 2$. Substituting into u gives

$$
\begin{aligned}
E &= (1+x) + \alpha_3(1 - x + x^2) + \alpha_4(x^3 + 2x - 4) \\
&= c_0 + c_1\alpha_3 + c_2\alpha_4
\end{aligned}
$$

using the notation of the programmed examples.

Hence
$$
a_{i,j} = \text{integrate}(f_{i,j}, 1, 2)
$$
where
$$
f_{i,j} = c_i * c_j
$$
which are defined above. The right-hand side vector b_i is

$$
b_i = \text{integrate}(c_0 * c_i, 1, 2).
$$

The outcome is the linear equations

$$
\begin{bmatrix} 0.7 & -2.283 \\ -2.283 & 8.276 \end{bmatrix} \begin{bmatrix} \alpha_2 \\ \alpha_3 \end{bmatrix} = \begin{bmatrix} -1.25 \\ 3.88 \end{bmatrix}
$$

with solution $\alpha_3 = -1.8068$ and $\alpha_4 = -0.29546$.

5.11 For the next set of examples, tables analogous to those in Exercises 5.3 to 5.7 will be generated, and hence direct comparisons can be made with the solutions there. In each case the solution will be sought in powers of x to utilise the function coef defined in Exercise 5.8.

Exercise 5.3

For this example
$$
a_{i,j} = \text{integrate}(f_{i,j}, 0, 1)
$$
where
$$
f_{i,j} = c_i * c_j
$$
and
$$
c_i = \text{coef}(2, i, x) + 4\text{coef}(1, i, x) + 3\text{coef}(0, i, x)
$$
and the right-hand side has

$$
b_i = \text{integrate}(g_i, 0, 1)
$$
with
$$
g_i = 10e^{-2x} * c_i.
$$

Increasing the terms in the series from 3 to 6 gives the results in Table A.42. In each of the five examples in this question the number of free

Table A.42.

$x \backslash k$	2	3	4	5	Analytic solution
0.0	0.00	0.00	0.00	0.00	0.00
0.1	0.14563	0.17574	0.17686	0.17755	0.17762
0.2	0.29286	0.35492	0.35854	0.35941	0.35940
0.3	0.43730	0.52452	0.52741	0.52745	0.52730
0.4	0.57452	0.67462	0.67290	0.67194	0.67182
0.5	0.70014	0.79839	0.79013	0.78890	0.78894
0.6	0.80975	0.89209	0.87845	0.87798	0.87814
0.7	0.89895	0.95509	0.94006	0.94085	0.94095
0.8	0.96332	0.98985	0.97862	0.98011	0.98005
0.9	0.99847	1.0019	0.99783	0.99873	0.99863
1.0	1.00	1.00	1.00	1.00	1.00
α_0	0.0	0.0	0.0	0.0	
α_1	1.4336	1.6890	1.6653	1.6663	
α_2	0.30033	0.96550	1.5046	1.6372	
α_3	−0.73393	−2.9451	−5.1146	−6.0563	
α_4		1.2906	4.1162	6.3197	
α_5			−1.1714	−3.2892	
α_6				0.72231	

α's will range from 2 to 5 with any extra ones required for the boundary conditions. In this example the extra equation is

$$\alpha_1 + \alpha_2 + \cdots = 1.$$

Exercise 5.4

For this example, simple changes are required to give

$$c_i = \mathrm{coef}(2, i, x) + 4\mathrm{coef}(1, i, x) + 4\mathrm{coef}(0, i, x)$$

and the right-hand side has

$$b_i = \mathrm{integrate}(g_i, 0, 1)$$

with

$$g_i = \cos x * c_i$$

and the same extra equation to force $y(1) = 1$; with the corresponding results in Table A.43.

Exercise 5.5

This is the third order example and will have the extra equation

$$\alpha_2 + \alpha_3 + \cdots = e^2 - 1$$

Table A.43.

$x\backslash k$	2	3	4	5	Analytic solution
0.0	0.00	0.00	0.00	0.00	0.00
0.1	0.60095	0.53296	0.53181	0.53128	0.53124
0.2	1.0203	0.88278	0.87873	0.87805	0.87085
0.3	1.2845	1.0931	1.0897	1.0896	1.0897
0.4	1.4197	1.2011	1.2028	1.2036	1.2036
0.5	1.4524	1.2379	1.2471	1.2480	1.2480
0.6	1.4089	1.2283	1.2436	1.2440	1.2439
0.7	1.3156	1.1986	1.2076	1.2071	1.2070
0.8	1.1988	1.1371	1.1500	1.1489	1.1489
0.9	1.0848	1.0737	1.0785	1.0779	1.0779
1.0	1.00	1.00	1.00	1.00	1.00
α_0	0.0	0.0	0.0	0.0	
α_1	7.0052	6.4062	6.4359	6.4351	
α_2	−10.396	−11.624	−12.254	−12.355	
α_3	4.3910	8.8345	11.327	12.039	
α_4		−2.6170	−5.8384	−7.4921	
α_5			1.3289	2.9102	
α_6				−0.53727	

as the trial series is

$$y = 1 + \alpha^2 x + \alpha_3 x^3 = \dots$$

to force the boundary conditions on the solution. The matrix elements will have

$$c_i = \text{coef}(3, i, x) - 6c_i\ \text{ef}(2, i, x) + 12\text{coef}(1, i, x) - 8\text{coef}(0, i, x)$$

and the right-hand side has

$$b_i = \text{integrate}(g_i, 0, 1)$$

with

$$g_i = -1.0 * c_i$$

and the results are shown in Table A.44.

Exercise 5.6

In this example, the coefficients are functions of x. This causes no particular problem. The integration range is from 1 to 2. The matrix elements will have the form

$$a_{i,j} = \text{integrate}(f_{i,j}, 1, 2)$$

Table A.44.

$x \backslash k$	2	3	4	5	Analytic solution
0.0	1.00	1.00	1.00	1.00	1.00
0.1	1.0147	1.0045	1.0010	1.0016	1.0016
0.2	1.0351	1.0335	1.0120	1.0146	1.0144
0.3	1.0515	1.1050	1.0525	1.0570	1.0568
0.4	1.0882	1.2395	1.1526	1.1573	1.1573
0.5	1.2036	1.4697	1.3559	1.3590	1.3591
0.6	1.4902	1.8511	1.7252	1.7264	1.7265
0.7	2.0744	2.4727	2.3520	2.3521	2.3520
0.8	3.1168	3.4674	3.3678	3.3682	3.3681
0.9	4.8121	5.0222	4.9599	4.9608	4.9607
1.0	7.3891	7.3891	7.3891	7.3891	7.3891
α_0	1.0	1.0	1.0	1.0	
α_1	0.0	0.0	0.0	0.0	
α_2	2.3379	-0.036464	0.0032118	$-0.30393(-5)$	
α_3	-10.145	5.5918	0.48753	1.4542	
α_4	14.196	-7.8796	5.6406	1.1778	
α_5		8.7133	-3.9952	3.6844	
α_6			4.2530	-1.5593	
α_7				1.6279	

where
$$f_{i,j} = c_i * c_j$$
and
$$c_i = x^2 \text{coef}(2, i, x) + 6x\text{coef}(1, i, x) + 6\text{coef}(0, i, x)$$
and the right-hand side has
$$b_i = \text{integrate}(g_i, 1, 2)$$
with
$$g_i = -1.0/x^2 * c_i.$$
There are two extra equations here, namely
$$\alpha_1 + \alpha_2 + \alpha_3 + \cdots = 1$$
and
$$2\alpha_1 + 4\alpha_2 + 8\alpha_3 + \cdots = 2$$
and the results are shown in Table A.45.

Exercise 5.7

In this example, the coefficients are again functions of x. The matrix elements will have the form
$$a_{i,j} = \text{integrate}(f_{i,j}, 0, 1)$$

Table A.45.

$x \backslash k$	2	3	4	5	Analytic solution
1.0	1.00	1.00	1.00	1.00	1.00
1.1	2.1265	1.8987	1.8665	1.8560	1.8529
1.2	2.8552	2.3733	2.2970	2.2815	2.2810
1.3	3.2495	2.5574	2.4698	2.4654	2.4694
1.4	3.3734	2.5615	2.5043	2.5167	2.5206
1.5	3.2903	2.4721	2.4721	2.4933	2.4933
1.6	3.0640	2.3522	2.4094	2.4252	2.4219
1.7	2.7581	2.2412	2.3288	2.3296	2.3368
1.8	2.4364	2.1547	2.2309	2.2200	2.2203
1.9	2.1625	2.0848	2.1170	2.1084	2.1099
2.0	2.000	2.0000	2.000	2.000	2.000
α_0	-46.159	-77.668	-137.27	-223.12	
α_1	91.463	193.64	412.17	784.83	
α_2	-54.916	-172.73	-485.62	-1147.9	
α_3	10.612	67.607	286.14	902.88	
α_4		-9.8467	-84.345	-401.85	
α_5			9.9331	95.659	
α_6				-9.4920	

where
$$f_{i,j} = c_i * c_j$$
and
$$c_i = (1 - x^2)\text{coef}(2, i, x) - x\,\text{coef}(1, i, x)$$
and the right-hand side has
$$b_i = 0.0.$$

The extra equation is
$$\alpha_1 + \alpha_2 + \alpha_3 + \cdots = \pi/2$$
and the results are shown in Table A.46.

5.12 Only very minor changes are required to implement this method using the least squares codes. Instead of the weighting function being c_i to represent $dE/d\alpha_i$, the weighting function is ϕ_i or in the notation used in these problems $\text{coef}(0, i, x)$. In the first example the basis functions are $\sin i\pi x$ and then
$$\text{coef}(0, i, x) = \sin i\pi x$$
$$\text{coef}(1, i, x) = i\pi \cos i\pi x$$
$$\text{coef}(2, i, x) = -i^2\pi^2 \sin i\pi x$$

Table A.46.

$x\backslash k$	2	3	4	5	Analytic solution
0.0	0.00	0.00	0.00	0.00	0.00
0.1	0.10446	0.11328	0.08399	0.11434	0.10017
0.2	0.18906	0.24168	0.17195	0.21135	0.20136
0.3	0.26563	0.36127	0.28790	0.29082	0.30469
0.4	0.34600	0.46063	0.42454	0.38638	0.41152
0.5	0.44201	0.54086	0.56010	0.51694	0.52360
0.6	0.56549	0.61559	0.67520	0.66843	0.64350
0.7	0.72827	0.71101	0.76970	0.80587	0.77540
0.8	0.94220	0.86578	0.87956	0.91563	0.92730
0.9	1.2191	1.1311	1.0937	1.0782	1.1198
1.0	1.5708	1.5708	1.5708	1.5708	1.5708
α_0	0.0	0.0	0.0	0.0	
α_1	1.1834	0.94598	1.0122	0.99731	
α_2	-1.5849	2.5279	-3.1408	3.8690	
α_3	1.9723	-7.1226	16.751	-32.116	
α_4		5.2196	-27.098	90.302	
α_5			14.046	-103.57	
α_6				42.073	

to give the first and second derivatives. The matrix coefficients are:

$$a_{i,j} = \text{integrate}(f_{i,j}, 0, 1)$$

where

$$f_{i,j} = c_j * \text{coef}(0, i, x)$$

and

$$c_i = -\text{coef}(2, i, x) + \text{coef}(0, i, x)$$

to mimic the equation coefficients. The right-hand side vector b_i is given by

$$b_i = \text{integrate}(g_i, 0, 1)$$

with

$$g_i = (-x) * \text{coef}(0, i, x).$$

The replacement of c_i by $\text{coef}(0, i, x)$ can be seen in the definitions of both f and g.

In this particular example there is much less work than might be expected as the set of functions $\sin i\pi x$ is orthogonal on $(0,1)$ and so all the non-diagonal coefficients in the matrix are zero. The use of such orthogonal functions is common to reduce the computational load in Galerkin computations.

The results are shown in tabular form as in the previous sections and are shown in Table A.47.

Table A.47.

$x \backslash k$	2	3	4	5	Analytic solution
0.0	0.00	0.00	0.00	0.00	0.00
0.1	0.013477	0.015388	0.014435	0.014949	0.014766
0.2	0.026947	0.029194	0.028605	0.028605	0.028680
0.3	0.039904	0.040634	0.041223	0.040709	0.040878
0.4	0.051080	0.049691	0.050644	0.050644	0.050483
0.5	0.058569	0.056206	0.056206	0.056720	0.056591
0.6	0.060324	0.058936	0.057983	0.057983	0.058620
0.7	0.054862	0.055592	0.055003	0.054489	0.054507
0.8	0.041905	0.044151	0.044740	0.044740	0.044295
0.9	0.022721	0.024632	0.025585	0.026099	0.026518
1.0	0.00	0.00	0.00	0.00	0.00
α_1	0.058569	0.058569	0.058569	0.0058569	
α_2	−0.0078637	−0.0078637	−0.0078637	−0.0078637	
α_3		0.0023624	0.0023624	0.0023624	
α_4			−0.0010015	−0.0010015	
α_5				0.00051394	

The effect of the orthogonal function sin the basis set is seen in the α coefficients which remain unchanged as the number of fitting terms increases.

5.13 For this problem, coef reverts to its definition as powers of x. There is also one extra equation to deal with the boundary condition at $x = 1$ which is

$$\alpha_1 + \alpha_2 + \alpha_3 + \cdots = \pi/2.$$

The coefficients c_i are

$$c_i = \mathrm{coef}(2, i, x) + 4\mathrm{coef}(1, i, x) + 3\mathrm{coef}(0, i, x)$$

from the equation coefficients. The right-hand side vector b_i is given by

$$b_i = \mathrm{integrate}(g_i, 0, 1)$$

with

$$g_i = -10e^{-2x} * \mathrm{coef}(0, i, x).$$

The results are shown in Table A.48.

Table A.48.

$x\backslash k$	2	3	4	5	Analytic solution
0.0	0.00	0.00	0.00	0.00	0.00
0.1	0.19189	0.18168	0.17791	0.17761	0.17762
0.2	0.36441	0.35920	0.35872	0.35931	0.35940
0.3	0.51718	0.52361	0.52662	0.52732	0.52730
0.4	0.64978	0.66807	0.67185	0.67191	0.67182
0.5	0.76184	0.78788	0.78955	0.78899	0.78894
0.6	0.85294	0.88043	0.87872	0.87809	0.87814
0.7	0.92271	0.94523	0.94100	0.94088	0.94095
0.8	0.97074	0.98391	0.97963	0.98004	0.98005
0.9	0.99663	1.0002	0.99829	0.99867	0.99863
1.0	1.00	1.00	1.00	1.00	1.00
α_0	0.0	0.0	0.0	0.0	
α_1	2.0144	1.8026	1.6954	1.6708	
α_2	−0.94857	0.33569	1.2598	1.5783	
α_3	−0.065848	−2.0191	−4.4747	−5.8124	
α_4		0.88087	3.4536	5.8813	
α_5			−0.93318	−2.9310	
α_6				0.6130	

5.14 With the same notation as in the previous example,

$$\alpha_1 + \alpha_2 + \alpha_3 + \cdots = e^2 - 1$$

is the extra equation and the coefficients c_i are

$$c_i = \text{coef}(3, i, x) - 6\text{coef}(2, i, x) + 12\text{coef}(1, i, x) - 8\text{coef}(0, i, x).$$

The right-hand side vector b_i is given by

$$b_i = \text{integrate}(g_i, 0, 1)$$

with

$$g_i = -10e^{-2x} * \text{coef}(0, i, x).$$

The results are shown in Table A.49.

It is of note here that the 3-term column is a poor representation of the solution. It appears that the natural signs of the α's are best represented in the even columns and poorer results appear in the odd columns. Nevertheless quite reasonable convergence does occur.

5.15 **Exercise 5.6**

The setup here is exactly as in Exercise 5.12, and the results are shown in Table A.50.

Table A.49.

$x \backslash k$	2	3	4	5	Analytic solution
0.0	1.00	1.00	1.00	1.00	1.00
0.1	1.0753	0.91248	1.0091	1.0007	1.0016
0.2	1.2733	0.73548	1.0350	1.0124	1.0144
0.3	1.5647	0.56154	1.0898	1.0538	1.0568
0.4	1.9378	0.45489	1.2017	1.1532	1.1573
0.5	2.3977	0.47583	1.4145	1.3539	1.3591
0.6	2.9670	0.70495	1.7915	1.7202	1.7265
0.7	3.6854	1.2675	2.4221	2.3453	2.3520
0.8	4.6096	2.3578	3.4337	3.3619	3.3681
0.9	5.8137	4.2633	5.0058	4.9565	4.9607
1.0	7.3891	7.3891	7.3891	7.3891	7.3891
α_0	1.0	1.0	1.0	1.0	
α_1	0.0	0.0	0.0	0.0	
α_2	8.3748	−1.1410	1.1353	−0.15481	
α_3	−9.1497	29.586	−3.3120	2.2718	
α_4	7.1640	−32.047	1.1511	−0.85708	
α_5		20.261	−8.2382	6.2603	
α_6			5.2934	−3.1531	
α_7				2.0220	

Exercise 5.7

This problem again has the same setup as Exercise 5.12, and now the results are shown in Table A.51.

5.16 As in the previous two sections, some fairly minor changes to the previous processes will yield the computations for this section.

The first example is in two parts. Using the basis set $\sin i\pi x$ requires defining the routine coef with both $\sin i\pi x$ itself and the first derivative $i\pi \cos i\pi x$. (This section will not need to have second derivatives defined because of the integration by parts.)

The matrix elements then follow along the same lines as the examples starting at equation (5.5.23). The matrix elements are:

$$a_{i,j} = \text{integrate}(f_{i,j}, 0, 1)$$

where

$$
\begin{aligned}
f_{i,j} = \ & -a * \text{coef}(1, j, x) * \text{coef}(1, i, x) \\
& + b * \text{coef}(1, j, x) * \text{coef}(0, i, x) + c * \text{coef}(0, j, x) * \text{coef}(0, i, x)
\end{aligned}
$$

Table A.50.

$x\backslash k$	2	3	4	5	Analytic solution
1.0	1.00	1.00	1.00	1.00	1.00
1.1	2.1265	1.8987	1.8665	1.8560	1.8529
1.2	2.8552	2.3733	2.2970	2.2815	2.2810
1.3	3.2495	2.5574	2.4698	2.4654	2.4694
1.4	3.3734	2.5615	2.5043	2.5167	2.5206
1.5	3.2903	2.4721	2.4721	2.4933	2.4933
1.6	3.0640	2.3522	2.4094	2.4252	2.4219
1.7	2.7581	2.2412	2.3288	2.3296	2.3368
1.8	2.4364	2.1547	2.2309	2.2200	2.2203
1.9	2.1625	2.0848	2.1170	2.1084	2.1099
2.0	2.000	2.0000	2.000	2.000	2.000
α_0	−46.159	−77.668	−137.27	−223.12	
α_1	91.463	193.64	412.17	784.83	
α_2	−54.916	−172.73	−485.62	−1147.9	
α_3	10.612	67.607	286.14	902.88	
α_4		−9.8467	−84.345	−401.85	
α_5			9.9331	95.659	
α_6				−9.4920	

Table A.51.

$x\backslash k$	2	3	4	5	Analytic solution
0.0	0.00	0.00	0.00	0.00	0.00
0.1	0.16452	0.057766	0.12043	0.097674	0.10017
0.2	0.27448	0.18598	0.18236	0.22661	0.20136
0.3	0.34971	0.33187	0.26969	0.31093	0.30469
0.4	0.41006	0.46244	0.39934	0.38672	0.41152
0.5	0.47537	0.56450	0.54897	0.49994	0.52360
0.6	0.56549	0.64465	0.68483	0.65439	0.64350
0.7	0.70024	0.72928	0.78951	0.80962	0.77540
0.8	0.89949	0.86457	0.88970	0.92899	0.92730
0.9	1.1831	1.1165	1.0840	1.0777	1.1198
1.0	1.5708	1.5708	1.5708	1.5708	1.5708
α_0	0.0	0.0	0.0	0.0	
α_1	1.9842	0.000	1.9995	0.000	
α_2	−3.7203	7.0686	−1.1349	16.647	
α_3	3.3069	−13.744	38.838	−86.052	
α_4		8.2467	−51.069	190.80	
α_5			23.151	−189.32	
α_6				69.491	

which gives the formula:

$$K_{i,j} = \int (-a\phi_i'\phi_j' + b\phi_i\phi_j' + c\phi_i\phi_j)dx$$

and the right-hand sides are:

$$b_i = \text{integrate}(g_i, 0, 1) - vu'|_0^1$$

with

$$g_i = (-x) * \text{coef}(0, i, x).$$

The basis functions each satisfy the boundary conditions and hence let $v(0) = 0$ and $v(1) = 0$ to leave just the integration term in b_i. As is now customary in this chapter the problem was solved for a range of fitting functions from 2 to 5 terms and the results are shown in Table A.52, together with analytic values from

$$u = A(e^x - e^{-x}) + x$$

with $A = e/(e^2 - 1)$.

Table A.52.

$x \backslash k$	2	3	4	5	Analytic solution
0.0	0.00	0.00	0.00	0.00	0.00
0.1	0.013477	0.015388	0.014435	0.014949	0.014766
0.2	0.026947	0.029194	0.028605	0.028605	0.028680
0.3	0.039904	0.040634	0.041223	0.040709	0.040878
0.4	0.051080	0.049691	0.050644	0.050644	0.050483
0.5	0.058569	0.056206	0.056206	0.056720	0.056591
0.6	0.060324	0.058936	0.057983	0.057983	0.058260
0.7	0.054862	0.055592	0.055003	0.054489	0.054507
0.8	0.041905	0.044151	0.044740	0.044740	0.044295
0.9	0.022721	0.024632	0.025585	0.026099	0.026518
1.0	0.00	0.00	0.00	0.00	0.00
α_0	0.0	0.0	0.0	0.0	
α_1	0.058569	0.058569	0.058569	0.058569	
α_2	−0.007864	−0.007864	−0.007864	−0.007864	
α_3		0.002362	0.002362	0.002362	
α_4			−0.001002	−0.001002	
α_5				5.1394(−4)	

Notice here that the basis functions form an orthogonal set on $(0, 1)$ and hence in the linear equations only the diagonal elements are non-zero, and hence the series of α coefficients repeats as the order increases.

In the second part polynomials are used as basis functions. The definition $\phi_i = x^i(1 - x)$ was employed, and the only change was to replace the routine coef with the new definition. The outcome is shown in Table A.53.

Table A.53.

$x\backslash k$	2	3	4	5	Analytic solution
0.0	0.00	0.00	0.00	0.00	0.00
0.1	0.014594	0.014782	0.014766	0.014766	0.014766
0.2	0.028550	0.028673	0.028680	0.028680	0.028680
0.3	0.040890	0.040856	0.040878	0.040878	0.040878
0.4	0.050638	0.050464	0.050483	0.050483	0.050483
0.5	0.056818	0.056590	0.056590	0.056591	0.056591
0.6	0.058452	0.058278	0.058260	0.058260	0.058260
0.7	0.054564	0.054530	0.054508	0.054507	0.054507
0.8	0.044178	0.044301	0.044295	0.044295	0.044295
0.9	0.026351	0.026503	0.026518	0.026518	0.026518
1.0	0.00	0.00	0.00	0.00	0.00
α_0	0.0	0.0	0.0	0.0	
α_1	0.14588	0.14965	0.14907	0.14908	
α_2	0.16279	0.14406	0.14926	0.14905	
α_3		0.018733	0.0066024	0.0074316	
α_4			−0.0080872	−0.0068437	
α_5				6.2179(−4)	

5.17 For this example, both types of non-homogeneous boundary conditions are exhibited. For case (a), a ϕ_0 is introduced which satisfies the boundary conditions. Hence $\phi_0 = 1 + x$, and then both $\phi_i = 0$ at both endpoints for $i > 1$. Again the routine coef is adjusted to $x^i(3 - x)$ and its derivative, and the matrix elements still follow as in the previous exercise, but with the new coef routine, and $a = 1$, $b = 2$ and $c = -1$, and the range from 1 to 3. The right-hand side vector will encompass the ϕ_0 term and the right-hand side x of the differential equation and will be

$$F_i = \int_0^3 (-\phi_0'\phi_i' + 2\phi_0'\phi_i - \phi_0\phi_i - x\phi_i)dx$$

which codes into

$$g_i = -\text{coef}(1, i, x) + 2\text{coef}(0, i, x) - (1 + x)\text{coef}(0, i, x) - x * \text{coef}(0, i, x).$$

These results are shown in Table A.54.

For boundary conditions (b), the integration by parts gives the term $u'v|_0^1$, and hence the error E will have the extra term $u'(1)v(1) -$

Table A.54.

$x \backslash k$	2	3	4	5	Analytic solution
0.0	1.00	1.00	1.00	1.00	1.00
0.2	0.89996	0.87672	0.87155	0.87031	0.87020
0.4	0.84516	0.81927	0.81764	0.81830	0.81882
0.6	0.83507	0.81855	0.82259	0.82456	0.82491
0.8	0.86912	0.86696	0.87496	0.87657	0.87635
1.0	0.94676	0.95846	0.96705	0.96707	0.96649
1.2	1.0675	1.0885	1.0944	1.0926	1.0922
1.4	1.2307	1.2541	1.2551	1.2525	1.2525
1.6	1.4358	1.4537	1.4496	1.4474	1.4479
1.8	1.6823	1.6874	1.6799	1.6793	1.6799
2.0	1.9697	1.9568	1.9491	1.9505	1.9508
2.2	2.2974	2.2649	2.2609	2.2635	2.2632
2.4	2.6648	2.6164	2.6191	2.6211	2.6205
2.6	3.0714	3.0174	3.0270	3.0266	3.0263
2.8	3.5167	3.4755	3.4870	3.4843	3.4847
3.0	4.00	4.00	4.00	4.00	4.00
α_0	0.0	0.0	0.0	0.0	
α_1	-0.53808	-0.59936	-0.62021	0.62852	
α_2	0.011466	0.11830	0.18732	0.22807	
α_3		-0.039709	-0.097198	-0.15425	
α_4			0.013615	0.043398	
α_5				$-5.1598(-3)$	

$u'(0)v(0)$, and this will become

$$[5 - 3u(1)]v(1) - 2v(0)$$

from the boundary condition. Hence the extra terms are:

$$\left[5 - 3\sum \alpha_j \phi_j(1)\right]\left(\sum \beta_j \phi_j(1)\right) - 2\sum \beta_i \phi_i(0)$$

which means that K_{ij} has added to it $-3\phi_j(1)\phi_i(1)$ and F_i has added $-5\phi_i(1) + 2\phi_i(0)$. Hence the matrix elements are:

$$a_{i,j} = \text{integrate}(f_{i,j}, 0, 1) - 3\text{coef}(0, j, 1.0) * \text{coef}(0, i, 1.0)$$

where

$$
\begin{aligned}
f_{i,j} &= -a * \text{coef}(1, j, x) * \text{coef}(1, i, x) \\
&+ b * \text{coef}(1, j, x) * \text{coef}(0, i, x) + c * \text{coef}(0, j, x) * \text{coef}(0, i, x)
\end{aligned}
$$

as before, but now with coef being x^{i-1} and its derivative.

The right-hand sides are:

$$b_i = \text{integrate}(g_i, 0, 1) - 5\text{coef}(0, i, 1) + 2\text{coef}(0, i, 0.0)$$

with

$$g_i = x * coef(0, i, x).$$

The results of this set up are shown in Table A.55.

Table A.55.

$x\backslash k$	2	3	4	5	Analytic solution
0.0	0.16495	0.16914	0.16771	0.16776	0.16776
0.1	0.28763	0.32918	0.34708	0.35000	0.35027
0.2	0.41031	0.47946	0.50240	0.50401	0.50381
0.3	0.53299	0.61999	0.63751	0.63637	0.63591
0.4	0.65567	0.75077	0.75622	0.75286	0.75256
0.5	0.77835	0.87179	0.86237	0.85843	0.85855
0.6	0.90103	0.98306	0.95978	0.95722	0.95766
0.7	1.0237	1.0846	1.0523	1.0525	1.0529
0.8	1.1464	1.1763	1.1437	1.1468	1.1469
0.9	1.2691	1.2583	1.2379	1.2418	1.2415
1.0	1.3918	1.3306	1.3386	1.3383	1.3383
α_0	0.0	0.0	0.0	0.0	
α_1	0.16495	0.16914	0.16771	0.16776	
α_2	1.2268	1.6492	1.9267	1.9877	
α_3		-0.48774	-1.3936	-1.7797	
α_4			0.63785	1.3052	
α_5				-0.34269	

5.18 The sequence of exercises in this chapter has allowed a steady incremental development to take place which culminates in this section with the finite element method for ordinary differential equations. The changes from the previous method are again small. The variational principle from Section 5.5 is used locally on separate line elements. For any typical element, there are two shape functions as in (5.6.3) and (5.6.4). Hence for the element from s_1 to s_2, the routine coef now computes $(s_2 - x)/h$ for $i = 1$ and $(x - s_1)/h$ for $i = 2$, with the first derivatives $-1/h$ and $1/h$ respectively. Hence (5.6.48) is now codable. The local (2×2) matrix is set up for each element, and these matrices are added into a full matrix ready for final solution.

Hence the (2×2) matrix for element e is:

$$k_{i,j} = \text{integrate}(f_{i,j}, s_1, s_2)$$

where

$$f_{i,j} = -a * \text{coef}(1, j, x) * \text{coef}(1, i, x)$$
$$+ b * \text{coef}(1, j, x) * \text{coef}(0, i, x) + c * \text{coef}(0, j, x) * \text{coef}(0, i, x)$$

just as in Section 5.5 but now for only one element. These local elements are added into the full matrix a_{ij} so that

$$a_{e+i-1,e+j-1} = a_{e+i-1,e+j-1} + k_{ij}.$$

The right-hand sides are equally computed locally to give the vector g_i from (5.6.49) and added into the full vector b say:

$$g_i = \text{integrate}(d_i, 0, 1) - vu'|_0^1$$

with

$$d_i = (1 - x^2) * \text{coef}(0, i, x)$$

in the case of the first example with right-hand side function $1 - x^2$. Because the basis functions are unity on a node and zero elsewhere as in (5.6.1), then the solution vector of α's gives the nodal values at each point.

The only variation from problem to problem is in the boundary conditions, and this follows either (5.6.65) and following with essential boundary conditions, or (5.6.58) with natural boundary conditions. This is again close to Section 5.5. In the first example $\alpha_1 = 0$ and $\alpha_N = 0$, to leave $N - 2$ equations to be solved. Hence for a 4-element solution the results are shown in Table A.56 and exact agreement is seen with the analytic values.

Table A.56.

$x \backslash N$	4	Analytic solution
0.0	0.00	0.00
0.25	0.073242	0.073242
0.5	0.088542	0.088542
0.75	0.057617	0.057617
1.0	0.00	0.00

5.19 In this example, the same procedure is re-used but the natural boundary conditions in (a) will cause element modification in the first and last rows. Case (b) is similar to Exercise 5.21, and case (c) is a mixture of the two.

Hence to deal with (a) the extra terms are

$$au'(s_2)\psi_i(s_2) - au'(s_1)\psi_i(s_1)$$

from the integration by parts (5.6.39). But by the construction of the basis functions, $\alpha_1 = u(s_1)$ and $\alpha_N = u(s_2)$. Hence for $u'(0) = 1$, subtract from g, $a(0)u'(0) = -1$, whereas for $u'(1) + u(1) = 2$, subtract $a(1)u'(1) = (-2)(2 - u(1)) = -4 + 2\alpha_N$. Hence here the final k_{22} has 2 subtracted and the final g_2 has -4 subtracted.

The results are shown in Table A.57 where the number of elements is increased from 4 to 10 in steps of 2.

Table A.57.

$x \backslash N$	4	6	8	10
0.0	−9.3408	−9.3550	−9.3600	−9.3624
0.1				−9.2231
0.125			−9.1746	
1/6		−9.0837		
0.2				−9.0175
0.25	−8.8679		−8.8913	
0.3				−8.7591
1/3		−8.6540		
0.375			−8.5344	
0.4				−8.4588
0.5	−8.0994	−8.1164	−8.1224	−8.1252
0.6				−7.7657
2/3		−7.5062		
0.625			−7.6698	
0.7				−7.3860
0.75	−7.1672		−7.1876	
0.8				−6.1705
5/6		−6.8492		
0.875			−6.6848	
0.9				−6.5847
1.0	−6.1518	−6.1641	−6.1685	−6.1705

The practical convergence is observed as the number of elements is increased.

Case (b) is like the previous exercise in which the first and last equations are removed by the boundary conditions. These elements then appear on the right-hand side in the vector b for equations 2 to $N - 1$. The results for $N = 10$ are are shown in Table A.58.

Case (c) has one equation eliminated because $u(1) = 3$ with the corresponding correction to b for the 1st to $(N - 1)$-th equations. The condition $u'(0) = 1$ just requires 1 to be subtracted from g_1. Again the results are shown alongside case (b) in Table A.58.

Table A.58.

$x\backslash N$	(b)	(c)
0.0	1.00	3.7500
0.1	1.1367	3.6503
0.2	1.2635	3.5525
0.3	1.3807	3.4583
0.4	1.4888	3.3692
0.5	1.5885	3.2865
0.6	1.6808	3.2112
0.7	1.7667	3.1443
0.8	1.8475	3.0865
0.9	1.9247	3.0383
1.0	2.00	0.00

Chapter 6

The solutions in this chapter generally require the use of computer code. The main tools needed are an integrator based on say an 8-point Gauss–Legendre rule using the given weights and abscissae from Abramowitz and Stegun (1964, p.916), and a linear equation solver, which for the purposes of these exercises need be no more than a conventional Gauss elimination procedure. The problems will not be so large that full sparse methods become imperative. Where two-dimensional integrals are required the simple integrator may be used recursively remembering to transform the intervals of integration to $(-1, 1)$ for each range.

A small (3×3) or (4×4) matrix is used to store the stiffness components, and these are assembled as they are computed into the larger matrix K, by adding elements into their correc positions in the larger matrix.

Some structures were employ d to allow easy specification of the problem and its boundary conditions. Hence a *node* consists of the fields:

```
STRUCTURE
    integer nodenumber;
    double x-coordinate;
    double y-cordinate;
    integer boundarytype;
END STRUCTURE node;
```

and an element:

```
STRUCTURE
    integer elementnumber;
    array [3] integers nodenumbers;
    array [3] integers boundaryconditions;
END STRUCTURE element;
```

Hence in the text the `nodenumbers` would be enclosed in circles and the `elementnumbers` in squares. The three-element array of integers gives the `nodenumbers` of the three nodes of the triangle. In the case of rectangular elements replace the 3 by 4. This makes for a very easy data set to define and input the elements.

The field `boundarytype` allows the boundary to be specified. If this is zero, there is a Dirichlet condition and the boundary value can be the next data element which can be used to set the relevant ϕ value. For values greater than zero, the number can indicate the number of sides across which the Neumann condition will be taken. This number can then be used to control the setting of `boundaryconditions` in the element fields, each value giving the side number (from 1 to 3 in the same order as the nodes numbers are defined) across which the Neumann conditions are required. These values are used to control the line integrals needed in these cases.

Hence the stages of the code are:

> read in the total number of nodes;
> read in the total number of elements;
> read in the node structure details, and
>> if bound is zero the corresponding ϕ value;
> read in the element details.
> for each element:
>> Use the stiffness matrix formulae to set
>> the local stiffness matrix.
>> Accumulate the result in the main K matrix.
>> The position of the elements in K follows
>> from the element structure which contains
>> the node numbers for that element.
> End of: for each element;
> Compute the force vectors by adding the
> double integral for the $f(x,y)$ part to the
> line integral for the $h(x,y)$ part.
> Assemble these components in the force vector Q.
> Remove the Dirichlet ϕ values to the RHS
> using the boundary values, and move the rows and columns
> of the remaining unknown elements to the top left-hand
> part of the matrix.
> Call Gauss elimination to solve.

The main pitfalls here include taking great care over the signs of the integrals. Remember the normal derivative is the outward derivative and the line integral differential ds is taken in the positive direction as indicated by its components (dx, dy). Care is also needed in compressing the matrix after putting the known ϕ values into the right-hand side. Use the field `boundarytype` to keep a tally on where the columns are which are being removed. Once the columns are sorted, the rows follow by an almost identical procedure.

It is advised that the worked examples are coded and the quoted results at each step can be used to get each part of the program working. It is then fairly easy to re-hash the data sets and the functions for $f(x,y)$ and $h(x,y)$ to tackle the problems in this exercise.

Hence for each exercise in this chapter, the data sets for initialising the node/element structures are given, followed by the compressed K matrix and the Q vector. Then the unknown ϕ values are quoted. Any additional detail is added at the end of the main results. The setting up of the matrices is more straightforward than the computation of the Q's, so in the interests of space saving the compressed matrix is usually quoted.

6.1 The node data set is

```
16
1    0.33333    0.33333    1
2    0.66667    0.33333    1
3    0.33333    0.66667    1
4    0.66667    0.66667    1
5    0.0        0.0        0    0.0
6    0.0        0.33333    0    0.33333
7    0.0        0.66667    0    0.66667
8    0.0        1.0        0    1.0
9    0.33333    1.0        0    1.0
10   0.666667   1.0        0    1.0
11   1.0        1.0        0    1.0
12   1.0        0.666667   0    1.0
13   1.0        0.33333    0    1.0
14   1.0        0.0        0    1.0
15   0.66667    0.0        0    0.66667
16   0.33333    0.0        0    0.333333
```

where the node number is followed by the node coordinates, the boundary type and the ϕ value where this type is zero for a Dirichlet point. The rectangular element set is

```
9
1    5     16    1     6     0
2    6     1     3     7     0
3    7     3     9     8     0
4    16    15    2     1     0
5    1     2     4     3     0
6    3     4     10    9     0
7    15    14    13    2     0
8    2     13    12    4     0
9    4     12    11    10    0
```

where the element number is followed by the node numbers of its corners counted anti-clockwise as in the worked example.

The resulting stiffness matrix is $K = [K_1 K_2]$ where

$$
K_1 = \begin{bmatrix}
2.67 & -0.333 & -0.333 & -0.333 & -0.333 & -0.333 & -0.333 & 0.0 \\
-0.333 & 2.67 & -0.333 & -0.333 & 0.0 & 0.0 & 0.0 & 0.0 \\
-0.333 & -0.333 & 2.67 & -0.333 & 0.0 & -0.333 & -0.333 & -0.333 \\
-0.333 & -0.333 & -0.333 & 2.67 & 0.0 & 0.0 & 0.0 & 0.0 \\
-0.333 & 0.0 & 0.0 & 0.0 & 0.67 & -1.67 & 0.0 & 0.0 \\
-0.333 & 0.0 & -0.333 & 0.0 & -1.67 & 1.33 & -1.67 & 0.0 \\
-0.333 & 0.0 & -0.333 & 0.0 & 0.0 & -1.67 & 1.33 & -1.67 \\
0.0 & 0.0 & -0.33 & 0.0 & 0.0 & 0.0 & -1.67 & 0.667 \\
0.0 & 0.0 & -0.333 & -0.333 & 0.0 & 0.0 & -0.333 & -1.67 \\
0.0 & 0.0 & -0.333 & -0.333 & 0.0 & 0.0 & 0.0 & 0.0 \\
0.0 & 0.0 & 0.0 & -0.333 & 0.0 & 0.0 & 0.0 & 0.0 \\
0.0 & -0.333 & 0.0 & -0.333 & 0.0 & 0.0 & 0.0 & 0.0 \\
0.0 & -0.333 & 0.0 & -0.333 & 0.0 & 0.0 & 0.0 & 0.0 \\
0.0 & -0.333 & 0.0 & 0.0 & 0.0 & 0.0 & 0.0 & 0.0 \\
-0.333 & -0.333 & 0.0 & 0.0 & 0.0 & 0.0 & 0.0 & 0.0 \\
-0.333 & -0.333 & 0.0 & 0.0 & -1.67 & -0.333 & 0.0 & 0.0
\end{bmatrix}
$$

and

$$
K_2 = \begin{bmatrix}
0.0 & 0.0 & 0.0 & 0.0 & 0.0 & 0.0 & -0.333 & -0.333 \\
0.0 & 0.0 & 0.0 & -0.333 & -0.333 & -0.333 & -0.333 & -0.333 \\
-0.333 & -0.333 & 0.0 & 0.0 & 0.0 & 0.0 & 0.0 & 0.0 \\
-0.333 & -0.333 & -0.333 & -0.333 & -0.333 & 0.0 & 0.0 & 0.0 \\
0.0 & 0.0 & 0.0 & 0.0 & 0.0 & 0.0 & 0.0 & -1.67 \\
0.0 & 0.0 & 0.0 & 0.0 & 0.0 & 0.0 & 0.0 & -0.333 \\
-0.333 & 0.0 & 0.0 & 0.0 & 0.0 & 0.0 & 0.0 & 0.0 \\
-1.67 & 0.0 & 0.0 & 0.0 & 0.0 & 0.0 & 0.0 & 0.0 \\
1.333 & -1.667 & 0.0 & 0.0 & 0.0 & 0.0 & 0.0 & 0.0 \\
-1.67 & 1.333 & -1.67 & -0.333 & 0.0 & 0.0 & 0.0 & 0.0 \\
0.0 & -1.67 & 0.67 & -1.67 & 0.0 & 0.0 & 0.0 & 0.0 \\
0.0 & -0.333 & -1.67 & 1.33 & -1.67 & 0.0 & 0.0 & 0.0 \\
0.0 & 0.0 & 0.0 & -1.67 & 1.33 & -1.67 & -0.333 & 0.0 \\
0.0 & 0.0 & 0.0 & 0.0 & -1.67 & 0.67 & -1.67 & 0.0 \\
0.0 & 0.0 & 0.0 & 0.0 & -0.333 & -1.67 & 1.33 & -1.67 \\
0.0 & 0.0 & 0.0 & 0.0 & 0.0 & 0.0 & -1.67 & 1.33
\end{bmatrix}
$$

and the full force vector has all zero elements as all the boundary conditions are Dirichlet. The node numbers have been chosen so that the unknowns are the first four rows and columns to leave the reduced matrix as

$$
\begin{bmatrix}
2.667 & -0.333 & -0.333 & -0.333 \\
-0.333 & 2.667 & -0.333 & -0.333 \\
-0.333 & -0.333 & 2.667 & -0.333 \\
-0.333 & -0.333 & -0.333 & 2.667
\end{bmatrix}
\begin{bmatrix}
\phi_1 \\ \phi_2 \\ \phi_3 \\ \phi_4
\end{bmatrix}
=
\begin{bmatrix}
0.6667 \\ 1.333 \\ 1.333 \\ 1.667
\end{bmatrix}
$$

with solution $\phi_1 = 0.55556$, $\phi_2 = 0.7778$, $\phi_3 = 0.7778$ and $\phi_4 = 0.8889$, which is in good agreement with the corresponding finite difference result in Exercise 4.1.

6.2 The data sets are:

16
1	0.33333	0.33333	1	
2	0.66667	0.33333	1	
3	0.33333	0.66667	1	
4	0.66667	0.66667	1	
5	0.0	0.0	0	0.0
6	0.0	0.33333	0	0.33333
7	0.0	0.66667	0	0.66667
8	0.0	1.0	0	1.0
9	0.33333	1.0	0	1.0
10	0.66667	1.0	0	1.0
11	1.0	1.0	0	1.0
12	1.0	0.66667	0	1.0
13	1.0	0.33333	0	1.0
14	1.0	0.0	0	1.0
15	0.66667	0.0	0	0.66667
16	0.33333	0.0	0	0.33333

and

18
1	5	16	1	0
2	5	1	6	0
3	6	1	3	0
4	6	3	7	0
5	7	3	9	0
6	7	9	8	0
7	16	15	2	0
8	16	2	1	0
9	1	2	4	0
10	1	4	3	0
11	3	4	10	0
12	3	10	9	0
13	15	14	13	0
14	15	13	2	0
15	2	13	12	0
16	2	12	4	0
17	4	12	11	0
18	4	11	10	0

The condensed matrix problem is then

$$
\begin{bmatrix}
4.0 & -1.0 & -1.0 & 0.0 \\
-1.0 & 4.0 & 0.0 & -1.0 \\
-1.0 & 0.0 & 4.0 & -1.0 \\
0.0 & -1.0 & -1.0 & 4.0
\end{bmatrix}
\begin{bmatrix}
\phi_1 \\
\phi_2 \\
\phi_3 \\
\phi_4
\end{bmatrix}
=
\begin{bmatrix}
0.6667 \\
0.6667 \\
1.6667 \\
2.0
\end{bmatrix}
$$

with solution $\phi_1 = 0.55556$, $\phi_2 = 0.7778$, $\phi_3 = 0.7778$ and $\phi_4 = 0.8889$, showing good agreement with previous calculations.

6.3 This example is a further Dirichlet problem with the grid defined by the data set

```
36
1     0.2   0.2   1
2     0.4   0.2   1
3     0.6   0.2   1
4     0.8   0.2   1
5     0.2   0.4   1
6     0.4   0.4   1
7     0.6   0.4   1
8     0.8   0.4   1
9     0.2   0.6   1
10    0.4   0.6   1
11    0.6   0.6   1
12    0.8   0.6   1
13    0.2   0.8   1
14    0.4   0.8   1
15    0.6   0.8   1
16    0.8   0.8   1
17    0.0   0.0   0   0.0
18    0.0   0.2   0   0.04
19    0.0   0.4   0   0.16
20    0.0   0.6   0   0.36
21    0.0   0.8   0   0.64
22    0.0   1.0   0   1.0
23    0.2   1.0   0   1.04
24    0.4   1.0   0   1.16
25    0.6   1.0   0   1.36
26    0.8   1.0   0   1.64
27    1.0   1.0   0   2.0
28    1.0   0.8   0   1.6
29    1.0   0.6   0   1.2
30    1.0   0.4   0   0.8
31    1.0   0.2   0   0.4
32    1.0   0.0   0   0.0
33    0.8   0.0   0   0.0
34    0.6   0.0   0   0.0
35    0.4   0.0   0   0.0
36    0.2   0.0   0   0.0
```

and

```
25
1    17   36   1    18   0
2    36   35   2    1    0
3    35   34   3    2    0
4    34   33   4    3    0
5    33   32   31   4    0
6    4    31   30   8    0
7    3    4    8    7    0
8    2    3    7    6    0
9    1    2    6    5    0
10   18   1    5    19   0
11   19   5    9    20   0
12   5    6    10   9    0
13   6    7    11   10   0
14   7    8    12   11   0
15   8    30   29   12   0
16   12   29   28   16   0
17   11   12   16   15   0
18   10   11   15   14   0
19   9    10   14   13   0
20   20   9    13   21   0
21   21   13   23   22   0
22   13   14   24   23   0
23   14   15   25   24   0
24   15   16   26   25   0
25   16   28   27   26   0
```

This problem is again a Dirichlet problem and follows closely the lines of the previous example. The condensed matrix is 16×16 and splits into $[K_1, K_2]$ which are given by:

$$
K_1 = \begin{bmatrix}
2.667 & -0.333 & 0.0 & 0.0 & -0.333 & -0.333 & 0.0 & 0.0 \\
-0.333 & 2.667 & -0.333 & 0.0 & -0.333 & -0.333 & -0.333 & 0.0 \\
0.0 & -0.333 & 2.667 & -0.333 & 0.0 & -0.333 & -0.333 & -0.333 \\
0.0 & 0.0 & -0.333 & 2.667 & 0.0 & 0.0 & -0.333 & -0.333 \\
-0.333 & -0.333 & 0.0 & 0.0 & 2.667 & -0.333 & 0.0 & 0.0 \\
-0.333 & -0.333 & -0.333 & 0.0 & -0.333 & 2.667 & -0.333 & 0.0 \\
0.0 & -0.333 & -0.333 & -0.333 & 0.0 & -0.333 & 2.667 & -0.333 \\
0.0 & 0.0 & -0.333 & -0.333 & 0.0 & 0.0 & -0.333 & 2.667 \\
0.0 & 0.0 & 0.0 & 0.0 & -0.333 & -0.333 & 0.0 & 0.0 \\
0.0 & 0.0 & 0.0 & 0.0 & -0.333 & -0.333 & -0.333 & 0.0 \\
0.0 & 0.0 & 0.0 & 0.0 & 0.0 & -0.333 & -0.333 & -0.333 \\
0.0 & 0.0 & 0.0 & 0.0 & 0.0 & 0.0 & -0.333 & -0.333 \\
0.0 & 0.0 & 0.0 & 0.0 & 0.0 & 0.0 & 0.0 & 0.0 \\
0.0 & 0.0 & 0.0 & 0.0 & 0.0 & 0.0 & 0.0 & 0.0 \\
0.0 & 0.0 & 0.0 & 0.0 & 0.0 & 0.0 & 0.0 & 0.0 \\
0.0 & 0.0 & 0.0 & 0.0 & 0.0 & 0.0 & 0.0 & 0.0
\end{bmatrix}
$$

and

$$K_2 = \begin{bmatrix}
0.0 & 0.0 & 0.0 & 0.0 & 0.0 & 0.0 & 0.0 & 0.0 \\
0.0 & 0.0 & 0.0 & 0.0 & 0.0 & 0.0 & 0.0 & 0.0 \\
0.0 & 0.0 & 0.0 & 0.0 & 0.0 & 0.0 & 0.0 & 0.0 \\
0.0 & 0.0 & 0.0 & 0.0 & 0.0 & 0.0 & 0.0 & 0.0 \\
-0.333 & -0.333 & 0.0 & 0.0 & 0.0 & 0.0 & 0.0 & 0.0 \\
-0.333 & -0.333 & -0.333 & 0.0 & 0.0 & 0.0 & 0.0 & 0.0 \\
0.0 & -0.333 & -0.333 & -0.333 & 0.0 & 0.0 & 0.0 & 0.0 \\
0.0 & 0.0 & -0.333 & -0.333 & 0.0 & 0.0 & 0.0 & 0.0 \\
2.667 & -0.333 & 0.0 & 0.0 & -0.333 & -0.333 & 0.0 & 0.0 \\
-0.333 & 2.667 & -0.333 & 0.0 & -0.333 & -0.333 & -0.333 & 0.0 \\
0.0 & -0.333 & 2.667 & -0.333 & 0.0 & -0.333 & -0.333 & -0.333 \\
0.0 & 0.0 & -0.333 & 2.667 & 0.0 & 0.0 & -0.333 & -0.333 \\
-0.333 & -0.333 & 0.0 & 0.0 & 2.667 & -0.333 & 0.0 & 0.0 \\
-0.333 & -0.333 & -0.333 & 0.0 & -0.333 & 2.667 & -0.333 & 0.0 \\
0.0 & -0.333 & -0.333 & -0.333 & 0.0 & -0.333 & 2.667 & -0.333 \\
0.0 & 0.0 & -0.333 & -0.333 & 0.0 & 0.0 & -0.333 & 2.667
\end{bmatrix}$$

with Q vector:

$$[0.0667 \quad 0.0 \quad 0.0 \quad 0.4 \quad 0.18667 \quad 0.0 \quad 0.0 \quad 0.8$$
$$0.38667 \quad 0.0 \quad 0.0 \quad 1.2 \quad 1.4 \quad 1.18667 \quad 1.38667 \quad 2.6]^T$$

and solution:

$$[0.1526 \quad 0.2262 \quad 0.2867 \quad 0.3436 \quad 0.3328 \quad 0.4621$$
$$0.5753 \quad 0.6867 \quad 0.5522 \quad 0.7089 \quad 0.8621 \quad 1.0262$$
$$0.8017 \quad 0.9522 \quad 1.1328 \quad 1.3526]^T.$$

6.4 This problem is very similar to the worked examples and uses the data
   set:

```
36
1     0.2   0.2   1
2     0.4   0.2   1
3     0.6   0.2   1
4     0.8   0.2   1
5     0.2   0.4   1
6     0.4   0.4   1
7     0.6   0.4   1
8     0.8   0.4   1
9     0.2   0.6   1
10    0.4   0.6   1
11    0.6   0.6   1
12    0.8   0.6   1
13    0.2   0.8   1
14    0.4   0.8   1
15    0.6   0.8   1
```

16	0.8	0.8	1	
17	0.0	0.0	0	0.0
18	0.0	0.2	1	
19	0.0	0.4	1	
20	0.0	0.6	1	
21	0.0	0.8	1	
22	0.0	1.0	0	1.0
23	0.2	1.0	0	1.04
24	0.4	1.0	0	1.16
25	0.6	1.0	0	1.36
26	0.8	1.0	0	1.64
27	1.0	1.0	0	2.0
28	1.0	0.8	1	
29	1.0	0.6	1	
30	1.0	0.4	1	
31	1.0	0.2	1	
32	1.0	0.0	0	0.0
33	0.8	0.0	0	0.0
34	0.6	0.0	0	0.0
35	0.4	0.0	0	0.0
36	0.2	0.0	0	0.0

and

25

1	17	36	1	18	1	4	
2	36	35	2	1	0		
3	35	34	3	2	0		
4	34	33	4	3	0		
5	33	32	31	4	1	2	
6	4	31	30	8	1	2	
7	3	4	7	8	0		
8	2	3	7	6	0		
9	1	2	6	5	0		
10		18	1	5	19	1	4
11	19	5	9	20	1	4	
12	5	6	10	9	0		
13	6	7	11	10	0		
14	7	8	12	11	0		
15	8	30	29	12	1	2	
16	12	29	28	16	1	2	
17	11	12	16	15	0		
18	10	11	15	14	0		
19	9	10	14	13	0		
20	20	9	13	21	1	4	

21	21	13	23	22	1	4
22	13	14	24	23	0	
23	14	15	25	24	0	
24	15	16	26	25	0	
25	16	28	27	26	1	2

In this data set the Neumann conditions are indicated using extra terms. Hence for example element 25 has nodes 16, 28, 27 and 26 and a Neumann condition along *one* side (hence the next one), namely side 2 which starts from the second node, 28, and runs to 27. The K matrices are now fairly standard and will not be quoted from now on. The force vector has all zero elements except for elements 17 to 21 with values $(-0.1, -0.2, -0.2, -0.1)$ and the solution:

$$\begin{bmatrix} 0.1287 & 0.1982 & 0.2383 & 0.2695 & 0.3029 & 0.4189 \\ 0.4965 & 0.5506 & 0.5378 & 0.6679 & 0.7807 & 0.8638 \\ 0.7982 & 0.9281 & 1.080 & 1.2400 \end{bmatrix}^T$$

for x_1 to x_{16}, the interior points, and then on the left-hand boundary x_{18} to x_{21} are $(-0.02529, 0.13454, 0.3627, 0.6894)$ and on the right-hand boundary x_{28} to x_{31} are $(1.298, 0.8967, 0.5675, 0.2760)$. On this boundary $\partial\phi/\partial x = 0$, and we see for example that x_8 is very close to x_{30}. In the corresponding finite difference solution these values are taken to be equal, which will give some error in this solution. Nevertheless the two solutions are comparable.

6.5 This is the same problem solved by triangles generated by drawing diagonals across every rectangle from bottom left to top right, and the resulting solution is:

$$\begin{bmatrix} 0.1438 & 0.2001 & 0.2403 & 0.2665 & 0.3126 & 0.4163 \\ 0.4954 & 0.5495 & 0.5278 & 0.6581 & 0.7719 & 0.8655 \\ 0.7785 & 0.9165 & 1.069 & 1.229 \end{bmatrix}^T$$

for x_1 to x_{16}, the interior points, and then on the left-hand boundary x_{18} to x_{21} are $(-0.0625, 0.1624, 0.3620, 0.6297)$ and on the right-hand boundary x_{28} to x_{31} are $(1.342, 0.9113, 0.5716, 0.2762)$. These values are in good agreement with the previous rectangular grid.

6.6 In this problem, which was previously solved by conversion to polars to deal with the main curvilinear sections, and then the one side which was irregular was treated by the curved boundary formula, a direct finite element triangularisation is used with the grid:

```
13
1    0.5          0.86603     1
2    1.0          1.732051    1
3    1.732051     1.0         1
4    0.86603      0.5         1
5    0.0          0.0         0    0.0
6    0.0          1.0         0    1.0
7    0.0          2.0         0    2.0
8    0.0          3.0         0    3.0
9    1.5          2.598076    0    3.0
10   2.25         1.299038    0    3.0
11   3.0          0.0         0    3.0
12   2.0          0.0         0    2.0
13   1.0          0.0         0    1.0
```

and

```
15
1    5    13   4    0
2    4    13   3    0
3    13   12   3    0
4    3    12   10   0
5    12   11   10   0
6    5    4    1    0
7    1    4    2    0
8    4    3    2    0
9    2    3    9    0
10   3    10   9    0
11   5    1    6    0
12   6    1    7    0
13   1    2    7    0
14   7    2    8    0
15   2    9    8    0
```

and using the codes developed for the previous examples gives the four unknowns $x_1 = 1.2538$, $x_2 = 2.2716$, $x_3 = 2.3855$ and $x_4 = 1.267$.

6.7 The last two cases are now very standard and give the solution

$$[0.7950 \quad 0.7327 \quad 0.7139 \quad 0.7327 \quad 0.7950$$
$$0.5207 \quad 0.4220 \quad 0.3900 \quad 0.4220 \quad 0.5207]^T$$

with the same numbering of nodes is in Exercise 4.11. Again there is good agreement and the symmetry is demonstrated in the solution.

6.8 This example requires the use of an f function but is otherwise straightforward to yield the vector

$$[0.3857 \quad 0.4821 \quad 0.3857 \quad 0.4821 \quad 0.6214$$
$$0.4821 \quad 0.3857 \quad 0.4821 \quad 0.3857]^T$$

with the same counting as in Exercese 4.1 with x_1 in the bottom left interior spot and the count continuing from left to right and bottom to top.

References and Further Reading

Abramowitch M and Stegun IA (1964) Handbook of mathematical functions. Dover Publications, New York.

Bramble JH (1993) Multigrid methods. Longman, Harlow.

Brebbia CA (1978) The boundary element method for engineers. Pantech Press, London.

Brebbia CA and Dominguez J (1992) Boundary elements: an introductory course. Computational Mechanics, Southampton.

Brebbia CA and Mackerle J (1988) The boundary element reference book. Computational Mechanics, Southampton.

Briggs WA (1987) A multi-grid tutorial. Society for Industrial and Applied Mathematics, Philadelphia.

Davies AJ (1980) Finite element method. Oxford University Press, Oxford.

Duff IS, Reid JK and Erisman AM (1986) Direct methods for sparse matrices. Cambridge University Press, Cambridge.

Evans GA (1993) Practical numerical integration. Wiley, New York.

Evans GA (1996) Practical numerical analysis. Wiley, New York.

Evans GA, Blackledge JM and Yardley PD (1999) Analytic methods for partial differential equations. Springer-Verlag, London.

Golumbic MC (1980) Algorithmic graph theory and perfect graphs. Academic Press, New York.

Hackbusch W (1985) Multigrid methods and applications. Springer-Verlag, Berlin.

Hildebrand FB (1974) Introduction to numerical analysis. McGraw-Hill, New York.

Iserles A (1996) A first course in the numerical analysis of differential equations. Cambridge University Press, Cambridge.

Lambert J (1990) Numerical solution of ordinary differential equations. Wiley, New York.

Richtmyer RD and Morton KW (1967) Difference methods for initial value problems. Interscience, New York.

Renardy M and Rogers RC (1992) An introduction to partial differential equations. Springer-Verlag, London.

Smith GD (1978) Numerical solution of partial differential equations. Oxford University Press, Oxford.

Sneddon IN (1957) Numerical solution of partial differential equations. McGraw-Hill, New York.

Varga RS (1962) Matrix iterative analysis. Prentice-Hall, Englewood Cliffs, NJ.

Vvedensky D (1993) Partial differential equations with Mathematica. Addison-Wesley, Reading, MA.

Wilkinson JH (1965) Algebraic eigenvalue problem. Oxford University Press, Oxford.

Young DM (1971) Iterative solution of large linear systems. Academic Press, London, New York.

Zienkiewicz OC (1977) The finite element method. McGraw-Hill, New York.

Index

Printed in the United States
78850LV00005B/16